International Conference on Computing and Precision Engineering 2015 (ICCPE 2015)

MATEC Web of Conferences Volume 71 (2016)

Sun-Moon Lake, Taiwan
27 - 29 November 2015

Editors:

Chien-Hung Liu
Cheng-Chi Wang

ISBN: 978-1-5108-2748-6

Printed from e-media with permission by:

Curran Associates, Inc.
57 Morehouse Lane
Red Hook, NY 12571

Some format issues inherent in the e-media version may also appear in this print version.

Printed by Curran Associates, Inc. (2016)

For additional information, please contact EDP Sciences – Web of Conferences
at the address below.

EDP Sciences – Web of Conferences
17, Avenue du Hoggar
Parc d'Activité de Courtabœuf
BP 112
F-91944 Les Ulis Cedex A
France

Phone: +33 (0) 1 69 18 75 75
Fax: +33 (0) 1 69 28 84 91

contact-edps@webofconferences.org

Additional copies of this publication are available from:

Curran Associates, Inc.
57 Morehouse Lane
Red Hook, NY 12571 USA
Phone: 845-758-0400
Fax: 845-758-2633
Email: curran@proceedings.com
Web: www.proceedings.com

TABLE OF CONTENTS

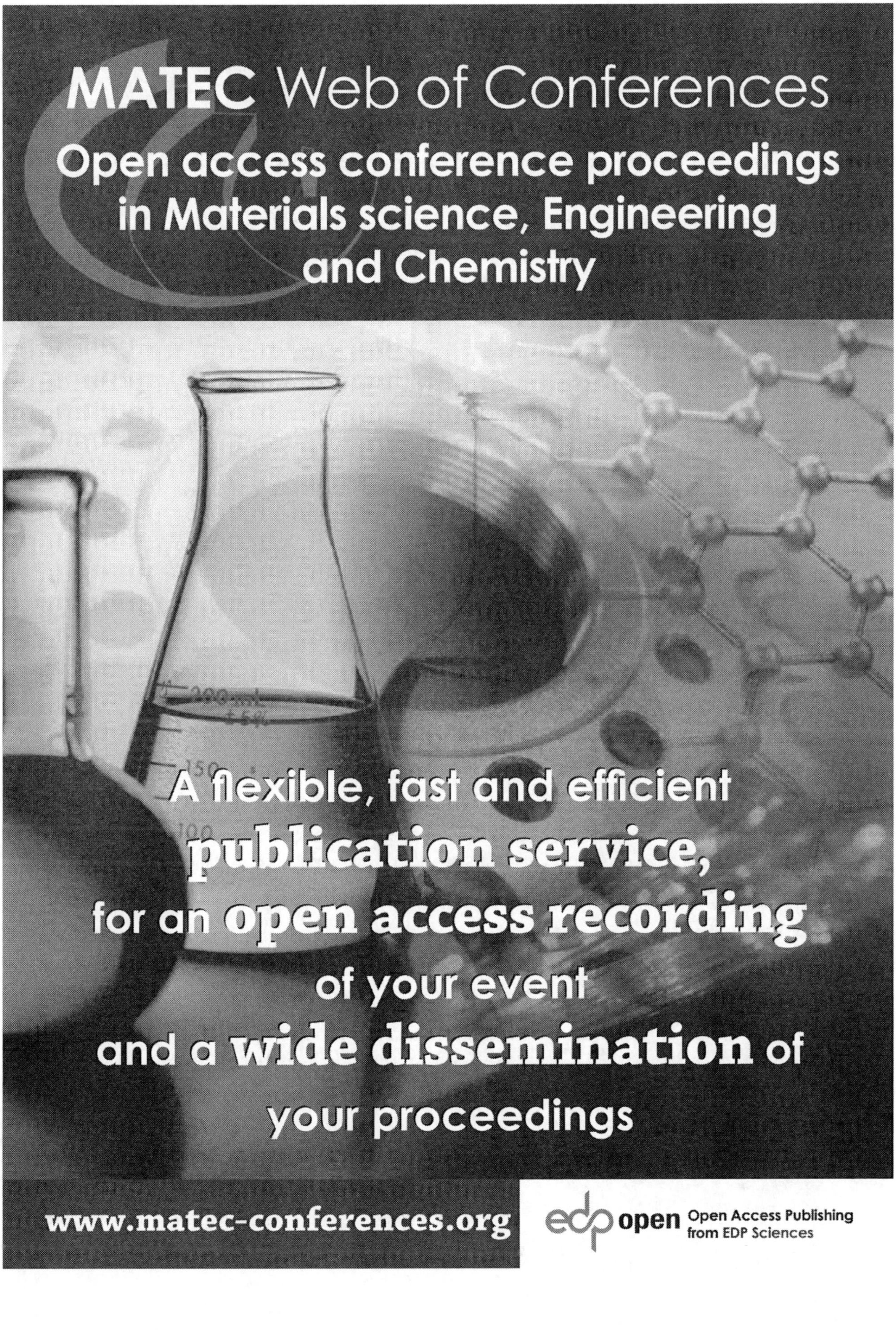
MATEC Web of Conferences
Open access conference proceedings in Materials science, Engineering and Chemistry
A flexible, fast and efficient publication service, for an open access recording of your event and a wide dissemination of your proceedings
www.matec-conferences.org
edp open Open Access Publishing from EDP Sciences

ISSN (electronic edition): 2261-236X

Committees

Program Chairmen

- Te-Hua Fang, National Kaohsiung University of Applied Sciences, Taiwan
- Cheng-Chi Wang, National Chin-Yi University of Technology, Taiwan
- Zheng-Rong Wang, Spintech Precision Machinery Co., Ltd., Taiwan
- Chien-Hung Liu, National Chung Hsiung University, Taiwan
- Sheng-Chung Tzeng, Chienkuo Technology University, Taiwan
- Yu-Ying Chiu, Chiuan Yan Technology Co., Ltd., Taiwan

Honorary Chairmen

- Cha'o-Kuang Chen, National Cheng Kung University, Taiwan
- Ralph Greif, University of California, Berkeley, CA, USA
- Costas P. Grigoropoulos, University of California, Berkeley, CA, USA

Program Co-Chairmen

- Chin-Chia Liu, National Changhua University of Education, Taiwan
- Yu-Fen Chen, National Formosa University, Taiwan
- Yi-Chang Wu, National Yunlin University of Science & Technology, Taiwan
- Ming-Tsang Lee, National Chung Hsing University, Taiwan
- Chi-Fan Liu, Feng Chia University, Taiwan
- Yu-Bin Chen, National Cheng Kung University, Taiwan
- Xin-gang Liang, Tsinghua University, Beijing, China
- Zhonghao Rao, China University of Mining and Technology, China
- Chang-Ying Zhao, School of Mechanical Engineering Shanghai JiaoTong University, China
- Jian-xing Ren, Shanghai University of Electric Power, China
- Liang Gong, China University of Petroleum (East China), China

Technical Committee Chairmen

- Chih-Jer Lin, National Taipei University of Technology, Taiwan
- Huei-Chu Weng, Chung Yuan Christian University, TaiwanG.Y. Tzou, Kao Yuan University of Sciences and Technology, Taiwan
- Her-TerngYau, National Chin-Yi University of Technology, Taiwan
- Hau-Wei Lee, Chiuan Yan Technology Co., Ltd., Taiwan
- Tao-Hsing Chen, National Kaohsiung University of Applied Sciences, Taiwan
- Yunhua Li, Beihang University, China

Invited Session Chairmen

- Cong-Hui Huang, Far East University, Taiwan
- Kuan-Chen Lin, Far East University, Taiwan
- Cheng-Chi Wang, National Chin-Yi University of Technology, Taiwan
- Chien-Hung Liu, National Chung Hsing University
- Chih-Jer Lin, National Taipei University of Technology, Taiwan
- Ming-Tsang Lee, National Chung Hsing University
- Jui-Feng Yeh, National Chiayi University, Taiwan
- Tao-Hsing Chen, National Kaohsiung University of Applied Sciences, Taiwan
- Young-Long Chen, National Taichung University of Science and Technology, Taiwan
- Hung-Jen Yang, National Kaohsiung Normal University, Taiwan

Program Committee

- Young-Long Chen, National Taichung University of Science and Technology, Taiwan
- Yunn-Lin Hwang, National Formosa University, Taiwan
- Chang-Tzuoh Wu, Kaohsiung Normal University, Taiwan
- Liang-Wen Ji, National Formosa University, Taiwan
- Siu-Tsen Shen, National Formosa University, Taiwan
- Walter Water, National Formosa University, Taiwan
- Jian-Ming Lu, National Center for High-Performance Computing,Taiwan
- Yu-Jen Hsiao, National Nano Device Laboratories, Taiwan
- Cheng-Che Hsu, National Taiwan University, Taiwan
- Si-Yu Li , National Chung Hsing University, Taiwan
- Jia-Yang Juang, National Taiwan University,Taiwan
- Ming-Chang Lu, National Chiao Tung University,Taiwan
- Seung Hwan Ko, Seoul National University, Korea
- HojeongJeon, Korea Institute of Science and Technology, Korea
- Ichiro Ueno, Tokyo University of Science, Japan
- Nico Hotz, Duke University, USA
- Renkun Chen, University of California, USA
- ChuanhuaDuan, Boston University, USA
- Heng Pan, Missouri University of Science and Technology, USA
- Coleman Kronawitter, Princeton University, USA
- IrfanJamil,Tsinghua University, Beijing, P.R. China
- Steven. D. Prior, University of Southampton, UK
- Hongying Meng, Brunel University, UK
- Liang Pan, Purdue University, USA
- David T.W. Lin, National University of Tainan, Taiwan
- Liman Yang, Beihang University, China

Preface

This proceeding presents thirty-six papers selected from the proceedings of the 1st International Conference on Computing and Precision Engineering(ICCPE), which was held from November 27 to 30, 2015, in Sun-Moon Lake, Nantou, Taiwan. The topics include Computational and mathematical sciences, Modeling, numerical simulation and optimization, Computer and information design and analysis technologies, Advanced manufacturing and analysis technology, Control and automation technology, mechatronics, robotics for manufacture and industry.

The conference particularly encouraged the interaction of research students and developing academics with the more established academic community to present and to discuss new and current work. Their contributions helped to make the Conference as outstanding as it has been. In addition to the contributed papers, four invited keynote presentations were given: by Prof. Oronzio Manca of the Second University of Naples of Italy, who spoke about comparative analysis in convective heat transfer enhancement by nanofluids, by Dr. Yunhua Li of the Beijing University of Aeronautics and Astronautics of China, who spoke about modeling, control, and scientific computation for complicated engineering problems, by Prof. Daeho Lee of Gachon University of Korea, who spoke about laser-based material processing and electronics fabricationandbyDr. Tien-Yuan Kuoof the President of Chengda Intellectual Property Tech. Ltd.Co. who spoke about systematization of technology valuation.

All the papers (about 130 papers) submitted to this conference were first subjected to peer review by two independent reviewers prior to the conference. There were only thirty-six papers of high quality recommended for submission to this proceeding and accepted with mandatory changes and final examination by the Chairmen. Lastly, I would like to thank ICCPE Program Chairman /Professor Chien-Hung Liu for his advice.

These Proceedings will furnish the scientists of the world with an excellent reference book. I trust also that this will be an impetus to stimulate further study and research in all these areas.
We thank all authors and participants for their contributions.

Statement of Peer review

In submitting conference proceedings to MATEC Web of Conferences, I certify to the Publisher that I adhere to the **Policy on Publishing Integrity** of the journal in order to safeguard good scientific practice in publishing.

1. All articles have been subjected to peer review administered by the proceedings editors.
2. Reviews have been conducted by expert referees, who have been requested to provide unbiased and constructive comments aimed, whenever possible, at improving the work.
3. Proceedings editors have taken all reasonable steps to ensure the quality of the materials they publish and their decision to accept or reject a paper for publication has been based only on the merits of the work and the relevance to the journal.

Title, date and place of the conference

The International Conference on Computing and Precision Engineering, 27th-29th November 2015, Sun Moon Lake Teacher's Hostel in Taiwan.

Proceedings editor(s):

*Prof.Chien-Hung Liu, National Chung Hsiung University
*Prof.Cheng-Chi Wang, National Chin-Yi University of Technology

Date and editor's signature

MATEC Web of Conferences 71, 01001 (2016) DOI: 10.1051/matecconf/20167101001
ICCPE 2015

A novel DEMATEL theory based on Liu's polytomous ordering theory

Hsiang-Chuan Liu[1], Yih-Chang Ou[2], Ben-Chang Shia[3] and Hsien-Chang Tsai[4,a]

[1]*Department of Informatics and Biomedical Engineering, Asia University, Taichung, Taiwan*
[2]*Department of Finance, Ling Tung University, Taichung, Taiwan*
[3]*Big Data Research Center & School of Management School of Health Care Administration, Taipei Medical University, Taipei, Taiwan*
[4]*Department of Biology, National Changhua University of Education, Changhua, Taiwan*

Abstract. The most important issue in DEMATEL theory is how to obtain a reliable initial direct relation matrix with order n, the traditional theory obtains it by using the pair-wise comparison method, in which, each respondent must answer n(n-1) times pair-wise comparisons of all of the direct influences, if n is a large number, the work of pair-wise comparing is becoming hard, time-consuming, and unreliable. In this paper, for overcoming above drawbacks, we replace the pair-wise comparison method with Liu's ordering theory to find the initial direct relation matrix. This new method without pair-wise comparing can be used for any order n, a simple example was also provided in this paper to illustrate the advantages of the proposed theory.

1 Introduction

Decision Making Trial and Evaluation Laboratory (DEMATEL) was developed by Gabus and Fontela [1], Fontela and Gabus [2], and it can be used to resolve complex and difficult problems in the world, up to now, it has been widely used as one of the best tools to solve the cause and effect relationship among the evaluation factors [3-8].

2 The Traditional DEMATEL

The procedure of the traditional DEMATEL method is briefly introduced below:

Step 1: Calculate the initial direct relation matrix Q

If n factors are asked by N experts to evaluate the degree of direct influence between two factors based on pair-wise comparison.

The degree to which the expert e perceived factor i affects on factor j is denoted as

$$q_{ij}^{(e)}, e=1,2,..,N,\ q_{ij}^{(e)}\in\{0,1,2,3,4\}, i,j=1,2,\cdots,n \quad (1)$$

For each expert e, an individual direct relation matrix is constructed as

$$Q_e=\left[q_{ij}^{(e)}\right]_{n\times n}, e=1,2,..,N,\ q_{ii}^{(e)}=0, i=1,2,...,n \quad (2)$$

and we can obtain their average direct relation matrix, called the initial direct relation matrix Q;

$$Q=\left[q_{ij}\right]_{n\times n}=\frac{1}{N}\sum_{e=1}^{N}Q_e,\ q_{ij}=\frac{1}{N}\sum_{e=1}^{N}q_{ij}^{(e)}, i,j=1,2,\cdots,n \quad (3)$$

Step 2: Calculate the direct relation matrix A

$$A=\left[a_{ij}\right]_{n\times n}=\lambda^{-1}Q,\ \lambda=\max_{1\le i,j\le n}\left\{\sum_{j=1}^{n}q_{ij},\sum_{i=1}^{n}q_{ij}\right\} \quad (4)$$

where

$$a_{ii}=0, i=1,2,...,n,\quad 0\le a_{ij}\le 1, i\ne j, i,j=1,2,...,n$$

$$\text{and}\quad 0\le\sum_{i=1}^{n}a_{ij},\sum_{j=1}^{n}a_{ij}\le 1,\ i,j=1,2,...,n \quad (5)$$

Step 3: Calculate the indirect relation matrix B and the total relation matrix T

Based on Markov chain theory, we have

$$\lim_{k\to\infty}A^k=0_{n\times n} \quad (6)$$

$$B=\left[b_{ij}\right]_{n\times n}=\lim_{k\to\infty}\left[A^2+A^3+...+A^k\right]=A^2(I-A)^{-1} \quad (7)$$

$$T=\left[t_{ij}\right]_{n\times n}=A+B=\left[\left(a_{ij}+b_{ij}\right)\right]_{n\times n} \quad (8)$$

Step 4: Calculate the relation degree and prominence degree of each factor

$$r_i=\sum_{j=1}^{n}t_{ij},\ c_i=\sum_{k=1}^{n}t_{ki},\ i=1,2,...,n \quad (9)$$

The value of r_i indicates the total dispatch both directly and indirectly effects, that factor i has on the other factors, and the value of c_i indicates the total receive both directly and indirectly effects, that factor i has on the other factors.

The relation degree of factor i is denoted as

$$x_i=r_i-c_i,\quad i=1,2,...,n \quad (10)$$

The prominence degree of factor i is denoted as

$$y_i=r_i+c_i,\quad i=1,2,...,n \quad (11)$$

$$\text{relation-prominence matrix;}\quad (x_i,y_i)_{i=1}^{n} \quad (12)$$

Step 5: Set the threshold value (α)

For eliminating some minor effects elements in matrix T to find the impact-relations map, Yang et al.[5] suggest their threshold value below;

[a] Corresponding author: bihft@cc.ncue.edu.tw

$$\alpha_Y = \frac{1}{n^2}\sum_{i=1}^{n}\sum_{j=1}^{n} t_{ij} \tag{13}$$

Lin and Tzeng [6] suggested a more information threshold value, α_M, based on their maximum mean de-entropy (MMDE) algorithm.

Step 6: Build a cause and effect relationship diagram

If $t_{ij} > \alpha_Y, or\ t_{ji} > \alpha_Y$, then factor i is a net dispatch node of factor j, and factor j is a net receive node of factor i, respectively, and denoted as

$$(x_i, y_i) \rightarrow (x_j, y_j), or\ (x_i, y_i) \leftarrow (x_j, y_j)$$

The graph of $(x_i, y_i)_{i=1}^{n}$ including the net direct edges can present a cause and effect relationship diagram.

3 New DEMATEL theory based on Liu's polytomous ordering theory

3.1. Liu's polytomous ordering theory

If n factors are asked by N experts to evaluate the degree of direct influence of each factor i to the rest, the degree to which the expert e perceived factor i affects on the rest is denoted as

Let $M = \{0,1,2,...,(m-1)\}$ and the data matrix is denoted below;

$$D = [d_{ie}]_{n \times N},\ d_{ie} \in M,\ i = 1,2,...,n,\ e = 1,2,...,N \tag{14}$$

Let $S_i(x) = \{d_{ie} \mid d_{ie} = x,\ e = 1,2,..,N\},\ x \in M,\ i = 1,2,...,n$ (15)

$$S_{ij}(x,y) = \{(d_{ie}, d_{je}) \mid (d_{ie}, d_{je}) = (x,y),\ e = 1,2,..,N\} \tag{16}$$

where $i, j = 1,2,...,n,\quad x, y \in M$,

then the probabilities of impact degree of factor are defined below;

$$P_i(x) = \frac{|S_i(x)|}{N}, P_j(y) = \frac{|S_j(y)|}{N}, P_{ij}(x,y) = \frac{|S_{ij}(x,y)|}{N} \tag{17}$$

Let $P_i = \sum_{x=0}^{m-1} P_i(x),\quad P_j = \sum_{y=0}^{m-1} P_j(y),$ (18)

$P_i > P_j$ means that the degree of direct influence of factor i to others is greater than which of factor j.

Liu's ordering coefficient from factor i to factor i [8] is defined below;

$$r_{ij}^{(LOT)} = 1 - V_{ij} \tag{19}$$

$$V_{ij} = \sum_{x=0}^{m-1}\sum_{y=0}^{m-1} P_{ij}(x > y) \tag{20}$$

where $P_{ij}(x > y) = \frac{1}{(m-1)}[x-y]^+ P_{ij}(x,y)$ (21)

$$[z]^+ = \begin{cases} z & if\ z > 0 \\ 0 & if\ z \le 0 \end{cases} \tag{22}$$

In (20) and (21), $P_{ij}(x > y)$ represents the impact degree of factor i is greater than which of factor j, hence V_{ij} is just the violated probability.

For all experts. Liu's ordering coefficient from factor i to factor j, $r_{ij}^{(LOT)}$, represents that the all probable probability minas the violated probability, it is reasonable and well defined.

3.2 Example for Liu's polytomous ordering theory

Example 1. Suppose the joint probability and marginal probability of the impact degree x of factor 1 and the impact degree y of factor 2 are listed in Table 1. below, from formula (20), we can get $V_{12} = 0.3050$, $V_{21} = 0.2700$, then using formula (19), we have $r_{12}^{(LOT)} = 0.6950$ and $r_{21}^{(LOT)} = 0.7300$.

Similarly, we can obtain the values of all Liu's ordering coefficients;

$r_{ij}^{(LOT)}, i \neq j,\ i, j = 1,2,...,n$

Table 1. The probabilities of $P_{12}(x, y)$

$P_{12}(x,y)$	$y=0$	$y=1$	$y=2$	$y=3$	$y=4$	$P_1(x)$
$x=0$	0.03	0.05	0.06	0.02	0.01	0.17
$x=1$	0.05	0.04	0.05	0.04	0.02	0.20
$x=2$	0.05	0.05	0.13	0.03	0.01	0.27
$x=3$	0.05	0.03	0.04	0.11	0.03	0.26
$x=4$	0.02	0.01	0.02	0.01	0.04	0.10
$P_2(y)$	0.20	0.18	0.30	0.21	0.11	1.00

3.3 New DEMATEL theory based on Liu's ordering theory

The procedure of the new DEMATEL theory based on Liu's ordering theory is briefly described below:

Step 1: Calculate the data matrix D

If n factors are asked by N experts to evaluate the degree of direct influence of each factor i to the rest, according to equation (14) to calculate the data matrix;

$$D = [d_{ie}]_{n \times N}, d_{ie} \in \{0,1,2,3,4\}.$$

Step 2: Calculate the probabilities of impact degree of each factor

According to equations (14)-(18) to calculate the probabilities; $P_i(x),\ P_j(y),\ P_{ij}(x,y)$.

Step 3: Calculate the Liu's ordering coefficient for each order-pair of all factors

According to equations (19)-(22) to calculate Liu's ordering coefficient from factor i to factor j;

$$r_{ij}^{(LOT)},\quad i, j = 1,2,...,n \tag{23}$$

Step 4: Calculate the initial direct relation matrix Q

$$Q=\left[q_{ij}\right]_{n\times n},\ q_{ii}=0, i=1,2,...,n, \qquad (24)$$
$$q_{ij}=r_{ij}^{(LOT)}, i\neq j, i,j=1,2,...,n$$

Step 5: Calculate the direct relation matrix A

According to equations (24) to calculate A

$$A=\left[a_{ij}\right]_{n\times n}=\lambda^{-1}Q,\ \lambda=\max_{1\le i,j\le n}\left\{\sum_{j=1}^{n}q_{ij},\sum_{i=1}^{n}q_{ij}\right\} \qquad (25)$$

The rest steps are the same as the traditional DEMATEL theory.

3.4 Example for New DEMATEL theory

Example 2 Suppose 4 factors are asked by 10 experts to evaluate the degree of direct influence of each factor i to the rest, according to the step 1, up to step 4 of the new method, we can obtain Liu's ordering coefficient of 4 factors and the initial direct relation matrix Q can be obtained, each expert does not need repeat the hard work by pair-wise comparison 4(4-1) times.

Step 1-Step 4.: Calculate Liu's ordering coefficient of 4 factors and the initial direct relation matrix Q According to equations (19)-(22), we can obtain Liu's ordering coefficient matrix of 4 factors and the initial direct relation matrix Q below;

$$Q=\left[q_{ij}\right]_{4\times4}=\left[r_{ij}^{(LOT)}\right]_{4\times4}=\begin{bmatrix}0&0.72&0.64&0.64\\0.64&0&0.68&0.60\\0.68&0.60&0&0.60\\0.56&0.56&0.60&0\end{bmatrix} \qquad (26)$$

Step 5: Calculate the direct relation matrix A

According to equations (4), we have

$$\lambda=\max_{1\le i,j\le n}\left\{\sum_{j=1}^{n}q_{ij},\sum_{i=1}^{n}q_{ij}\right\}=2$$
$$A=\left[a_{ij}\right]_{4\times4}=\lambda^{-1}Q=\begin{bmatrix}0&0.36&0.32&0.32\\0.32&0&0.34&0.30\\0.34&0.30&0&0.30\\0.28&0.28&0.30&0\end{bmatrix} \qquad (27)$$

Step 6: Calculate the indirect relation matrix B

$$B=\left[b_{ij}\right]_{4\times4}=A^2\left(I-A\right)^{-1}=\begin{bmatrix}3.9448&3.8483&3.9290&3.7995\\3.7483&3.8237&3.7991&3.6907\\3.6807&3.6963&3.8253&3.6331\\3.4499&3.4478&3.4963&3.4508\end{bmatrix} \qquad (28)$$

Step 7: Calculate the total relation matrix T

$$T=\left[t_{ij}\right]_{4\times4}=A+B=\left[\left(a_{ij}+b_{ij}\right)\right]_{4\times4}=\begin{bmatrix}3.9448&4.2083&4.2490&4.1195\\4.0683&3.8237&4.1391&3.9907\\4.0207&3.9963&3.8253&3.9331\\3.7299&3.7278&3.7963&3.4508\end{bmatrix} \qquad (29)$$

Step 8: Calculate the relation degree and prominence degree of each factor and relation prominence matrix;

$$x_i=r_i-c_i=\sum_{j=1}^{4}t_{ij}-\sum_{k=1}^{4}t_{ki},\quad i=1,2,...,n \qquad (30)$$
$$y_i=r_i+c_i=\sum_{j=1}^{4}t_{ij}+\sum_{k=1}^{4}t_{ki},\quad i=1,2,...,n$$
$$(x_i,y_i)_{i=1}^{4}=\begin{bmatrix}(x_1,y_1)\\(x_2,y_2)\\(x_3,y_3)\\(x_4,y_4)\end{bmatrix}=\begin{bmatrix}(0.7579,&32.2851)\\(0.2658,&31.7781)\\(-0.2344,&31.7850)\\(-0.7893,&30.1989)\end{bmatrix} \qquad (31)$$

Step 9: Set the threshold value and find the significant values using Liu's threshold value below;

$$\alpha_L=\max\left\{\alpha_Y,\max_{i=1,2,...n}t_{ii}\right\}=3.9448 \qquad (32)$$

$$\text{Where}\quad \alpha_Y=\frac{1}{n^2}\sum_{i=1}^{n}\sum_{j=1}^{n}t_{ij}=3.9390 \qquad (33)$$

If $t_{ij}>\alpha_Y=3.9390$, then t_{ij} is a significant value, denoted $t_{ij}^{*},\quad i,j=1,2,...,n$,

$$T^{*}=\begin{bmatrix}3.9448&4.2083^{*}&4.2490^{*}&4.1195^{*}\\4.0683^{*}&3.8237&4.1391^{*}&3.9907^{*}\\4.0207^{*}&3.9963^{*}&3.8253&3.9331\\3.7299&3.7278&3.7963&3.4508\end{bmatrix} \qquad (34)$$

Step 10: Build a cause and effect relationship diagram according to the reduced T^{*} matrix

The formal definition of the reduced T^{*} matrix is defined as follows;

If $t_{ij}>\alpha_Y$, then factor i is significantly a dispatch of factor j , denoted as $x_i\to x_j,\quad i,j=1,2,...,n$, the reduced T^{*} matrix is denoted as;

$$T_{red}^{*}=\left[z_{ij}\right]_{n\times n},\ \text{where}\ z_{ij}=\begin{cases}1 & if\ x_i\to x_j\\0 & if\ x_i\not\to x_j\end{cases} \qquad (35)$$

For our example, we can obtain the reduced T^{*} matrix below;

$$T_{red}^{*}=\left[z_{ij}\right]_{4\times4}=\begin{bmatrix}0&1&1&1\\1&0&1&1\\1&1&0&0\\0&0&0&0\end{bmatrix} \qquad (36)$$

and then, the cause and effect relationship diagram can easily be constructed.

4 Conclusion

In this paper, we pointed out that each respondent answering the questionnaire of the traditional DEMATEL by pair-wise comparison is a hard, time-consuming, and unreliable work, if the number of factors is large. For overcoming this drawback, we replaced the pair-wise comparison method with Liu's ordering theory to find a more reliable initial direct relation matrix without pair-wise comparison, and this new method can be used for any number of factors. A simple example was also provided to illustrate the advantages of the proposed theory.

This paper is partially supported by the National Science Council grant (NSC 100-2511-S-468 -001).

References

1. A. Gabus, E. Fontela, *Perceptions of the World Problematique: Communication Procedure, Communicating with those Bearing Collective Responsibility* (Switzerland Geneva, 1973)
2. E. Fontela, A. Gabus, *The DEMATEL observer*, (Switzerland Geneva, 1976)
3. G.H. Tzeng, C.H. Chiang, C.W. Li, Expert. Syst. Appl **32**, 4 (2007)
4. J.H. Liou, L. Yen, G.H. Tzeng, J. Air. Transp. Manag **14**, 1 (2008)
5. Y.P.O. Yang, H.M. Shieh, J.D. Leu, G.H. Tzeng, Int. J. Oper. Res **5**, 3 (2008)
6. C.L. Lin, G.H. Tzeng, Expert. Syst. Appl **36**, 6 (2009)
7. C.W. Li, G.H. Tzeng, Expert. Syst. Appl **36**, 6 (2009)
8. H.C. Liu, W.S. Chen, C.M. Chiang, B.C. Shia, J.M. Ju, ICIC Express Lett Part B Appl **6**, 2 (2015)

Towards Shift Tolerant Visual Secret Sharing Schemes without Pixel Expansion

Justie Su-Tzu Juan[1,a], Yung-Chang Chen[1] and Song Guo[2]

[1]*Department of Computer Science & Information Engineering, National Chi Nan University, 54561 Nantou, Taiwan*
[2]*School of Computer Science and Engineering, The University of Aizu, 965-8580 Fukushima, Japan.*

Abstract: Naor and Shamir proposed the visual cryptography in 1995, they encrypted secret image into two meaningless random images, called shares, and it can be decrypted by human vision without any calculations. However, there would be problems in alignment when these two shares are staked by hand in practical. Therefore, this paper presents the fault-tolerant schemes of stacking two shares which are acquired from secret image encryption without pixel expansion. The main idea of these schemes is combining several pixels to be a unit, then encrypting every unit into a specific combination of pixels. It makes visual secret sharing scheme more practical.

1 Introduction

In 1995, Noar and Shamir proposed *visual cryptography* (VC for short), which is a way to encrypt one secret image and it can decoded by human vision without any calculation [11]. The concept is encrypting secret image S into two meaningless random images, called *share* (also called *sheet* or *random grid*), G_1 and G_2, one can be seen as a cipher text and the other is key to it. Stacking them is the only way to restore the hidden secret. *Random grid visual secret sharing* (VSS for short) which is made by [6] receives more attention in recent years, such as references [3, 4, 5, 9, 12, 14] are visual cryptography algorithms, designed on the basis of random grid. In the method proposed by Kafri and Keren, taking each *pixel* as a grid on the image and apply the concept of random variables to encrypt images.

For VSS, the secret image can be visually reconstructed with shares, printed on transparencies, and stacked precisely on an overhead projector. A slight misalignment between the shares could dramatically degrade the visual quality of the reconstructed image. If the size of shares are small, the alignment will be difficult. Therefore, some literatures study in this problem, such as [1, 2, 6, 8, 11, 13, 15]. Nakajima and Yamaguchi propose proposed an extended VSS scheme which enhanced registration tolerance when stacked shares are not aligned perfectly in 2004 [10]. Transfer the secret image into black and white values with half-tone technique, and then encrypt into two random images, the difference with other methods is that one of the random images becomes large diamond pattern and the other one is smaller so there will be some space for fault tolerance when stacking the shares. However, their scheme will make the pixel expansion. We will refer to their ideas to design the visual secret sharing scheme to achieve fault tolerance mechanism without pixel expansion.

The rest of this paper is organized as follows. Section 2 will detail techniques mentioned above. Section 3 discuss the major findings, we have designed three VSS schemes that achieve better fault tolerance. In Section 4, the fault tolerance result acquired from computer simulation of various schemes are mentioned in previous section will be presented. Finally, our conclusions and references.

2 Related works

2.1 Visual cryptography concepts

Difference with previous secret sharing scheme technique, when the shares are stacked, the confidential content can be interpreted with human vision directly. That is, a VC scheme can restore the secret without additional computational calculation. Figure 1 shows the encryption and decryption process model of a VC scheme.

[a] Corresponding author: jsjuan@ncnu.edu.tw

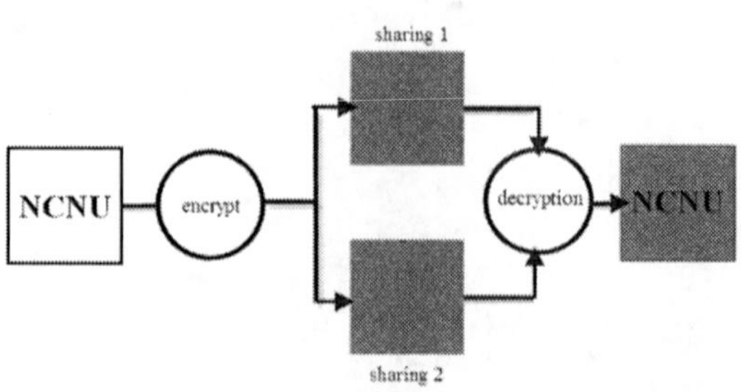

Figure 1. VC encryption and decryption process model.

For understanding the following sections, we have to understand some important notations about random grid which listed in this section in advance. In general, we define S is a secret image with size w pixels by h pixels for positive integers w and h.

Definition 1: Let $S(i, j)$ denote a *pixel value* of the secret image S at the position (i, j), where

$$S(i,j)=\begin{cases}0, & \text{if } S(i,j) \text{ is white;}\\ 1, & \text{if } S(i,j) \text{ is black.}\end{cases}$$

Actually, 1 is opaque and 0 is transparent when S is printed on transparencies.

Definition 2: Let $S(i, j)$ denote a pixel value of the secret image S at the position (i ,j), then

$$\overline{S(i,j)}=\begin{cases}0, & \text{if } S(i,j)=1;\\ 1, & \text{if } S(i,j)=0,\end{cases}=1-S(i,j).$$

Definition 3: *Transmittance* (T) is defined as proportion of white pixels to total pixels.

A secret image S is encrypted into two shares, one is G_1 and the other is G_2. r_i is a pixel in G_i; $\oplus$ stands for Boolean "OR" operation, so the output of $r_1 \oplus r_2$ means the result pixel of pixel r_1 overlap pixel r_2. All results when stacking any two pixels together are shown as Table 1.

Table 1: Results for Stacking Two Different Pixels Together

r_1	r_2	$r_1 \oplus r_2$
0	0	0
0	1	1
1	0	1
1	1	1

2.2 Random grid encryption algorithm

In 1987, Kafri and Keren proposed three encryption algorithms for halftone images. Transparent (white) is defined as 0 and opaque (black) is 1, the *random grid* obtained from determining each pixel by flipping a coin, it means that the probability of getting black or white pixel is the same. Therefore, the transmittance of random image is 1/2. Kafri and Keren [6] proposed three different algorithms to encrypt a secret image S with the size $w \times h$ pixels for some positive integers w and h into two shares, G_1 and G_2 with the same size. We list the most important one as follow:

Algorithm KK:

```
Generate a w × h random grid G1
For (i = 0; i < w; i ++)
    For (j = 0; j < h; j ++)
        If (S[i][j] == 0)
            G2[i][j] = G1[i][j];
        else
            G2[i][j] = 1 – G1[i][j];
Output (G1, G2)
```

Table 2. The Transmittance of KK Algorithm

S	Probability	G_1	G_2	$G_1 \oplus G_2$	$T(G_1 \oplus G_2)$
□	1/2	□	□	□	1/2
	1/2	■	■	■	
■	1/2	□	■	■	0
	1/2	■	□	■	

2.3 The concept of fault tolerance

Naor and Shamir proposed an extended visual secret sharing scheme in 1995 [11]. When they encrypt the secret image, each pixel on the secret image will be expand into m subpixels. Nakajima and Yamaguchi proposed an extended visual secret sharing scheme to show that the fault tolerance mechanism can be achieved in 2004 [10]. They encrypted the secret image into two expand shares. Each pixel in the secret image will produce a 7 x 7 subpixels which become a diamond-like pattern, one is small and the other is big, as shown in Figure 2. Even though there are slight deviation when stacking, primary color (black or white) still can be restore. Their design will allow some space for fault tolerance. However, due to the expansion, the size of the restored image will be 49 times of the original secret image. Hence, this paper will focus on halftone secret image, designing the VSS scheme using the idea of the method mentioned above, for towards shift tolerance without pixel expansion.

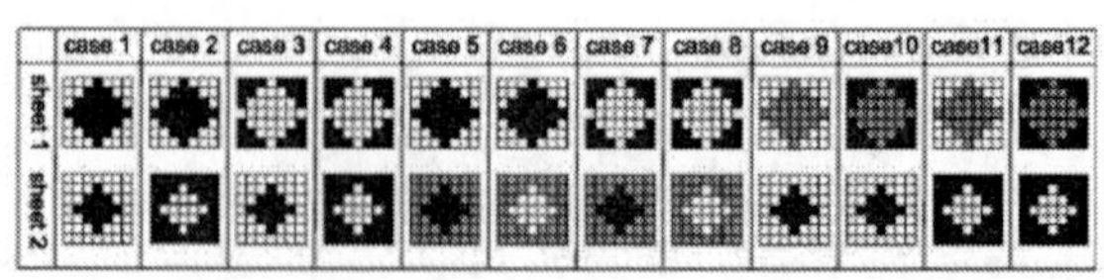

Figure 2. The patterns in Nakajima and Yamaguchi's scheme. This figure come from [10].

3 Main result

The main concept of our algorithm is as follows. First, taking $n \times n$ pixels as a unit, the image is divided into several units. Next, basis on the pattern we designed,

DOI: 10.1051/matecconf/20167101002

randomly choose one pattern for each unit, and generate the first share as combine those patterns. For the generation of second share, the amount of black and white pixel in each unit on original secret image need to be counted individually. Then identify which one is more than the other, selecting the suitable pattern according to the pattern of first share for the same unit. Run the steps sequentially and repeatedly, and second share will be generated.

Take the idea of Nakajima and Yamaguchi's scheme as a reference [10], we design the special patterns for the main encryption scheme and apply the KK algorithm [6]. With $n \times n$ pixels as a unit, we design the fault-tolerant VSS schemes n is 3, 4 or 5. After computing, analysis and screening, the final design of the patterns for encryption schemes as shown in Tables 3, 4 and 5, for n is 3, 4 and 5, respectively.

Table 3. The Designed Patterns for $n = 3$

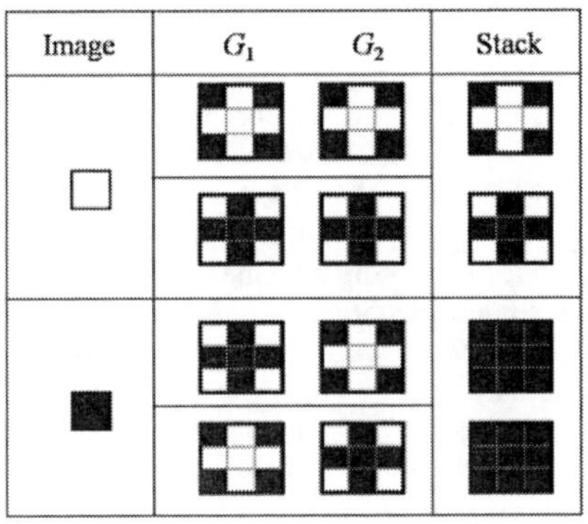

Table 4. The Designed Patterns for n =4

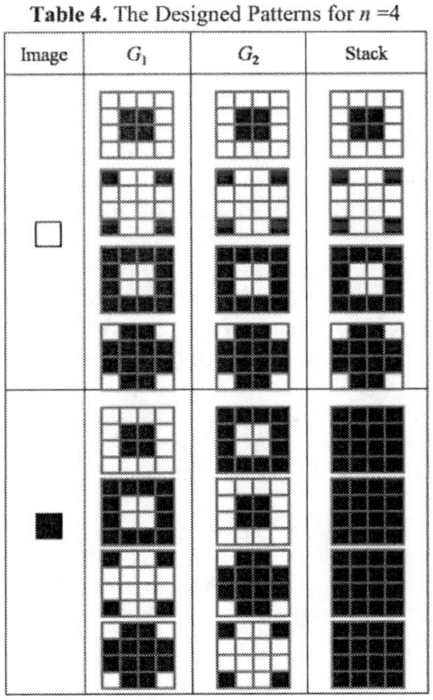

Table 5. The Designed Patterns for $n = 5$

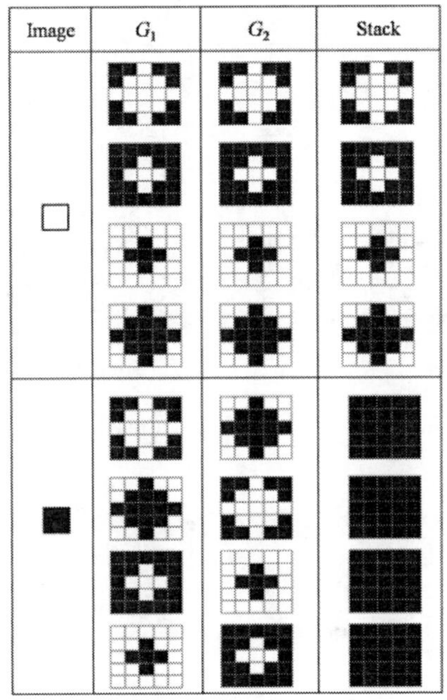

4 Experimental results

In this section, we use the scheme proposed in previous section to perform the simulation, the experimental results are shown as Figures 3-5 for n is 3, 4, or 5, respectively. The secret image we used in the experiment is 300 x 300 pixels halftone image. Because there is no pixel expansion, the size of two shares are also 300 x 300 pixel after encryption. Information on the restored image can be identified clearly after stacking the resulting two shares directly, or one of the share with 1 or 2 pixel shift. It shows that the effectiveness that we analysis in last section is correct as we expect.

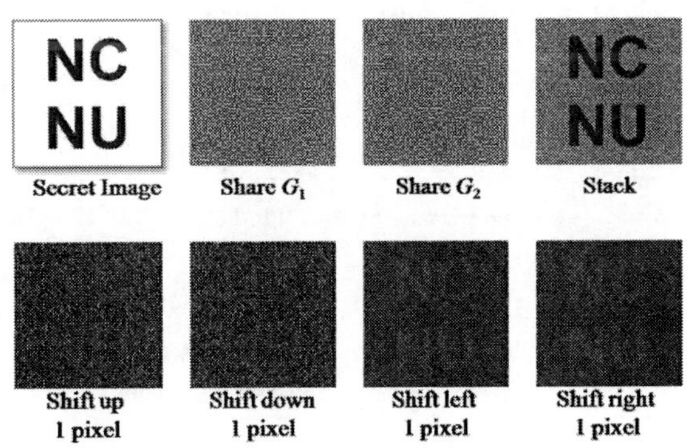

Figure 3. The experimental results for $n = 3$.

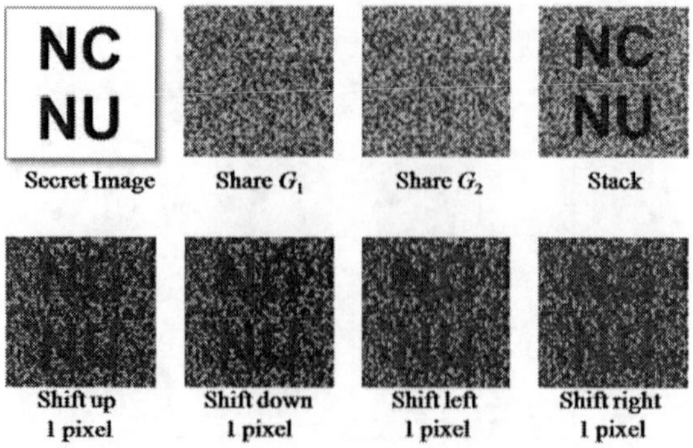

Figure 4. The experimental results for $n = 4$.

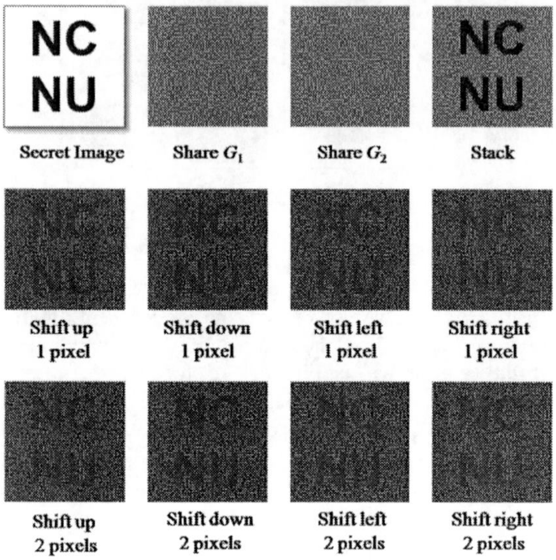

Figure 5. The experimental results for $n = 5$.

5 Conclusion

This paper presents that a visual secret sharing scheme can be designed with fault tolerance mechanism and without pixel expansion; the original information can be identified even if we stack the two resulting shares not perfectly. We expect further study for designing the encryption scheme with n is 6 or more. However, due to along the value of n is greater, the distortion of the graph will be much more. So the value of n has its upper limitation. Besides, improve the current algorithms for increasing the difference of the transmittance between black and white areas and reducing the distortion. So that the restored image could be more clearly. That is our goal as well.

Acknowledgments

The authors would like to thank the Ministry of Science and Technology of the Republic of China for financially supporting this research under Contract No. MOST 104-2918-I-260-004.

References

1. C. Blundo, A. Santis, Visual cryptography schemes with perfect reconstruction of black pixels. *Computer Graph* **22**(4): 449-455 (1998).
2. C. Blundo, A. Bonis, A. Santis, Improved schemes for visual cryptography. *Designs, Codes and Cryptograph* **24**: 255-278 (2001).
3. J. J.-Y. Chang, J. S.-T. Juan, Multi-VSS scheme by shifting random grids. In *Proc. World Academy of Science, Engineering and Technology*, **65**, Tokyo, Japan, 29-30 May 2012: 1277-1283 (2012).
4. L.-C. Chen, *Multi-Secret Images Sharing Schemes*, Master Thesis: National Chi Nan University (2014).
5. T.-H. Chen, K.-H. Tsao, K.-C. Wei, Multiple-image encryption by rotating random grids. In *Intelligent Systems Design and Applications, 2008. ISDA'08. Eighth International Conference on* **3**: 252-256 (2008): IEEE.
6. O. Kafri, E. Keren, Encryption of pictures and shapes by random grids. *Optics Letters* **12**(6): 377-379 (1987).
7. K. Kobara, H. Imai, Limiting the visible space visual secret sharing schemes and their application to human identification. In *Advances in Cryptology—ASIACRYPT'96* 185-195 (1996): Springer Berlin Heidelberg.
8. F. Liu, C. K. Wu, X. J. Lin, The alignment problem of visual cryptography schemes. *Designs, Codes and Cryptography* **50**(2): 215–227 (2009).
9. M. Nakajima, Y. Yamaguchi, Extended visual cryptography for natural images. *Journal of WSCG* **10**(2): 303-310 (2002).
10. M. Nakajima, Y. Yamaguchi, Enhancing registration tolerance of extended visual cryptography for natural images. *Journal of Electronic Imaging* **13**(3): 654-662 (2004).
11. M. Naor, A. Shamir, Visual cryptography. In *Advances in Cryptology—EUROCRYPT'94* 1-12 (1995): Springer Berlin Heidelberg.
12. S. J. Shyu, Image encryption by random grids. *Pattern Recognition* **40**: 1014- 1031(2007).
13. E. R.Verheul, H. C. Van Tilborg, Constructions and properties of k out of n visual secret sharing schemes. *Designs, Codes and Cryptography* **11**(2): 179-19(1997).
14. D. S. Wang, L. Dong, X. Li, Towards shift tolerant visual secret sharing schemes. *IEEE Transactions on Information Forensics and Security* **6**(2): 323-337 (2011).
15. C. N. Yang, A. G. Peng, T. S. Chen, MTVSS: Misalignment tolerant visual secret sharing on resolving alignment difficulty. *Signal Process* **89**: 1602-1624 (2009).

MATEC Web of Conferences **71**, 01003 (2016) DOI: 10.1051/matecconf/20167101003
ICCPE 2015

Optimal Improvement Ratios of Multi-Secret Sharing Schemes Can Be Achieved

Justie Su-Tzu Juan [1,a], Jennifer Hui-Chan Tsai [1] and Yi-Chun Wang [1]

[1]*Department of Computer Science & Information Engineering, National Chi Nan University, Nantou, Taiwan*

Abstract. Secret sharing schemes (SSS) deal with activities relative to the secure distribution of a secret among a group of participants who then securely reconstruct the secret by collecting the shares which are held by individuals in qualified sets. A multi-secret sharing scheme (MSSS) is an extension of this where multiple secrets are shared simultaneously and where the performance is estimated using both the maximum improvement and average improvement ratios. In 2003, Crescenzo calculated a lower bound of maximum and average improvement ratios for a MSSS and proposed some special MSSS which achieve the lower bound of the maximum improvement ratio. Wang and Juan deployed a more flexible MSSS to achieve the optimal maxi-mum improvement ratio 2006. This paper provides evidence such that the optimal maximum improvement ratio and the optimal average improvement ratio may both be achieved through even simpler and more flexible MSSS through even simpler and more flexible multi-secret sharing schemes for any arbitrary n and m.

1 Introduction

Secret sharing was invented by Shamir [7] and Blakely [1] independently. A secret sharing scheme (SSS, for short) includes two algorithms (D, R); D refers to the *distribution algorithm* and R refers to the *reconstruction algorithm*. Formally, given a group of participants $P = \{P_1, P_2, \ldots, P_n\}$, the distribution algorithm is executed by a *dealer* who has been given a secret. The dealer then calculates the *shares* S_i and distributes them to each participant P_i. The reconstruction algorithm is executed by authorized subsets of participants who combine their shares to reconstruct the secret. A subset A of P is called a *qualified subset* and a secret can be reconstructed if every participant in A uses their shares and executes the reconstruction algorithm R. $\Gamma \subseteq 2^P$ is an *access structure* which is the set of all qualified subsets.

Let a secret K be chosen from K with the uniform distribution. p_K denotes a probability distribution on K and $p_{S(A)}$ denotes a probability distribution on the shares $S(A)$ given to a subset $A \subseteq P$. A SSS is *perfect* if

$$H(K|A) = \begin{cases} 0 & \text{if } A \in \Gamma \\ H(K) & \text{if } A \notin \Gamma \end{cases}$$

where $H(K)$ and $H(K|A)$ denote the entropy of p_K and the conditional entropy of the joint probability distribution $p_{K\times S(A)}$, respectively. In other words, when a dealer distributes a shared S_i to every participant P_i, only qualified participants can reconstruct the secret K by using their shares. If we collect all the shares owned by nonqualified subsets, K cannot be reconstructed. Then we call this scheme is *perfect* which means that the nonqualified participants receive no information about the original secret.

Let $S(P_i)$ denote the set of possible shares that P_i might receive. $\rho_i = \log|K| \,/\, \log|S(P_i)|$ denotes the *information rate* for P_i. In general, the efficiency of a SSS (D, R) is measured by the information rate ρ defined as $\rho = \min\{\rho_i : 1 \le i \le n\}$ [9]. A SSS (D, R) is *ideal* if $\rho((D, R)) = 1$.

A multi-secret sharing scheme (MSSS, for short) deals with m secrets $s_0, s_1, \ldots, s_{m-1}$ at the same time. The most efficient algorithm of each single SSS can be used for distributing shares and reconstructing the secret. In this case, it is called a *basic MSSS* (b-D, b-R) in which the size of given shares will increase according to the secrets. Notably, the issue of reducing the size of shares given to every participant becomes an important one. Generally, an *Improvement Ratio* (*IR*) is used to estimate the performance of a MSSS. Crescenzo [3] announced a MSSS and studied this issue in 2003. After that, Wang and Juan [10] gave the instances of MSSS to show that the optimal maximum improvement ratio can be achieved for any $n > m$ in 2006.

This paper focuses on the optimal maximum improvement ratios and optimal average improvement ratios of MSSSs such that they can both be achieved in more flexible cases, most notably, for any arbitrary n and m. The next section will lay out Crescenzo's definitions and Wang and Juan's results. Section 3 proposed a MSSS, which extends the results of Crescenzo and Wang and Juan. The proposed scheme is proved to be secure and perfect in Section 4, conclusions are given.

[a] Corresponding author: jsjuan@ncnu.edu.tw

2 Preliminaries

Maximum improvement ratio and *average improvement ratio* are two criteria which are used to estimate the performance of a MSSS. In a MSSS (D, R), the dealer distributes m secrets $s_0, s_1, \ldots, s_{m-1}$ with associated access structures $A_0, A_1, \ldots, A_{m-1}$. The maximum improvement ratio (denoted by MaxIR) and the average improvement ratio (denoted by AvIR) are defined as follows Crescenzo [3]:

$$\text{MaxIR}(A_0, A_1, \ldots, A_{m-1}, (D, R)))$$
$$= \frac{\text{MaxShSize}(A_0, A_1, \ldots, A_{m-1}, (D, R))}{\text{MaxShSize}(A_0, A_1, \ldots, A_{m-1}, (\text{b-}D, \text{b-}R))};$$
$$\text{AvIR}(A_0, A_1, \ldots, A_{m-1}, (D, R)))$$
$$= \frac{\text{AvShSize}(A_0, A_1, \ldots, A_{m-1}, (D, R))}{\text{AvShSize}(A_0, A_1, \ldots, A_{m-1}, (\text{b-}D, \text{b-}R))}.$$

where MaxShSize(A_0, A_1, …, A_{m-1}, (D, R)) and AvShSize(A_0, A_1, …, A_{m-1}, (D, R)) denote, respectively, the maximum size of shares and the total size of shares obtained by the qualified participants when executing the (D, R) algorithm.

Next, the *optimal maximum improvement ratio* (denoted by OpMaxIR) and the *optimal average improvement ratio* (denoted by OpAvIR) are defined for the access structures $A_0, A_1, \ldots, A_{m-1}$ as follows:

$$\text{OpMaxIR}(A_0, A_1, \ldots, A_{m-1})$$
$$= \min_{(D,R)} \text{MaxIR}(A_0, A_1, \ldots, A_{m-1})$$
$$\text{OpAvIR}(A_0, A_1, \ldots, A_{m-1})$$
$$= \min_{(D,R)} \text{AvIR}(A_0, A_1, \ldots, A_{m-1})$$

The following theorems show the lower bounds and upper bounds of OpMaxIR and OpAvIR which are given by Crescenzo [3].

Theorem 1 [3] Let m, n be positive integers and let A_0, $A_1, \ldots, A_{m-1}$ be access structures over a set of size n. It holds that

$$\text{OpMaxIR}(A_0, A_1, \ldots, A_{m-1}) \geq 1 / m,$$
$$\text{OpAvIR}(A_0, A_1, \ldots, A_{m-1}) \geq 1 / m.$$

Theorem 2 [3] Let m be an integer. For any $\varepsilon > 0$, there exist access structures $A_0, A_1, \ldots, A_{m-1}$ such that $\text{OpAvIR}(A_0, A_1, \ldots, A_{m-1}) \geq 1 / m + \varepsilon$.

Theorem 3 [3] Let m be an integer. There exists an integer n and access structures $A_0, A_1, \ldots, A_{m-1}$ over a set of size n such that

$$\text{OpMaxIR}(A_0, A_1, \ldots, A_{m-1}) = 1 / m.$$

In order to prove Theorem 3, Crescenzo provides an instance of MSSS in which the max improvement ratio is equal to $1/m$. He uses one particular access structure and an associative algorithm which are overly complicated and is forced to use some previous results in the literature. In this instance, for any positive integer m, the number of participant $n = m + 6^{d-1}$, where $d = 2(1 + \lceil \log m \rceil) - 1$. For example, if $m = 5$, n will be 46661. Also, to understand this scheme, one must be familiar with the results of Blundo et al. [2], Dijk [4], and Stinson [8]. Furthermore, the other access structures in [2, 5, 6, 8] are not intuitive enough either. Wang and Juan [10] solved this problem and proposed two simple schemes which can directly achieve the lower bound of the OpMaxIR. Additionally, their result shows that the OpMaxIR can be achieved for all cases such that m only needs to be greater than n.

Theorem 4 [10] For any two positive integers $n > m$, there exist access structures $A_0, A_1, \ldots, A_{m-1}$ over a set of size n such that

$$\text{OpAvIR}(A_0, A_1, \ldots, A_{m-1}) = 1 / m.$$

The OpMaxIR is shown to be achievable in Theorem 3 and Theorem **4**. Additionally, the OpAvIR is shown to be nearly achievable in Theorem 2. However, those existing schemes only obtained in some special cases. That is, for any two arbitrary positive integers n (the number of secrets) and m(the number of participants), it is still unknown whether the OpMaxIR and OpAvIR are achievable. Specifically, this paper solves this dilemma such that for any two arbitrary positive integers n and m, and builds a set of access structures with a pair of associative algorithms which obtains both the lower bounds of OpMaxIR and OpAvgIR at the same time.

3 The Proposed Scheme

Let m denote the number of secrets and n denote the number of participants. For convenience, the operations at the footnote are calculated in Z_{n-1} space (in modulo $n - 1$) for the whole of this work. The m secrets $s_0, s_1, \ldots, s_{m-1}$ are selected from K. Note that $\oplus$ is the "exclusive-or" operation and this paper defines the "general exclusive-or" operation $\bigoplus$ as follows:

Definition 1 *Let I be a nonempty set and U be a universe. For each $i \in I$ let $A_i \subseteq U$. Then I is called an index set (or set of indices), and each $i \in I$ is called an index.*
Under these conditions, say $I = \{i_1, i_2, \ldots, i_k\}$

$$\bigoplus_{i \in I} A_i = A_{i_1} \oplus A_{i_2} \oplus \ldots \oplus A_{i_k}.$$

A particular access structure is defined as follows. Let the participants $P = \{X_0, X_1, \ldots, X_{n-2}, Y\}$. For $i = 0, 1, \ldots, m - 1$, there are two cases:

1) when $n > m$, access structure A_i of secret s_i is defined as: $A_i = \{A_{i1}, A_{i2}\}$, where $A_{i1} = \{X_i, Y\}$, $A_{i2} = P - A_{i1}$;

2) when $m \geq n$, access structure A_i of secret s_i is defined as: $A_i = \{A_{i1}, A_{i2}\}$, where $A_{i1} = \{X_i, Y\}$, $A_{i2} = P - A_{i1}$ for $0 \leq i \leq n - 2$; $A_i = \{A_{i1}, A_{i2}\}$, and $A_{i1} = \{Y\}$, $A_{i2} = P - A_{i1}$, for $n - 1 \leq i \leq m - 1$.

When sharing secret s_i for any $0 \leq i \leq m - 1$, two ideal schemes are proposed.At first, a basic MSSS is given.

An ideal SSS (b-D, b-R) for $A_0, A_1, \ldots, A_{m-1}$:
Distribution algorithm (b-D):
When $n > m$:
(1)Select random numbers $a_{i,1}, a_{i,2}, \ldots, a_{i,n-3}$ from K. Let $S_{i,i+j} = a_{i,j}$ for $1 \leq j \leq n - 3$ and $0 \leq i \leq m - 1$.
(2)Select random number $a_{i,n-2}$ and let $S_{i,i} = a_{i,n-2}$ for $0 \leq i \leq m - 1$.

DOI: 10.1051/matecconf/20167101003

(3)Let $S_{i,i-1} = (\oplus_{j=1}^{n-3} a_{i,j}) \oplus s_i$ for $0 \le i \le m-1$.

(4)Let $S_{i,y} = s_i \oplus a_{i,n-2}$ for $0 \le i \le m-1$.

(5)For $0 \le j \le n-2$, send $S_j = \{S_{0,j}, S_{1,j}, \ldots, S_{m-1,j}\}$ to X_j and send $S_y = \{S_{0,y}, S_{1,y}, \ldots, S_{m-1,y}\}$ to Y.

When $n \le m$:

For $0 \le i \le n-2$, the share S_i of A_i is given as same as the case of $n > m$; and for $n-1 \le i \le m-1$, the share S_i is given as follows:

(1)Select random numbers $a_{i,1}, a_{i,2}, \ldots, a_{i,n-1}$ from K. Let $S_{i,i+j} = a_{i,j}$ for $1 \le j \le n-1$.

(2)Let $S_{i,0} = (\oplus_{j=1}^{n-1} a_{i,j}) \oplus s_i$ and $S_{i,y} = s_i$.

(3)For $n-1 \le i \le m-1$, $0 \le j \le n-2$, send $S_j = \{S_{0,j}, S_{1,j}, \ldots, S_{m-1,j}\}$ to X_j and send $S_y = \{S_{0,y}, S_{1,y}, \ldots, S_{m-1,y}\}$ to Y.

Reconstruction algorithm (b-*R*):

(1)If the shares in A_{i1} are collected, for $0 \le i \le m-1$, s_i can be reconstructed by using $S_{i,i} \oplus S_y$.

(2)If the shares in A_{i2} are collected, for $0 \le i \le m-1$, s_i can be reconstructed by using $\oplus$ operations to $S_{i,k}$ for all X_k in A_{i2}.

According to the above algorithms, the results are as follows:

AvShSize($A_0, \ldots, A_{m-1}$, (b-*D*, b-*R*)) = mn,
MaxShSize($A_0, \ldots, A_{m-1}$,(b-*D*, b-*R*)) = m;

Actually, for the same access structures $A_0, \ldots, A_{m-1}$ of multi-secret $s_0, \ldots, s_{m-1}$, there exists an efficient MSSS (G-*D*, G-*R*) which is given:

An efficient MSSS (G-*D*, G-*R*) for $A_0, A_1, \ldots, A_{m-1}$:
Distribution algorithm (b-*D*):

When $n > m$:

(1)Select random numbers $Q_1, Q_2, \ldots, Q_k$ from K, where $k = n - m$ if n is even, otherwise $k = n - m - 1$. Let $s_{m+l-1} = Q_l$ for $1 \le l \le k$ and $m' = m + k$.

(2)For $0 \le j \le n-2$, calculate $S_j = \oplus_{0 \le i \le m'-1, i \ne j} s_i$ and send S_j to X_j.

(3) Calculate $S_y = \oplus_{i=0}^{m'-1} s_i$ and send S_y to Y.

(4) If n is even, add one unreal person U_1 and publish the value of $\oplus_{i=0}^{n-2} s_i$ to U_1 as the share.

When $n \le m$:

(1)Add unreal persons $U_1, U_2, \ldots, U_k$ into P where $k = m + 1 - n$ if m is even, otherwise $k = m + 2 - n$. Let $n' = n + k$ and $X_{n-2+l} = U_l$ for $1 \le l \le k$.

(2)If m is odd, select a random number Q_1 and let $s_m = s_{n'-2} = Q_1$.

(3)For $0 \le j \le n'-2$, calculate $S_j = \oplus_{0 \le i \le n'-2, i \ne j} s_i$ and send S_j to X_j.

(4)Calculate $S_y = \oplus_{i=0}^{n'-2} s_i$ and send S_y to Y.

Note that the shares sent to unreal persons are actually published.

Reconstruction algorithm (G-*R*):

(1)If the shares in A_{i1} are collected, for $0 \le i \le m-1$, s_i can be reconstructed by $S_i \oplus S_y \oplus$ published shares if necessary.

(2)If the shares in A_{i2} are collected, for $0 \le i \le m-1$, s_i can be reconstructed by $\oplus_{0 \le i \le n-2, j \ne i} S_i \oplus$ published shares if necessary.

In algorithm G-*R* (2), note that in the equation when reconstructing s_i, if $n > m$, then let n' equal to n if n is odd; n' will be equal to $n + 1$ otherwise. The secret s_i appears exactly $n' - 2$ times and for any $j \ne i$, s_j appears exactly $n' - 3$ times. Since n' is always an odd number, after exclusive-or calculations, it results in s_i.

According to the above algorithms, the results are as follows:

AvShSize($A_0, \ldots, A_{m-1}$, (G-*D*, G-*R*)) = n,
MaxShSize($A_0, \ldots, A_{m-1}$,(G-*D*, G-*R*)) = 1.

The published shares are over K^{α}, where α shows as Table 1. Table 2 shows two examples. In (a), $m = 4$ and $n = 5$; in (b), $m = 3$ and $n = 4$.

Table 1. The The number of published shares α

	$n > m$	$n \le m$
max{n,m} is even	1	$m - n + 1$
max{n,m} is odd	0	$m - n + 2$

Table 2. Shares in two examples.

(a) for $m = 4$ and $n = 5$.

	(b-*D*, b-*R*)	(G-*D*, G-*R*)
X_0	$a_{0,3}, s_1 \oplus a_{1,1} \oplus a_{1,2}, a_{2,2}, a_{3,1}$	$s_1 \oplus s_2 \oplus s_3$
X_1	$a_{0,1}, a_{1,3}, s_2 \oplus a_{2,1} \oplus a_{1,2}, a_{3,2}$	$s_0 \oplus s_2 \oplus s_3$
X_2	$a_{0,2}, a_{1,1}, a_{2,3}, s_3 \oplus a_{3,1} \oplus a_{3,2}$	$s_0 \oplus s_1 \oplus s_3$
X_3	$s_0 \oplus a_{0,1} \oplus a_{0,2}, a_{1,2}, a_{2,1}, a_{3,3}$	$s_0 \oplus s_1 \oplus s_2$
Y	$s_0 \oplus a_{0,3}, s_1 \oplus a_{1,3}, s_2 \oplus a_{2,3}, s_3 \oplus a_{3,3}$	$s_0 \oplus s_1 \oplus s_2 \oplus s_3$

(b) for $m = 3$ and $n = 4$.

	(b-*D*, b-*R*)	(G-*D*, G-*R*)
X_0	$a_{0,2}, s_1 \oplus a_{1,1}, a_{2,1}$	$s_1 \oplus s_2 \oplus Q_1$
X_1	$a_{0,1}, a_{1,2}, s_2 \oplus a_{2,1}$	$s_0 \oplus s_2 \oplus Q_1$
X_2	$s_0 \oplus a_{0,1}, a_{1,1}, a_{2,2}$	$s_0 \oplus s_1 \oplus Q_1$
U_1	No unreal participant	$s_0 \oplus s_1 \oplus s_2$
Y	$s_0 \oplus a_{0,2}, s_1 \oplus a_{1,2}, s_2 \oplus a_{2,2}$	$s_0 \oplus s_1 \oplus s_2 \oplus Q_1$

4 Security Analysis

Whether the efficient MSSS can be secure and perfect must be proven. This study proves that, in addition to the fact that the subset belongs to the specific access structure, when collecting any other shares there is no way to reconstruct the original secret or even glean any information about the secret.

Theorem 5 *After executing the algorithm G-D, collect all the shares belong to the participants in any subset A of P that does not include A_{i1} and A_{i2}, there is no way to reconstruct the secret s_i*, and $H(s_i|A) = H(s_i)$.

Proof. According to the properties of $\oplus$, it is not satisfied with a combination rule. Consequently, there is no way in which four fundamental operations of arithmetic can cancel the information excluding s_i but also the calculation is made more complicated. Notably, the inverse operation of $\oplus$ is itself. Therefore, by using $\oplus$,

whether one can obtain any information about s_i when one collects all the shares in any subset A of P that does not include any qualified subset of A_i. In this proof, we take into account the added unreal persons $U_1, U_2, \ldots, U_k$ and random numbers $Q_1, Q_2, \ldots, Q_k$ as real persons $X_{n-2+l} = P_l$ and real secrets $s_{m+l} = Q_l$ for $1 \le l \le k$. Note that $P = \{X_0, X_1, \ldots, X_{n-1}, Y\}$, and let n' and m' be the new number of participants and secrets, respectively. Now, $n' = m' + 1$ if m' is even and $n' = m' + 2$ otherwise. For any subset $A \subseteq P$, A does not include all sets in A_{i1} or A_{i2}. According to the associative and commutative properties of $\oplus$, a series of exclusive-or operations can be taken when executing multiple exclusive-or operations. Therefore, we only have to prove that there is no way to reconstruct s_i when executing exclusive-or operations for any subset of A. In other words, for any subset A of P, let

$$T = \begin{cases} \oplus_{\forall X_j \in A} S_j \oplus S_y & \text{if } Y \in S \\ \oplus_{\forall X_j \in A} S_j & \text{if } Y \notin S \end{cases}$$

If A_{i1} and $A_{i2} \not\subset A$, we must show that $T \neq s_i$. If $|A| \geq n' - 1$, A is definitely includes A_{i1} or A_{i2}. On the other hand, if $|A| = 1$, besides $m \geq n$ and $n - 1 \leq i \leq m - 1$, $Y \in A_{i1}$, implies Y can reconstruct the secret alone, in another subset A which does not include A_{i1} or A_{i2}; because there is only one participant in A, it is easy to verify that $T \neq s_i$. Therefore, only $2 \leq |A| \leq n' - 2$ needs to be considered. Note that $A_{i1} = \{X_i, Y\} \not\subset A$, so we have the following three cases.

Case 1. When $|A|$ is even and $X_i \notin A$ or when $|A|$ is odd and $X_i \in A$: Since each share of A contains s_i except S_i, there are even numbers of s_i in A in this situation. Therefore, the s_i will be cancelled after executing exclusive-or operations. Hence, it results $T \neq s_i$.

Case 2. When $|A|$ is even and $X_i \in A$ (hence $Y \notin A$): For any $X_j \in A$, each S_k contains s_j for any $X_k \in A$ except S_j. Thus, there is an odd number of s_j. Since $|A| \geq 2$, then $T = \oplus_{\forall X_j \in A} S_j = s_i \oplus s_{j1} \oplus s_{j2} \oplus \ldots \oplus s_{|A|-1} \neq s_i$, for $X_i, X_{j1}, X_{j1}, \ldots, X_{|A|-1} = A$

Case 3. When $|A|$ is odd, $X_i \notin A$: Since $A_{i2} = P - A_{i2} \not\subset A$, there exists a number $k \in \{0, 1, \ldots, n' - 1\}$ such that $k \neq i$ and $X_k \neq A$. For any $X_j \in A$, each S_j contains s_k and there is an odd number of s_k when calculating T. Therefore, it results $T \neq s_i$.

According to these cases, if we take unreal persons and secrets to be real ones, we still have proven that under $\oplus$, excluding the specific access structures A_{i1} and A_{i2}, there is no possibility of reconstructing s_i and no information has been revealed regarding the secret s_i. Hence, it results $H(s_i|A) = H(s_i)$ for any secret s_i.

Therefore, we have proven that the proposed MSSS is perfect and secure.

The values of MaxIR and AvIR of (G-D, G-R) for the specific access structures $A_0, A_1, \ldots, A_{m-1}$ are given, and according to Theorem 1, Theorem 6 is concluded.

Theorem 6 For any two positive integers m and n, there exist access structures $A_0, A_1, \ldots, A_{m-1}$ over a set of size n such that

$$\text{OpMaxIR}(A_0, A_1, \ldots, A_{m-1}) = 1/m,$$
$$\text{OpAvIR}(A_0, A_1, \ldots, A_{m-1}) = 1/m.$$

5 Conclusions

This paper has successfully proven that there exists a set of access structures and an associated MSSS which can achieve:

$$\text{OpAvIR}(A_0, \ldots, A_{m-1}) = 1/m,$$
$$\text{OpMaxIR}(A_0, \ldots, A_{m-1}) = 1/m.$$

Although this result may be largely of theoretical interest and the access structure will likely seldom be used in practical, it encourages further inquiry for any one that may attempt to design a MSSS. Specifically, when designing an efficient MSSS, the maximum and average improvement ratios should approach $1/m$. Since the results in [3]. are a special case (m is decided according to n), this work has extended (for both cases) and improved (for the case of OpAvIR) the results of [3] and [10]. Therefore, this scheme is considered flexible, and thereby generally applicable, because the number of participants and secrets are independent. In conclusion, given a set of access structures for any number of participants and secrets, finding an associated MSSS which can achieve two optimal improvement ratios at the same time is possible.

References

1. G.R. Blakley, Safeguarding cryptographic keys. *Proceedings of the National Computer Conference*, 48 of American Federation of Information Processing Societies, pp. 313-317 (1979)
2. C. Blundo, A.D. Santis, R.D. Simone, U. Vaccaro, Tight bounds on the information rate of secret sharing schemes. *Designs, Codes and Cryptography* **11**, 2, pp. 107-122 (1997)
3. G.D. Crescenzo, Sharing one secret vs. sharing many secrets. *Theoretical Computer Science*, **295**, pp. 123-140 (2003)
4. M.V. Dijk, On the information rate of perfect secret sharing schemes. *Designs, Codes and Cryptography* **6**, pp. 143-169 (1995)
5. W.A. Jackson, K.M. Martin, C.M. O'Keefe, A construction for multi-secret threshold schemes. *Designs, Codes and Cryptography* **9**, pp. 287-303(1996)
6. W.A. Jackson, K.M. Martin, C.M. O'Keefe, Ideal secret sharing schemes with multiple secrets. *Journal of Cryptology* **9**, pp. 233-250 (1996)
7. A. Shamir, How to share a secret. *Communications of the ACM* **22**, 11, pp. 612-613 (1979)
8. D.R. Stinson, An explication of secret sharing schemes. *Designs, Codes and Cryptography* **2**, pp. 357-390 (1992)
9. D.R. Stinson, *Cryptography: Theory and practice.* CRC Press (1995)
10. Y.-C. Wang, J.S.-T. Juan, Simple multi-secret sharing schemes to achieve the optimal maximum improvement ratio. *Proc. of the IEEE International Conference on Sensor Networks, Ubiquitous, and Trustworthy Computing*, pp. 58-63 (2006)

MATEC Web of Conferences **71**, 01004 (2016) DOI: 10.1051/matecconf/20167101004
ICCPE 2015

Comprehensive Weighted Clique Degree Ranking Algorithms and Evolutionary Model of Complex Network

Jie Xu[1,*], Zhen Liu[1], and Jun Xu[1]

[1] *Key Lab of Optical Fiber Sensing and Communications (Ministry of Education), School of Communication and Information Engineering, University of Electronic Science and Technology of China, No.2006, Xiyuan Ave, Chengdu 622731,China*

Abstract. This paper analyses the degree ranking (DR) algorithm, and proposes a new comprehensive weighted clique degree ranking (CWCDR) algorithms for ranking importance of nodes in complex network. Simulation results show that CWCDR algorithms not only can overcome the limitation of degree ranking algorithm, but also can find important nodes in complex networks more precisely and effectively. To the shortage of small-world model and BA model, this paper proposes an evolutionary model of complex network based on CWCDR algorithms, named CWCDR model. Simulation results show that the CWCDR model accords with power-law distribution. And compare with the BA model, this model has better average shortest path length, and clustering coefficient. Therefore, the CWCDR model is more consistent with the real network.

1 Introduction

In recent years, research of complex networks[1-2] become the hot topic in network research. More and more fields involve the research of complex network, such as biology, economics, sociology, etc[3]. Research of important nodes ranking[4-5] and evolutionary model of complex network have become more and more important in network research. Important nodes ranking is always based on degree, betweenness, closeness, etc. These algorithms have different preference characteristics, so different algorithms may reach different results of importance ranking of the nodes[6]. These typical nodes ranking algorithms have their own defect. ZHOU proposes the concept of clique degree in his paper[7]. This concept considers the close degree between neighbor nodes, and provides a new way for nodes ranking.

Clique degree can be a characterization which can represent the close degree between the nodes, and the clique equals the concept complete subgraph in graph theory. That means *m*-order equal a fully connected network with *m* nodes and m(m-1)/2 edges. Define the *m*-clique degree of a node *i* as the number of defferent *m*-cliques containing *i*, denoted by k_i^m. Obviously, a 2-clique is an edge and k_i^2 equals the degree k_i. So clique degree can be considered as an extension of degree. We calculated the clique degree of node-4 from 2-order to 5-order for representative networks, $k_4^2=7, k_4^3=5, k_4^4=1$, $k_4^5=0$, as shown in Figure 1.

Clique degree partly reflects the node degree, and it considers the node location in the network. So we propose a new important node ranking algorithms in complex networks base on clique degree concept, named comprehensive weighted clique degree ranking (CWCDR) algorithm.

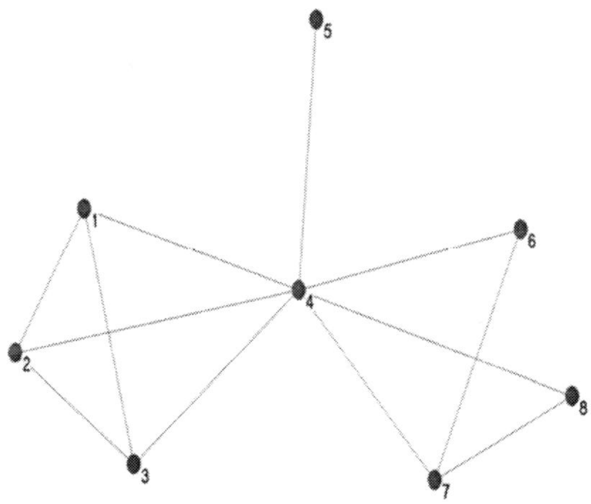

Figure 1. Clique degree of network

2 Comprehensive weighted clique degree algorithm

Comprehensive weighted clique degree algorithm introduces four parameters: $\alpha, \beta, \gamma, \lambda$, and four characteristic quantity: 2-order to 5-order clique degree. ZHOU indicates that the higher order in clique degree no longer has the strict power-law distribution in his paper[7]. According to the theory of six degrees of separation, people can find anyone through six sides in human social

* Corresponding author: xuj@uestc.edu.cn

networks. So we do not consider above 6-order clique degrees. Thus, the equation of algorithm is as follows:

$$N(i) = \alpha N^2(i) + \beta N^3(i) + \gamma N^4(i) + \lambda N^5(i) \quad (1)$$

Where $N^j(\mathrm{i})$ represents j-order clique degree, $N(\mathrm{i})$ represents multi-order comprehensive weighted degree, which is the measurement of nodes ranking. Then we talk about how to get value of $\alpha, \beta, \gamma, \lambda$.

$$N(i) = \begin{cases} N^2(i)(\alpha = 1, \beta = 0, \gamma = 0, \lambda = 0) \\ N^3(i)(\alpha = 0, \beta = 1, \gamma = 0, \lambda = 0) \\ N^4(i)(\alpha = 0, \beta = 0, \gamma = 1, \lambda = 0) \\ N^5(i)(\alpha = 0, \beta = 0, \gamma = 0, \lambda = 1) \\ \alpha N^2(i) + \beta N^3(i) + \gamma N^4(i) + \lambda N^5(i), others \end{cases} \quad (2)$$

Where $\alpha + \beta + \gamma + \lambda = 1$, assuming m as the 2-order clique degree, n as the 3-order clique degree, k as the 4-order clique degree, and t as the 5-order clique degree. By normalization of m, n, k, t, we get $p(m), p(n), p(k), p(t)$. Their reciprocals are as follows:

$$\frac{1}{p(m)}, \frac{1}{p(n)}, \frac{1}{p(k)}, \frac{1}{p(t)} \quad (3)$$

Comprehensive weighted clique degree considers the closeness level among nodes in the network. It measures the effect of each order clique degree in important nodes ranking by normalization. The higher order clique degree in the network is more important. So we get the results of $\alpha, \beta, \gamma, \lambda$ as follows:

$$\begin{cases} \alpha = \dfrac{\frac{1}{p(m)}}{\frac{1}{p(m)} + \frac{1}{p(n)} + \frac{1}{p(k)} + \frac{1}{p(t)}} \\ \beta = \dfrac{\frac{1}{p(n)}}{\frac{1}{p(m)} + \frac{1}{p(n)} + \frac{1}{p(k)} + \frac{1}{p(t)}} \\ \gamma = \dfrac{\frac{1}{p(k)}}{\frac{1}{p(m)} + \frac{1}{p(n)} + \frac{1}{p(k)} + \frac{1}{p(t)}} \\ \lambda = \dfrac{\frac{1}{p(t)}}{\frac{1}{p(m)} + \frac{1}{p(n)} + \frac{1}{p(k)} + \frac{1}{p(t)}} \end{cases} \quad (4)$$

3 Analysis of algorithm

Karate club network is the most popular real network data model in complex network research. This network is composed of 34 people. Then this network is abstracted to the graph theory model. This model contains 34 nodes, and every edge between two nodes denotes the two members have strong social relationships in real world. This network model is as follows in Figure 2.

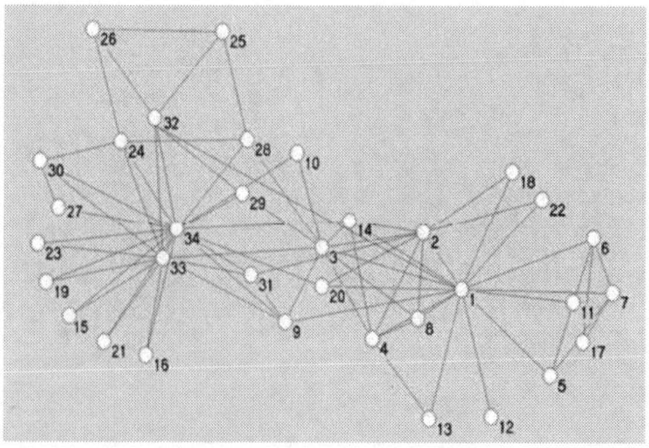

Figure 2. Karate club network

We know that the importance of the nodes in the margin is relatively low through Figure 2. In order to verify that our algorithm is better than degree ranking (DR) algorithm, closeness degree ranking (CDR) algorithm, betweenness ranking (BR) algorithm and PageRank ranking (PR) algorithm. We get the ranking results through the comprehensive weighted clique degree ranking (CWCDR) algorithm, and then we compare the results with these typical algorithms in Table 1.

Table 1. The comparison between the algorithms

ID	DR	ID	BR	ID	CDR	ID	PR	ID	CW CDR
34	17	1	0.8238	1	0.0172	34	0.1009	1	3.67
1	16	34	0.5724	3	0.0169	1	0.0970	3	3.37
33	12	33	0.2734	34	0.0167	33	0.0717	2	3.33
3	10	3	0.2704	32	0.0164	3	0.0571	4	3.11
2	9	32	0.2603	33	0.0156	2	0.0529	8	2.01
32	6	9	0.1053	14	0.0156	32	0.0372	34	1.87
4	6	2	0.1015	9	0.0156	4	0.0359	33	1.49
24	5	14	0.0863	20	0.0152	24	0.0315	14	0.44
9	5	20	0.0611	2	0.0147	9	0.0298	30	0.44
14	5	6	0.0564	4	0.0141	14	0.0295	9	0.38

In the Table 1, we can see in BR algorithm ranking, Node-6 is the tenth in the top 10 ranking, and Node-4 cannot rank in top 10. It's mean that Node-6 is more important than Node-4 in this algorithm. But it is easy to see in Figure 2 that Node-4 is in the central position of the network, and it is the bridge for Node-1, Node-2, Node-3, and Node-14, so we think in fact Node-4 is more important than Node-6. Thus, the CWCDR presents this result better as shown in Table 1. On the other hand, CDR algorithm cannot distinguish the importance of the nodes in the central position of the network, such as Node-32 and Node-8. The CDR value of Node-32 is much bigger than Node-8 (Node-8 cannot rank in top 10 in Table 2), but we find that the network cannot operate without Node-8 in reality and can operate without Node-32. So we think Node-8 is more important than Node-32.

In Table 1, PR algorithm obtains Node-32 is more important than Node-4. However, Node-4 is the core node of right sub-network, and it plays the role bridge connecting two networks. Node-4 also has close relationship with Node-1, Node-2, Node-14, which are at very important positions in the network. Although Node-32 has the function bridge connecting, it is far from core positions of two sub-networks. So we obtain Node-4 is more important than Node-32, and CWCDR presents this result well.

4 The evolutionary network model based on CWCDR algorithm

4.1 The evolutionary network model

Real networks have scale-free network characteristics, and BA is a typical model of scale-free networks. So we choose growth properties and preferential connection properties of BA evolutionary network model to evolve the model based on CWCDR algorithms[8-10], the evolutionary algorithms is as follows:

i) Original Network: the original network has n nodes, and we calculate the CWCDR value through CWCDR algorithm.

ii) Growth: a new node will join in the network every once in a while, we choose m nodes of the network to connect the new node, where $m < n$.

iii) Preferential connection: the new node i choose m nodes which the CWCDR value is highest in the network to connect. The connection probability p_i is:

$$p_i = \frac{N_i}{\sum_j N_j} \tag{5}$$

iv) Iteration and update: the new node changes the topological connection of the network, so we update the CWCDR value of the nodes.

v) Back: if it has new node to join in the network, back to ii); if not, the algorithm is over. Then the network has $N = n + t$ nodes and $m * t$ edges, where t is the number of new nodes.

4.2 Simulation and Analysis

The preferential connection mechanism of BA network evolution model is based on the DR algorithm. But CWCDR algorithm reconstruct the importance ranking of the nodes, and calculate the connection probability based on these ranking values. By the simulation in Matlab, we compare the evolutionary network model based on CWCDR algorithm with the BA model, and analyze degree distribution, average shortest path, and clustering coefficient.The results are in the Figure 3.

We assume the network has 2000 nodes after a period of time, and the first new node connects all the nodes in the network when it joins the network. The original networks have m_0 nodes, and m_0 are 2, 4, 6, and 8 respectively. When the new nodes join in the network, we choose $m = m_0$.

In the Figure 3(a), we know that the evolutionary network model based on CWCDR algorithms obeys power-law distribution. When the number of initial nodes is different, the value of power-law distribution function is the same. We can see that the small degree values nodes are in the majority. Degree value distribution is scattered, and degree distribution in the network is scale-free distribution.

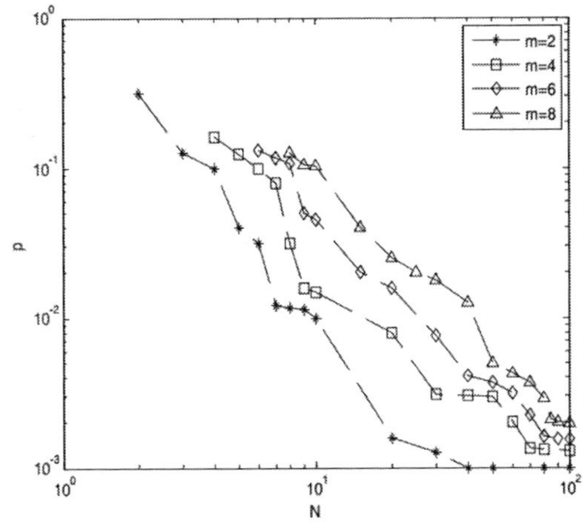

(a) Degree distribution

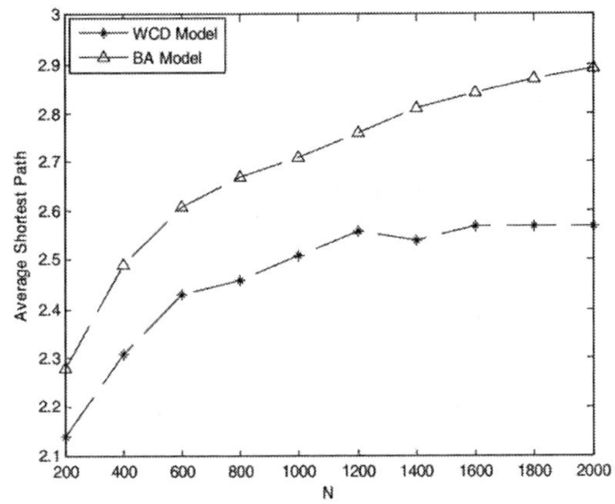

(b) Average shortest path

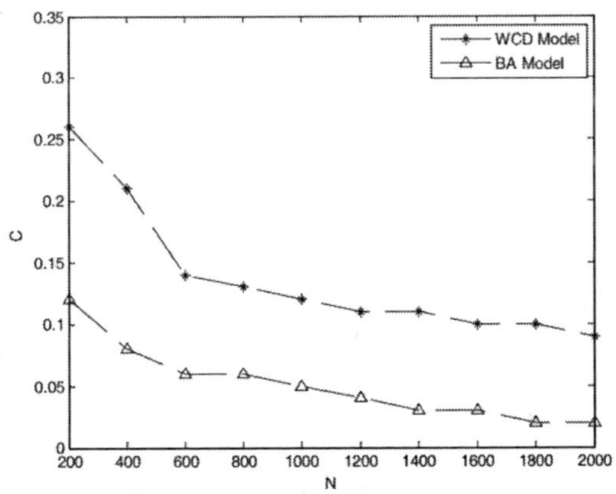

(c) Clustering coefficient

Figure 3. The comparison between our model and BA model

In the Figure 3(b), the average shortest path of our model is smaller than BA model, and it grows slowly with the growth of the network size. When the network has more than 2000 nodes, the average shortest path of the mode will remain unchanged. So the model is consistent with the small world network features.

In the Figure 3(c), the clustering coefficient of the model is bigger than BA model. When the network has more than N_t nodes, the clustering coefficient of the two models will remain unchanged.

The network in reality always is power-law distribution, and it has big clustering coefficient and small average shortest path. We know BA model consistent with the real network features and the model based on CWCDR algorithms is better than BA model in degree distribution, average shortest path, and clustering coefficient.

5 Conclusions

This paper proposes the comprehensive weighted clique degree algorithm based on the clique degree. The algorithm considers not only degree, but also the closeness of nodes with their neighbor nodes. Analyzed some famous real network models, we prove it more accurate than degree ranking algorithm, closeness degree ranking algorithm, betweenness ranking algorithm and PageRank ranking algorithm. Then according to the growth and preferential connection steps of evolutionary model, we propose a new evolutionary network model based on comprehensive weighted clique degree algorithm. Calculate the degree distribution, clustering coefficient and average shortest path of the model, and we compare the result with BA model. The result is better than BA model, which proves it better than BA model to consistent with the real network features.

Acknowledgments

This work was supported by NSFC Foundation (61201128), Sichuan Province Scientific and Technological Support Project (2014GZ0017, 2016GZ0093), Program for Changjiang Scholars and Innovative Research Team (PCSIRT) in University and the 111 Project B14039, the Fundamental Research Funds for the Central Universities (ZYGX2015J009).

References

1. S.H. Strogatz, Exploring complex networks. Nature, **410**, 268-276 (2001)
2. R. Albert, A.L. Barabasi, Statistical mechanic of complex networks. Rev.Mod.Phys, **74**, 1-54 (2002)
3. G.X. Wang, Y. Shen, E.J. Luan, A measure of centrality based on modularity matrix, Progress in Nature Science, **18**, 1043-1047 (2008)
4. J. Wu, Y.J. Tan, Finding the Most Vital Node by Node Contraction in Communication Networks. International Conference on Communications, Circuits and Systems, **2** (2005)
5. D.C. Huang, H.C. Qi, PageRank Algorithm Research. Computer Engineering, **32**, 145-146 (2006)
6. X.P. Zhang, Y.S. Li, Evaluation Method of Importance for Nodes in Complex Networks Based on Importance Contribution. Comlplex Systems And Complexity Science, **11**, 25-26 (2014)
7. W.K. Xiao, T. Zhou, Empirical study on clique-degree distribution of networks. PHYSICAL REVIEW E, **76**, 1-4 (2007)
8. J.R. Li, A Node Centrality Evaluation Model for Weighted Social Networks. Journal of University of Electronic Science and Technology of China, **43** (2014)
9. X. Li, G.R. Chen, A local-world evolving network model. Physical A: Theoretical and Statistical Physics, **328**, 274-286 (2003)
10. X.L. Wang, P2P network node importance measurement method based on multi-attribute. Journal of Computer Applications, **34**, 7-10 (2014)

The Design of a Modified PSO Guidance Law Using Predictor and LOS Rate Evaluation

Yung-Lung Lee[1], Kuei-Yi Chen[2,a], Yi-Wei Chen[3], Yen-Bin Chen[1] and Sheng-Ju Liao[2]

[1]*Department of Power Vehicle and Systems Engineering, C.C.I.T., National Defense University, Taoyuan County 33551, Taiwan, ROC.*
[2]*School of Defense Science, C.C.I.T., National Defense University, Taoyuan County 33551, Taiwan, ROC.*
[3]*Department of Aircraft Engineering, Air Force Institute of Technology, Kaohsiung 82047, Taiwan, ROC.*

Abstract. This paper proposes a modified particle swarm optimization guidance (MPSOG) for the pursuit-evasion optimization problem. If the line-of-sight (LOS) rate equals zero, the curvature of the missile's trajectory would be smaller and the probability of intercept greater. Thus, we propose using a MPSOG to improve the guidance performance. The MPSOG uses a Kalman filter to predict a target's dynamic. The lateral acceleration commands of the y- and z-axis are optimized by the particle swarm optimization (PSO) algorithm, respectively. Numerical simulation results show that the MPSOG have better guidance performance than proportional navigation guidance and particle swarm optimization guidance in miss distance, time-to-go, and final lateral acceleration commands.

1 Submitting the manuscript

Since the inception of missile technology, the research of guidance laws, which guide a missile to attack its target effectively, has attracted a lot of attention. Many research achievements have been made, including proportional navigation guidance (PNG) [1], augmented PNG [2], pure PNG [3], and command to line-of-sight (CLOS) guidance [4]. With the development of control theory, optimal control theory has also been utilized in missile guidance design. This includes optimal intercept guidance [5], optimal missile guidance [6], and modified receding horizon guidance [7].

In recent years, artificial intelligence algorithms have successfully been applied in many fields which started the research into intelligent guidance. For example, Rajasekhar et al. [8] proposed a fuzzy logic controller to adjust the navigational constant of the PNG. Deskovski et al. [9] developed a fuzzy-rule-based system which can calculate the closing velocity and lead angle in order to compute the navigation constant of the PNG. Song et al. [10] proposed an on-line suboptimal midcourse guidance law which used neural networks for air-to-air medium-range missiles. Choi et al. [11] presented a differential game missile guidance which used neural networks.

Since the inception of a particle swarm optimization (PSO) algorithm, its many advantages which include its few parameters, fast convergence, simple scheme and easy implementation have caused concern. Some studies, such as Parsopoulos et al. [12] and Kuo et al. [13] have also pointed out that the capability of a PSO is similar to other AI algorithms. In particular, some studies, such as Kung et al. [14], have been used in missile guidance.

In this paper, we propose a modified PSO guidance (MPSOG) to improve the guidance performance. The MPSOG uses a Kalman filter to predict the target's dynamic at the first step, and the line-of-sight (LOS) rates of the y- and z-axis are taken as the objective functions. Then the fitness functions are evaluated according to the defined objective function. Therefore, the lateral acceleration commands of the y- and z-axis can be optimized by the MPSOG, respectively.

In the remainder of this paper, Section 2 briefly describes the background of the PSO, and describes the MPSOG in detail. After presenting the experimental design and discussing the results in Section 3, finally, Section 4 gives conclusions.

2 PSO algorithm and MPSOG

2.1 PSO algorithm

The PSO algorithm was proposed in 1995, by Kennedy et al. [15]. Since the proposal of the PSO algorithm, a number of more advanced research papers have been published. The improved formula, described in Eqs. (1)-(2), was proposed by Shi et al. [16, 17], and is now called the standard PSO.

$$V_{ij}^{new} = wV_{ij}^{old} + c_1 rand_1\left(Pbest_{ij} - X_{ij}^{old}\right) + c_2 rand_2\left(Gbest_i - X_{ij}^{old}\right) \tag{1}$$

$$X_{ij}^{new} = X_{ij}^{old} + V_{ij}^{new} \tag{2}$$

where V_{ij}^{old} is the previous velocity of a particle; V_{ij}^{new} is the new velocity of a particle after updating; $Pbest_{ij}$ is the self-experience of an individual particle; $Gbest_i$ is the neighbouring experience of particle swarm; w is the

[a] Corresponding author: KueiYi.Chen@gmail.com

inertial weight factor; Both c_1 and c_2 are learning factors; Both $rand_1$ and $rand_2$ are random numbers between [0, 1]; X_{ij}^{old} is a current value of a particle; X_{ij}^{new} is a new value of a particle after updating; the subscript i is the iteration index, and the subscript j is the particle index.

The PSO algorithm implementation initializes all parameters which the PSO algorithm needs at first. Second, objective values are calculated by the objective function, and the fitness function is evaluated according to the defined objective function. The algorithm uses Eqs. (1)-(2) to update the states of the particle swarm after the fitness function. Finally, the algorithm should stop and output a solution if the stop condition is satisfied, otherwise goes to the second step.

2.2 MPSOG

At each time-step, the MPSOG decides the best guidance commands. The inputs are states of the missile and the target, and the outputs are the best guidance commands that are the y- and z-axis lateral acceleration commands. According to the symbol definitions of the PSO algorithm described in the previous section, the details of the MPSOG can be described in the following:

Step 1. ***(Prediction):*** In the first step, the Kalman filter is used to predict the target's dynamic.

Step 2. ***(Initialization):*** Preset the w, c_1, c_2, maximum iteration index $i_{\max}$, swarm size n, global best fitness value $GbestJ_{\phi i}$ and $GbestJ_{\theta i}$, respectively. According to n, preset the velocities of particle $V_{\phi ij}^{old}$ and $V_{\theta ij}^{old}$, and the local best fitness values $PbestJ_{\phi ij}$ and $PbestJ_{\theta ij}$. Let particle X_{yij}^{old} and X_{zij}^{old} be the sets of the y- and z-axis lateral acceleration commands, respectively. We initialize the particles within a random distribution between the limits of lateral acceleration command.

Step 3. ***(Calculating objective values and fitness values):*** We assume that the missile is modeled as a point-mass, and the motion equations refer to Imado [18]. Let the a_{pc} equal a particle X_{yij}^{old}, and a_{yc} equal a particle X_{zij}^{old}. We can then predict the missile's behavior when we utilize the Runge-Kutta numerical method to solve the point-mass model of the missile. Also, we assume that the target is modeled as a point-mass. Therefore, LOS can be defined as a virtual straight line between the missile and target. The LOS rate is the angular rate of the angle between LOS and horizontal reference line in each time-step. Let LOS rate be the objective value. The Eqs. (3)-(4) of the three-dimensional LOS rate refer to Yang [19], and the objective values $J_{\phi ij}$ and $J_{\theta ij}$ for each particle are calculated by Eq (5).

$$\dot{\phi}_{ij} = \frac{\dot{z}_{ij}\left(x_{ij}^2+y_{ij}^2+z_{ij}^2\right)-z_{ij}\left(x_{ij}\dot{x}_{ij}+y_{ij}\dot{y}_{ij}+z_{ij}\dot{z}_{ij}\right)}{\sqrt{x_{ij}^2+y_{ij}^2}\left(x_{ij}^2+y_{ij}^2+z_{ij}^2\right)} \tag{3}$$

$$\dot{\theta}_{ij} = \frac{x_{ij}\dot{y}_{ij}-\dot{x}_{ij}y_{ij}}{x_{ij}^2+y_{ij}^2} \tag{4}$$

$$\begin{cases} J_{\phi ij} = \sqrt{\dot{\phi}_{ij}^2} \\ J_{\theta ij} = \sqrt{\dot{\theta}_{ij}^2} \end{cases} \tag{5}$$

where x, y, and z are the relative distance of x-, y-, and z-axis between the missile and the target, respectively; $\dot{\theta}$ and $\dot{\phi}$ are the horizontal and vertical LOS rate, respectively.

Step 4. ***(Individual best and swarm best evaluating):*** In a pursuit-evasion game, the variation of the LOS rate will vary with the dynamic of the missile and target. If the LOS rate equals zero at each step-time, namely the LOS parallels with each other, the curvature of the missile's trajectory will be smaller. The probability of intercept will be greater. Therefore, the two fitness functions, Eqs. (6)-(7), are based on these concepts.

$$\begin{cases} Pbest_{\phi ij} = X_{yij}^{old}\left(\min(J_{\phi ij})\right) \& \; PbestJ_{\phi ij} = \min(J_{\phi ij}),\ if\ PbestJ_{\phi ij} > \min(J_{\phi ij}) \\ Pbest_{\theta ij} = X_{zij}^{old}\left(\min(J_{\theta ij})\right) \& \; PbestJ_{\theta ij} = \min(J_{\theta ij}),\ if\ PbestJ_{\theta ij} > \min(J_{\theta ij}) \end{cases} \tag{6}$$

$$\begin{cases} Gbest_{\phi i} = Pbest_{\phi ij}\left(\min(PbestJ_{\phi ij})\right) \& \; GbestJ_{\phi i} = \min(PbestJ_{\phi ij}),\ if\ Gbest_{\phi i} > Pbest_{\phi ij}\left(\min(PbestJ_{\phi ij})\right) \\ Gbest_{\theta i} = Pbest_{\theta ij}\left(\min(PbestJ_{\theta ij})\right) \& \; GbestJ_{\theta i} = \min(PbestJ_{\theta ij}),\ if\ Gbest_{\theta i} > Pbest_{\theta ij}\left(\min(PbestJ_{\theta ij})\right) \end{cases} \tag{7}$$

$$\begin{cases} V_{\phi ij}^{new} = wV_{\phi ij}^{old} + c_1 rand_1\left(Pbest_{\phi ij} - X_{yij}^{old}\right) + c_2 rand_2\left(Gbest_{\phi i} - X_{yij}^{old}\right) \\ V_{\theta ij}^{new} = wV_{\theta ij}^{old} + c_1 rand_1\left(Pbest_{\theta ij} - X_{zij}^{old}\right) + c_2 rand_2\left(Gbest_{\theta i} - X_{zij}^{old}\right) \end{cases} \tag{8}$$

where $PbestJ_{\phi ij}$ and $PbestJ_{\theta ij}$ are the local best fitness values, respectively; $Pbest_{\phi ij}$ and $Pbest_{\theta ij}$ are the local best y- and z-axis lateral acceleration commands, respectively; $GbestJ_{\phi i}$ and $GbestJ_{\theta i}$ are the global best fitness values, respectively; $Gbest_{\phi i}$ and $Gbest_{\theta i}$ are the global best y- and z-axis lateral acceleration commands, respectively.

Step 5. ***(Velocity and Position updating):*** We use Eq. (8) to update the $V_{\phi ij}^{new}$ and $V_{\theta ij}^{new}$, and use Eq. (9) to update the X_{yij}^{old} and X_{zij}^{old}, respectively.

$$\begin{cases} X_{yij}^{new} = X_{yij}^{old} + V_{\phi ij}^{new} \\ X_{zij}^{new} = X_{zij}^{old} + V_{\theta ij}^{new} \end{cases} \tag{9}$$

Step 6. ***(Stop condition checking):*** Calculate the standard deviations of the X_{yij}^{new} and X_{zij}^{new}, respectively. If the standard deviations are both greater than the convergent tolerance, set $V_{\phi ij}^{old}$ and $V_{\theta ij}^{old}$ equal to $V_{\phi ij}^{new}$ and $V_{\theta ij}^{new}$, respectively. We also set $X_{\phi ij}^{old}$ and $X_{\theta ij}^{old}$ equal to $X_{\phi ij}^{new}$ and $X_{\theta ij}^{new}$, respectively, and then return to Step 3. If the standard deviations are both less than the convergent tolerance or i equals to $i_{\max}$, and output the $Gbest_{\phi i}$ and $Gbest_{\theta i}$.

3 Numerical simulation and results

The maximum g-force left turn flight is used to evaluate the guidance performance of the MPSOG. The results would compare with the PNG and the PSO guidance (PSOG). [14].

The parameter set of the MPSOG are as follows: n is set as 100; $i_{\max}$ is set as 700; w is set as 0.5; Both c_1 and c_2 are set as 1; The convergent tolerance is set as 10^{-2}; The $X_{\phi ij}^{old}$, $X_{\theta ij}^{old}$, $V_{\phi ij}^{old}$, and $V_{\theta ij}^{old}$ are initialized randomly $\pm 25\,g$. The simulation stops at the minimal miss distance (MD), otherwise, the initial conditions of missile and target are shown in Table 1.

Table 1. The initial conditions of missile and target.

Item	Missile	Target
x-axis (m)	0	4,500
y-axis (m)	0	3,500
z-axis (m)	3,000	3,100
Initial speed (*Mach*)	2	0.8
Time-step (*sec*)	0.02	0.02

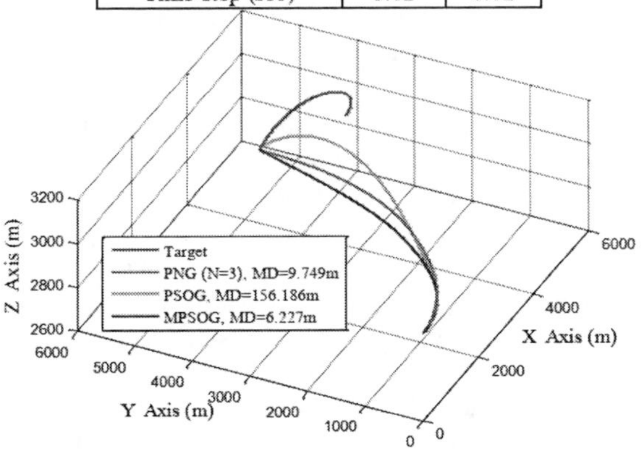

Figure 1. The simulation results where the maneuver of the target is a maximum g-force left turn flight.

The simulation trajectories are illustrated in Figure 1, and Table 2 collates the simulation results. According to the simulation results, the guidance performance of the MD makes it quite clear that the MPSOG is better than other methods. Also, the guidance performance of the time-to-go (T_{go}) of the MPSOG is better than other methods. These results revealed that the MPSOG has a higher tracking ability and accuracy. As a brief summary, the simulation results revealed that the MPSOG had a better guidance performance than the PNG and the PSOG.

Table 2. The simulation results of the evasion scenario.

Item		PNG	PSOG	MPSOG
MD (m)		9.75	156.19	6.23
T_{go} (*sec*)		8.84	8.96	8.82
The lateral acceleration commands at the final stage (g)	y-axis	-0.65	-25.00	0.48
	z-axis	-6.15	25.00	-5.33

4 Conclusion

In this study, we propose a MPSOG and demonstrate its validity. We use a Kalman filter to predict the target's dynamic at first. Then the LOS rate of the y- and z-axis are taken as the objective functions, and the lateral acceleration commands of the y- and z-axis are optimized by a PSO algorithm, respectively. Simulation results show that the MPSOG had a better guidance performance in MD, T_{go}, and final lateral acceleration commands.

References

1. S. A. Murtaugh and H. E. Criel, IEEE Spectrum **3**, 75 (1966)
2. G. M. Siouris, NTZ-Commun J. **27**, 278 (1974)
3. K. Becker, IEEE T. Aero. Elec. Sys. **26**, 526 (1990)
4. I.J. Ha and S. Chong, IEEE T. Aero. Elec. Sys. **28**, 51 (1992)
5. R. G. Cottrell, AIAA J. 9, 1414 (1971)
6. R. K. Aggarwal, IEEE Decis. Contr. P. **3**, 2775 (1996)
7. K. B. Kim, M. J. Kim, and J. W. Choi, IEEE T. Aero. Elec. Sys. **36**, 725 (2000)
8. V. Rajasekhar and A. G. Sreenatha, Acta Astronaut. **46**, 17 (2000)
9. S. Deskovski, Z. Gacovski, and G. Dimirovski, RAST, 680 (2003)
10. E.J. Song and M.J. Tahk, Control Eng. Pract. **6**, 957 (1998)
11. H.L. Choi, Y. Park, H.G. Lee, and M.J. Tahk, AIAA GNC., 1 (2001)
12. K. E. Parsopoulos and M. N. Vrahatis, *Intelligent Technologies–Theory and Application: New Trends in Intelligent Technologies* **76**, 214 (2002)
13. H.C. Kuo, J.R. Chang, and C.H. Liu, J. Mar. Sci. Technol. **14**, 170 (2006)
14. C.C. Kung and K.Y. Chen, T. Can. Soc. Mech. Eng. **37**, 971 (2013)
15. J. Kennedy and R. Eberhart, IEEE ICNN **4**, 1942 (1995)
16. Y. Shi and R. Eberhart, IEEE ICEC, 69 (1998)
17. Y. Shi and R. C. Eberhart, IEEE CEC **3**, 1945 (1999)
18. F. Imado, J. Guid. Control Dynam. **16**, 289 (1993)
19. C.C. Yang, Master Thesis, Tainan, (1993)

The Effect of mechanical resistive loading on optimal respiratory signals and breathing patterns under added dead space and CO2 breathing

Shyan-Lung Lin[1,2,a] and Hsing-Cheng Chang[1]

[1]*Department of Automatic Control Engineering, Feng Chia University, 40724, Taichung, Taiwan*
[2]*Department of Information Communication, MingDao University, 52345, Changhua, Taiwan*

Abstract. Current study aims to investigate how the respiratory resistive loading affects the behaviour of the optimal chemical-mechanical respiratory control model, the respiratory signals and breathing pattern are optimized under external dead space loading and CO2 breathing. The respiratory control was modelled to include a neuro-muscular drive as the control output to derive the waveshapes of instantaneous airflow, lung volume profiles, and breathing pattern, including total/alveolar ventilation, breathing frequency, tidal volume, inspiratory/expiratory duration, duty cycle, and arterial CO2 pressure. The simulations were performed under various respiratory resistive loads, including no load, inspiratory resistive load, expiratory resistive load, and continuous resistive load. The dead space measurement was described with Gray's derivation, and simulation results were studied and compared with experimental findings.

1 Introduction

Among recently developed respiratory control models [1-6], possible optimality principle has found to be existed in the modeling of the respiratory control [6-8]. It was suggested that the normal ventilatory responses to chemical stimuli, exercise inputs, and mechanical loadings are predicted by the minimization of a controller objective function that consists of the total chemical and mechanical costs of breathing [8]. The hypothesis was later extended to model the integrative control of ventilatory responses and respiratory patterns, by expressing respiratory work rate of breathing in terms of the isometric respiratory driving pressure (instead of ventilation) [9]. Based on the optimal chemical-mechanical respiratory control model [8-9], a simulation platform initially implemented under MATLAB [10], was later revised using LabVIEW [11], to provide a real-time simulation tool for respiratory control, and a signal monitoring platform for instantaneous profiles and breathing pattern.

The problem concerning the nature and the function of the dead space is of basic importance for the full comprehension of the respiratory physiology and pathophysiology [12-18]. Extended studies [19-23] on the concept of dead space have been made in the last half-century. Gray [22] presented an analysis of the respiratory dead space problem with careful distinction drawn between measured anatomical dead space and calculated physiological (virtual) dead space. When a subject inhales a CO_2 and air mixture, the ventilation increases, as does the concentration of CO_2 in his blood and tissues. Tracheostomized ponies were utilized to determine the effect of reducing anatomical dead space on P_{aCO_2} (arterial CO_2 pressure) during CO_2 inhalation. It was concluded [23] that the relationships of ventilatory responses did not differ between normal and decreased dead space cases with tracheostomy breathing. Experiments were conducted on healthy men [24] by adding dead space and CO_2 inhalation to determine the ventilatory responses between rapid-eye-movement sleep and slow-wave sleep. The ventilatory response during external dead space breathing and CO_2 inhalation were studied [25] for a given increase in P_{ETCO_2} with different levels of P_{ETO_2} in human subjects. The effect of positive and negative inspiratory pressure on respiration was investigated [26] both without and with an added external dead space, using either added inspiratory flow resistances or by providing inspiratory pressure support. The effects of different modes of airway CO_2 load on the ventilation-CO_2 output relationship were examined [27] during exercise with and without an external dead space.

Acute lung injury usually characterized with alveolar and capillary endothelial cell injuries, and that also results in compromised adequate pulmonary ventilation and blood flow across the lungs and increased V_D/V_T [28]. This high dead space fraction generally symbolizes impaired ability to excrete CO_2 and an increased risk of death in patients diagnosed with acute

[a]Corresponding author: sllin@fcu.edu.tw, sllin@mdu.edu.tw

DOI: 10.1051/matecconf/20167102002

respiratory distress syndrome [29]. The experiment [30], performed in hyperoxia with (IRL) and without (NL) an inspiratory resistive load (15 cm-$H_2O \cdot l^{-1} \cdot s$ at 1l/s) in healthy subjects, also depicted a similar $\dot{V}_E$ vs. P_{ETCO_2} relationship.

Respiratory-mechanical factors [31] have been postulated to provide a further source of control for ventilatory response to moderate exercise. Experimental studies on the influence of inspiratory assistance on respiratory control have been conducted [31,30]. Poon et al. [33] examined the steady-state effects of expiratory resistive loading on the time course of inspiratory and postinspiratory muscle activities and ventilatory pattern during quiet breathing from five conscious human subjects. To identify the neural changes responsible for ventilatory compensation, the respiratory driving pressure waveform was calculated [34] during steady-state unloaded and loaded breathing in eight conscious normal subjects. The simulation of IRL can be performed by increasing R_{in} (inspiratory airway) [35] from control (NL) with R_{ex} (expiratory airway resistance) at no load, while ERL can be performed by increasing resistance in R_{ex} with R_{in} at no load [33, 36, 37]. To perform continuous resistive loading (CRL), the inspiratory and expiratory airway resistances (R_{in} and R_{ex}) are increased to the intended levels of loading. Brack [38] studied healthy subjects breathing at rest and with continuous inspiratory resistive loads applied randomly through flow resistors. With eight subjects, flow was recorded at rest [39], while breathing against inspiratory threshold loads and resistive loads, and while inhaling 7% CO_2.

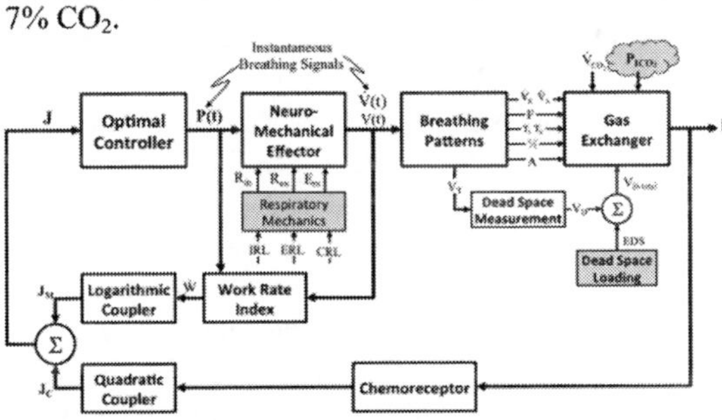

Fig. 1. The optimal chemical-mechanical respiratory control and simulation prospect.

2 Optimal Respiratory Control Model

Previous researches have successfully proposed and implemented a simulation strategy based upon a mathematical model of the optimal respiratory control with MATLAB and LabVIEW [11]. The model of the optimal chemical-mechanical respiratory control and simulation prospect of current study are illustrated in Fig. 1. The controller is driven by both chemical and neuro-mechanical feedback signals. In the figure, this is demonstrated by the coupling of chemical cost J_C and mechanical cost J_M, represented by the chemical feedback signal quadratic coupler and the work rate logarithmic coupler, respectively. The fundamental hypothesis of the model is that a total cost function can be formulated to reflect the balance of a combined challenge due to the chemical and mechanical cost of breathing [8]. The mathematical descriptions of these functional blocks depicted in Fig. 1 have been detailed in earlier reports [9-11] and are outlined below Eqs. (1)~(7):

$$J = J_C + J_M \tag{1}$$

$$J_C = \alpha^2 \cdot \left(P_{aco_2} - \beta\right)^2 \tag{2}$$

$$J_M = \ln \dot{W} \tag{3}$$

$$\begin{gathered} \dot{W} = \dot{W}_I + \dot{W}_E \\ \dot{W}_I = \frac{1}{T}\int_0^{T_I} \frac{P(t)\cdot \dot{V}(t)}{\xi_1^n \cdot \xi_2^n} dt \\ \begin{cases} \xi_1 = 1 - \frac{P(t)}{P_{max}} \\ \xi_1 = 1 - \frac{\dot{P}(t)}{P_{max}} \end{cases} \\ \dot{W}_E = \frac{1}{T}\int_{T_I}^{T} P(t)\cdot \dot{V}(t)\, dt \end{gathered} \tag{4}$$

In Eq. (4), the total mechanical index is assumed to be a weighted sum of the inspiratory and expiratory indexes, with a weighting parameter λ, and where $\dot{W}$, $\dot{W}_I$, $\dot{W}_E$, represent the total respiratory, inspiratory, and expiratory work rate in kg·m/sec, respectively. Efficiency factors ξ_1 and ξ_2 account for the effects of respiratory-mechanical limitation and the decrease in neuro-mechanical efficiency with increasing effort. The overall efficiency is dependent on two factors: the maximum isometric pressure (P_{max}) and the maximum rate of rise in the isometric pressure ($\dot{P}_{max}$). The respiratory period of a breathing cycle (T = t_1+t_2, sec) accounts for the inspiratory (t_1) and expiratory (t_2) durations of neural activity, however, the parameter T_I describes the duration of mechanical inspiratory, where the airflow flows into lung for inspiratory work of breathing.

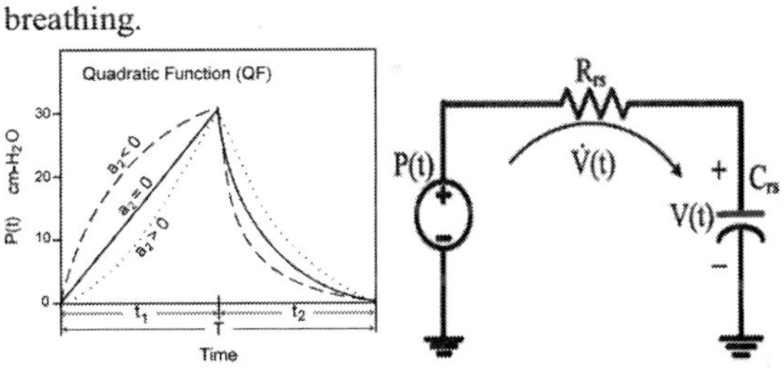

Fig. 2. Left: pressure waveshape over a complete respiratory cycle; Right: RC model of neuro-mechanical effector.

The neuro-mechanical effector relates the neural respiratory output to the resultant mechanical airflow and is required to optimize the neural input to the respiratory muscles for the optimization of the ventilatory airflow ($\dot{V}(t)$, l/min). The effector is described by the electrical RC model of Fig. 2(right) based on a lumped-parameter model for the relation between respiratory neural and mechanical outputs. In this model, the equation of motion is given by the following dynamic equation:

$$P(t) = \dot{V}(t) \cdot R_{rs} + V(t) \cdot E_{rs} \quad (5)$$

The parameters R_{rs} (cm-$H_2O \cdot l^{-1} \cdot$sec) and E_{rs} (=1/C_{rs}, cm-H_2O/l) in Eq. (5) represent the total flow-resistive and volume-elastic components, respectively. These include the passive resistances and elastance of the lung, chest wall, and airways. The neuro-muscular driving pressure waveform (P(t), cm-H_2O) is modeled as inspiratory and expiratory phases, as were depicted in Fig. 2(left). The inspiratory pressure is approximated by a quadratic function, and the expiratory pressure is represented by an exponential discharge function of the form

$$P(t) = a_0 + a_0 t + a_0 t^2 \quad 0 \le t \le t_1 \quad (6)$$

$$P(t) = P(t_1) \cdot e^{\frac{t-t_1}{\tau}} \quad t_1 \le t \le t_1 + t_2 \quad (7)$$

where the parameters a_0 and a_1 in Eq. (6) represent the net driving pressure and its rate of increase at the onset of the neural inspiratory phase respectively, and the parameter a_2 describes the shape of the wave; t_1 (sec) and t_2 (sec) represent the neural inspiratory and expiratory durations, respectively. In Eq. (7), P(t_1) is the peak inspiratory pressure in units of cm-H_2O at the end of inspiration (t_1), and τ denotes the rate of decline of inspiratory activity. With the mathematical description of Eqs. (5)~(7) over a complete respiratory cycle, $T = t_1 + t_2$, the analytical solution for the instantaneous profile of the lung volume (V(t), l) over the functional residual capacity (FRC, l) and airflow ($\dot{V}(t)$, l/s) within a breath can be obtained.

The plant of Gas Exchanger in Fig. 2 describes the events of the pulmonary exchange subject to the control signal $\dot{V}_E$ (total ventilation, l/min), disturbances in the inhaled and metabolic CO_2 (P_{ICO2}, Torr, $\dot{V}_{CO_2}$, l/min). The system's output is the pressures of arterial CO_2 (P_{aCO2}, Torr). The gas exchanger equation describes the dependence of P_{aCO2} on the total ventilation and other disturbances:

$$P_{aco_2} = P_{aco_2} + \frac{863 \cdot \dot{V}_{CO_2}}{\dot{V}_E\left(1 - \frac{V_D}{V_T}\right)} \quad (8)$$

where P_{aCO2} is assumed to be identical to the mean alveolar P_{CO2}. Equation (8) describes the steady-state effect of ventilation on P_{aCO2} subject to any disturbances in the inhaled and metabolic production of CO_2. To account for the changes in anatomic dead space with an airway caliber, the empirical relation [22] of Eq. (9) is employed:

$$V_D = 0.037 \cdot VC \cdot \left(1 + \frac{V_T}{8}\right) \quad (9)$$

$$V_{D-total} = V_D + EDS \quad (10)$$

where VC is the vital capacity (l) and V_T is the tidal volume (l), EDS is the imposed external dead space (l). Based on the optimized pressure profile of Eqs. (6) and (7), for any given respiratory resistance and elastance, the breathing patterns, including tidal volume (V_T, l), breathing frequency (f, bpm), total ventilation ($\dot{V}_E$, l/min), and alveolar ventilation ($\dot{V}_A$, l/min) can be obtained [11].

3 Results and Discussion

It is assumed that the parameters R_{rs} and E_{rs} of Eq. (5) represent the active resistance and active elastance, respectively, which include the effective impedance of the respiratory muscles. The normal load (NL) is defined to have respiratory resistance and elastance of R_{rs} = 3.02 (cm-$H_2O \cdot l^{-1} \cdot$s) and E_{rs} = 21.9 (cm-H_2O/l). The inspiratory and expiratory resistances (R_{in} and R_{ex}) can be adjusted from the normal load condition to the intended values independently. In current study, the simulations were performed under various respiratory resistive loads, including no load, inspiratory resistive load (IRL, R_{in} = 8 cm-$H_2O \cdot l^{-1} \cdot$s), expiratory resistive load (ERL, R_{ex} = 8 cm-$H_2O \cdot l^{-1} \cdot$s), and continuous resistive load (CRL, $R_{in} = R_{ex}$ = 8 cm-$H_2O \cdot l^{-1} \cdot$s).

In the resting state (**Rest**), P_{ICO2} is set to be 0% and $\dot{V}_{CO_2}$ is set to be 0.2 (l/min). In simulation **Case-1**, the CO_2 inhalation was performed under various CO_2 concentrations (P_{ICO2}=3 and 5%). To investigate the effect of the imposed dead space on resultant respiratory signals and breathing patterns, the simulations of dead space loading were performed under with and without CO_2 inhalation. Without CO_2 inhalation, an external dead space (EDS) of 0.4, and 0.8 (l) were added for **Case-2**. With 5% of CO_2 inhalation, **Case-3** was performed with an additional dead space loading of 0.4 (l). The respiratory parameter settings for the performed simulations are concluded in Table 1.

To examine the effect of various types of mechanical resistive loads on the optimized instantaneous respiratory waveforms under dead space loading and CO_2 inhalation, Figure 3 depicts the profiles with **Rest** (upper left of Figure 3), **Case-3** (upper right of

Table 1. Respiratory parameter setting for simulations

Simulations		$\dot{V}_{CO_2}$ (l/min)	P_{ICO2} (%)	EDS (l)	Respiratory Mechanical Resistive Loadings: NL	IRL	ERL	CRL
					(cm-$H_2O \cdot l^{-1} \cdot$s)			
Rest		0.2	0%	0				
Case-1	Case-1-3%	0.2	3%					
	Case-1-5%	0.2	5%		R_{in}=3.02	R_{in}=8	R_{in}=3.02	R_{in}=8
Case-2	Case-2-0.4	0.2	0%	0.4	R_{ex}=3.02	R_{ex}=3.02	R_{ex}=8	R_{ex}=8
	Case-2-0.8	0.2	0%	0.8				
Case-3		0.2	5%	0.4				

Figure 3), **Case-1-3%** (middle left of Figure 3), **Case-1-5%** (middle right of Figure 3), **Case-2-0.4** (lower left of Figure 3), and **Case-2-0.8** (lower right of Figure 3). In the waveforms of each simulation case, three instantaneous respiratory signals, neural muscular driving pressure P(t) (first row), airflow $\dot{V}(t)$ (second row), and lung volume V(t) (third row), were showed. On the other hand, three types of mechanical resistive loadings, IRL (first column), ERL (first column), and CRL (first column), were displayed to demonstrate the comparisons on optimized waveforms of loaded (dotted line) and normal load (solid line).

We found IRL generally resulted in higher amplitude of p(t) and V(t) with more concave upward, prolonged inspiratory duration (t_1), and higher duty cycle (T_I/T) for the waveforms in **Rest**, **Case-1**s, and **Case-3**.

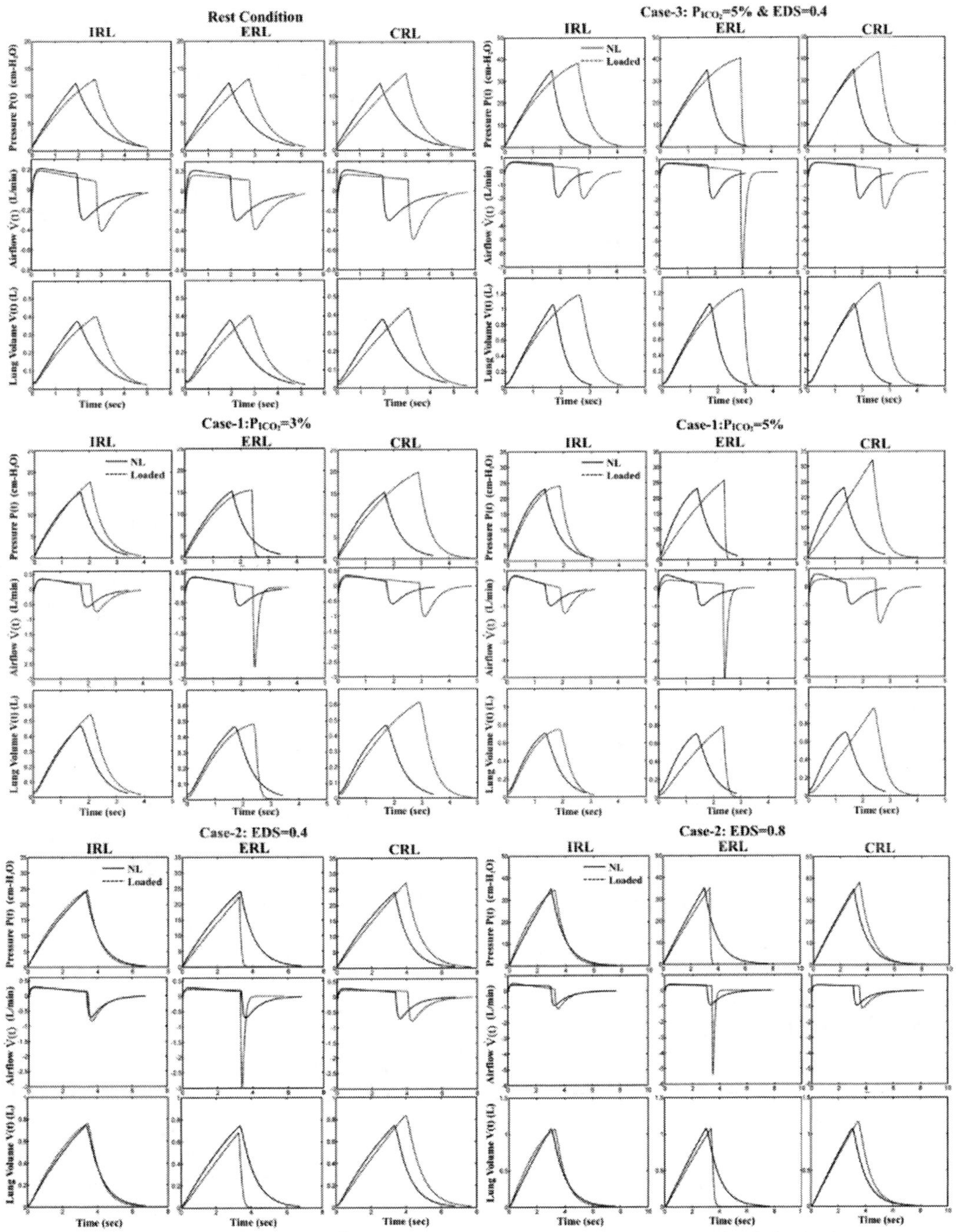

Fig. 3. Instantaneous respiratory signals, P(t), $\dot{V}(t)$, and V(t), under various types of mechanical resistive loads.

DOI: 10.1051/matecconf/20167102002

However, the imposed IRL did not show much effect on **Case-2** as dead space loading was imposed. In resting condition, the mechanical loadings of ERL and CRL exhibited similar effect on the waveshapes of respiratory signals but with a elongated t_1 and consequently a higher peak amplitude in P(t) and V(t). During CO_2 inhalation (**Case-1-3%** and **Case-1-5%**), EDS loadings (**Case-2-0.4** and **Case-2-0.8**), and a combination of both (**Case-3**), an abruptly fall in the beginning of expiratory phases of P(t) were found and the peak flows in expiratory phase were significantly boosted during ERL in comparison with those of **Rest**.

As EDS loadings were imposed alone in **Case-2-0.4** and **Case-2-0.8**, both IRL and ERL did not have much effect on the peaks of P(t) and V(t) at the end of inspiration as they did to the other simulation cases. However, in **Case-3** with P_{ICO_2}=0% and EDS=0.4 (l), three types of mechanical resistive loadings attained largest effect on the increasing of peak amplitudes on the neural muscular driving pressure, lung volume, and expiratory peak flow. As long as the waveforms of the P(t) and V(t) are concerned, the loading of CRL in all cases resulted in the most notable rises in amplitude in comparison with IRL and CRL did. Nevertheless, CRL appeared not to have much influence on the waveshapes of airflow except in **Case-1-5%**, as higher concentration of CO_2 was inhaled.

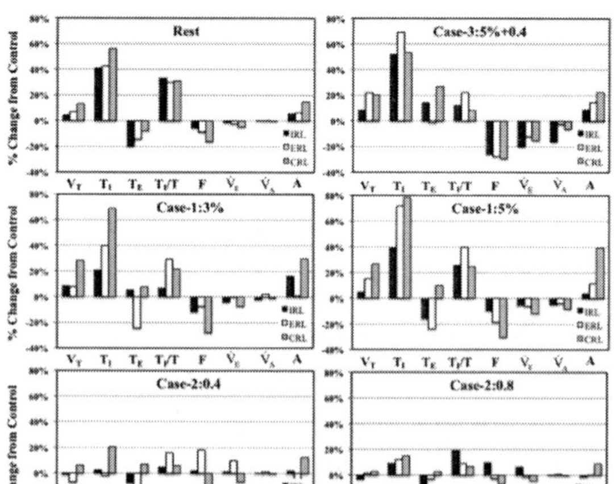

Fig. 4. Changes in breathing patterns under various types of mechanical resistive loads relative to control (normal load). V_T: tidal volume; T_I , T_E: inspiratory time, and expiratory time; Ti/T: duty cycle; F: breathing frequency; $\dot{V}_E$: minute ventilation; $\dot{V}_E$: alveolar ventilation; A: peak amplitude of pressure.

The relative percentage changes of corresponding pattern variables were concluded in Figure 4 to examine the effect of various types of mechanical resistive loadings on model behavior. As shown in Figure 4, mechanical resistive loadings (IRL in dark; ERL in blank; CRL in gray) caused increases in V_T during **Rest** and CO_2 inhalation (**Case-1-3%**, **Case-1-5%**, and **Case-3**). The effect of EDS loading on V_T was not significant in both cases of **Case-2**. All mechanical loadings resulted in an increase in T_I, which is consistent with the experimental data [33,34,38]. The expiratory duration T_E decreased in all cases of simulations during ERL while the effect of IRL remained ambiguous. However, T_E for ERL exhibited more substantial increase in **Case-1-3%**, **Case-1-5%**, and **Case-2-0.4%**. As a result of elongated T_I during mechanical loading, the duty cycle generally increased in all cases, however, the effect of CO_2 inhalation showed larger extent of increase in T_I/T than EDS loading did. In contrast to the increase in tidal volume, the breathing frequency F decreased during **Rest** and CO_2 inhalation (**Case-1-3%**, **Case-1-5%**, and **Case-3**), and effect of EDS loading on F was not clear in both cases of **Case-2**. Consequently, the decrease in total ventilation $\dot{V}_E$ displayed similar behavior as in breathing frequency for all simulations. No matter what type of mechanical loading was imposed, the alveolar ventilation $\dot{V}_A$ maintained nearly constant as in **Rest** during CO_2 inhalation or EDS loading. In comparison with IRL and ERL, the CRL exhibited substantial increases on the peak amplitude (A) of neural muscular drive P(t) during all cases of simulation.

4 Conclusions

With a unified prediction of exercise and chemical responses was shown entirely in terms of conventional feedback-mechanisms based on the optimal chemical-mechanical respiratory control model, instead of using a separate stimulus signal, we use a human respiratory control model to simulate the effect of mechanical resistive loading on the optimized respiratory signals and breathing patterns. The simulations are performed in resting state, CO_2 inhalation (P_{ICO_2}=3 and 5%), external dead space loading (EDS=0.4 and 0.8 l), and a combination of both (P_{ICO_2}=5% and EDS=0.4 l). Various types of resistive loading, IRL, ERL, and CRL are imposed for the study of model behavior in comparisons with normal load. An earlier research [40] with the same model has explored the effect of the imposed dead space loading on the optimized respiratory profiles under increased levels of CO_2 inhalation. It was concluded that higher amplitude with prolonged inspiratory duration and higher duty cycle were attained with EDS=0.4 (l). The increase of EDS =0.8 (l) showed opposite effects on the inspiratory duration and duty cycle under P_{ICO_2}=0% and 7%. No significant change in the waveshapes of these respiratory profiles was found with the imposed dead space loading under CO_2 inhalation.

In current study, the simulation results of successfully provided extended view into the respiratory signals, including neural muscular driving pressure, airflow, and lung volume, and breathing patterns. We found IRL generally resulted in higher amplitude of p(t) and V(t) with more concave waveshape, prolonged inspiratory duration and higher duty cycle (T_I/T) for the waveforms. In comparison with those of resting state, an abruptly fall in the beginning of expiratory phases of P(t) were displayed and the peak flows in expiratory phase were significantly boosted with ERL during CO_2 inhalation, EDS loadings, or a combination of both. The results of CRL in all cases exhibited the most notable rises in amplitude of pressure and lung volume in comparison with IRL and CRL, nevertheless, they also appeared not to have much influence on the waveshapes of airflow except in the case as higher concentration of CO_2 inhalation.

We attained valuable results in simulations of mechanical resistive loadings on respiratory control with the optimal control model during imposed external dead space and CO_2 inhalation. Further work can be extended to include the effects of mechanical elastic loading along with increased metabolic CO_2 output. Recent evidences indicated that the increase in dyspnea and decrease in exercise tolerance is generally associated with the result of increases in physiological dead space and/or ventilator-perfusion mismatching [12-14]. The study of relative behavior of respiratory signals and comparative relationship of the ventilator responses between dead space loading and unloading during rest, CO_2 inhalation, and exercise will provide further understanding of increases in central respiratory motor command output of human respiratory control, which is also associated with Dyspnea (respiratory discomfort) on exertion.

Acknowledgement

This study was granted by the Ministry of Science and Technology, Taiwan. (MOST 104-2221-E-451-005)

References

1. M.C. Khoo, R.E. Kronauer, K.P. Strohl, A.S. Slutsky, J. Appl. Physiol., **53** (1982).
2. M. Modarreszadeh, E.N. Bruce, J. Appl. Physiol., 76 (1994).
3. A. Rybak, M. Pottmann, B.A. Ogunnaike, J.S. Schwaber, Proc. of Am. Control Conf. (1995).
4. S. Mesic, R. Babuska, H.C. Hoogsteden, A.F.M. Verbraak, IEEE Trans. Biomed. Eng., 50 (2003).
5. G.D. Mitsis, Proc. of 9th International Conf. on Infor. Tech. and Appl. in Biomed. (2009).
6. H. Hirayama, Y. Shimizu, Y. Fukuyama, Proc. of 34th SICE Annual Conf., Sapporo (1995).
7. F.T. Tehrani, Proc. 20th Annu. International Conf. of the IEEE Eng. in Med. and Biol. Soc., 20 (1998).
8. C.S. Poon, J. Appl. Physiol. **62** (1987).
9. C.S. Poon, S.L. Lin, O.B. Knudson, J. Appl. Physiol. **72** (1992).
10. S.L. Lin, C.C. Chiu, J Biomed Eng. Appl. Basis Comm. **6** (1994).
11. S.L. Lin, N.R. Guo and C.C. Chiu, Modelling and simulation of respiratory control with LabVIEW, J. Med. Biol. Eng. **32** (2012): 51-60.
12. D.E. O'Donnell, R.B. Banzett, V. Carrieri-Kohlman, R. Casaburi, P.W. Davenport, S.C. Gandevia, A.F. Gelb, D.A. Mahler, K.A. Webb, Proc. Am. Thorac. Soc. **4** (2006).
13. D. Sajkov, N. Petrovsky, P. Palange, Curr. Opin. Support Palliat. Care, **4** (2010).
14. D. Jensen, D.E. O'Donnell, R. Li, Y.M. Luo, Respi. Physio. & Neuro. **179** (2011).
15. M.N. Van der Plas, H.J. Reesnik, C.M. Roos, R.P. Van Steenwijk, J.J. Kloek, P. Bresser, Ann. Thorac. Surg. **89** (2010).
16. H.E. Wood, G.S. Mitchell, T.G. Babb, J. Appl. Physiol. **104** (2008).
17. C.S. Poon, J. Appl. Physiol. **105** (2008).
18. R.A. Klocke, J. Appl. Physiol. **100** (2006).
19. J.R. Pappenheimer, A.P. Fishman, L.M. Borrero, J. Appl. Physiol. **4** (1952).
20. R.L. Coon, J.P. Kampine, J. Appl. Physiol. **51** (1981).
21. C.B. Wolff, R.C. Garratt, J. Appl. Physiol. **53** (1982).
22. J.S. Gray, F.S. Groadins, E.T. Carter, J. Appl. Physiol. **9** (1956).
23. H.V. Forster, J. Appl. Physiol. **61** (1986).
24. A. Warley, Respir. Physiol. **75** (1989).
25. R. Maruyama, Jpn. J. Physiol. **38** (1988).
26. F. Lofaso, D. Isabey, H. Lorino, P. Scheid, Physiol. **89** (1992).
27. C.S. Poon, J. Appl. Physiol. **73** (1992).
28. U. Lucangelo, L. Blanch, Intensive Care Med. **30** (2004).
29. T.J. Nuckton, J.A. Alonso, R.H. Kallet, B.M. Daniel, J.F. Pittet, M.D. Eisner, Matthay A, N. Eng. J. Med. **346** (2002).
30. D.A. Sidney, C.S. Poon, J. Appl. Physiol. **78** (1995).
31. C.S. Poon, S.A. Ward, B.J. Whipp, J. Appl. Physiol. **62** (1987).
32. C.G. Gallagher, R. Sanii, M. Younes, J. Appl. Physiol. **66** (1989).
33. C.S. Poon, M. Younes, C.G. Gallagher, J. Appl. Physiol. **63** (1987).
34. V.I. Hof, P. West, M. Younes, J. Appl. Physiol. **60** (1986).
35. C.S. Poon, J. Appl. Physiol. **66** (1989).
36. L. Fee, R.M. Smith, M.B. English, J. Appl. Physiol. **83** (1997).
37. W.H. Thompson, P. Carvalho, J.P. Souza, N.B. Charan, J. Appl. Physiol. **89** (2000).
38. T. Brack, A. Jubran, M.J. Tobin, Am. J. Respir. Crit. Care Med. **157** (1998).
39. Z. Samara, M. Raux, M.N. Fiamma , A. Gharbi, S.B. Gottfried, C.S. Poon, T. Similowski, C. Straus, Respi. Physiol. & Neuro. **165** (2009).
40. S.L. Lin, T.C. Chen, H.C. Chang, Int. Appl. Sci. and Prec. Eng. Conf. (ASPEC2013), (2013).

Parametric Study of Defect Detection in Pipes with Bend Using Guided Ultrasonic Waves

Jin Jack Tan[1], Xin Wang[1a], Ningqun Guo[1] and Jee Hou Ho[2]

[1]School of Engineering, Monash University Malaysia, Jalan Lagoon Selatan, 47500 Selangor, Malaysia
[2] Department of Mechanical, Materials and Manufacturing Engineering, University of Nottingham Malaysia Campus, Jalan Broga, 43500 Semenyih, Malaysia

Abstract. The propagation behaviour of guided ultrasonic waves in a steel pipe with welded bend is studied by finite element simulation. The effectiveness of the longitudinal L(0,2) and torsional T(0,1) guided waves in detecting circumferential cut near the weld is investigated. In order to identify the presence of the defect, the reflection strength due to the cut is studied. The geometry of the weld is constructed based on common V-bevel butt joints and the anisotropy of the 316L stainless steel weld is included to correctly predict the scattering of ultrasonic waves. The finite element model is built to allow high accuracy. Detection of small circumferential cut (up to 60° circumferential extent) can be achieved with longitudinal L(0,2) mode. Detection of moderate to large circumferential cut can be achieved by torsional T(0,1) or longitudinal L(0,2) modes, with T(0,1) mode preferred due to its less mode conversion to higher order modes.

1 Introduction

Ultrasonic guided wave testing is an attractive non-destructive testing (NDT) technique as it allows rapid and long range inspection from a single location. It also allows inspection of otherwise inaccessible regions of structure. Compared to conventional bulk waves, excited ultrasonic guided wave propagates along a waveguide like plates or pipes and has continual interaction with the boundaries of the waveguide. Therefore, the wave is confined and subsequently propagated over a long distance. Upon meeting a defect or anomaly, guided wave interacts with the defect which results in a reflected wave. The reflected wave is detected by the excitation transducer where the time difference between the initial and reflected wave can be used to locate the anomaly given known velocities of the wave. Thus guided wave is very attractive for industries involving large and complex pipelines structures.

Guided wave was applied to detect various kinds of defects, including circumferential notches [1], sludge [2], axisymmetric defects with varying depth [3], multiple circular holes [4], defect beyond weld [5], complex defect profiles [6] and corrosion defects [7]. More complex cases like pipes with fluid loading [8-10] and coated pipes [11-13] are also pursued by various authors.

Pipe with bend, on the other hand, receives relatively less attentions due to several difficulties. Defect detection based on the reflection signals scattered by the defects can be interfered by or confused with the reflection signals due to welds, which is commonly used to join pipe bend to straight pipes. In addition to that, mode conversion and dispersion of guided wave occur as it propagates through bend and deteriorates the reflection signals. Several methods have been used to derive the dispersion relations and to describe the propagation and scattering behaviour across the bend, including semi-analytical finite element formulation [14], analytical [15], numerical [16-17] and experimental approach [18]. There are also several works on detecting defect in a pipe with bend with defect at the bend [19-21] and beyond the bend [22-23].

In this paper, we investigate the guided wave detection ability of defects that exist near to weld and bend using finite element simulation. In the study, two steel pipes are joined by a 90-degree bend of same material via V-bevel butt joint welds. A thin circumferential cut is artificially introduced near to the weld to simulate a crack. The circumferential extent of the cut is varied and its effect to the reflection coefficient of the guided wave is monitored. Reflection coefficient is defined as the ratio of the amplitude of the reflection wave due to the defect to the amplitude of the reflection wave due to the end of the pipe in a defect-free pipe.

[a] Corresponding author: wang.xin@monash.edu

2 Finite element simulation

In the simulation, two 2-inch schedule 40 steel pipes (external diameter of 60mm and wall thickness of 4mm) with length of 0.3m are joined together with a bend by welds as shown in Figure 1. The bend has same diameter and wall thickness and has a bend radius of 76mm. The welds are V-bevel butt joint weld and have the geometry as shown in Figure 2(b) and both welds are identical. Root opening and root height is 0.8mm and included angle is 60°. The width of the welds at the outer surface is 4mm. The welds are 316L industrial austenitic stainless steel welds and their anisotropy is described accordingly [35]. Thin circumferential cut E is introduced 5mm after the weld D and is centered at the extrados of the pipe as shown in Figure 2(a).

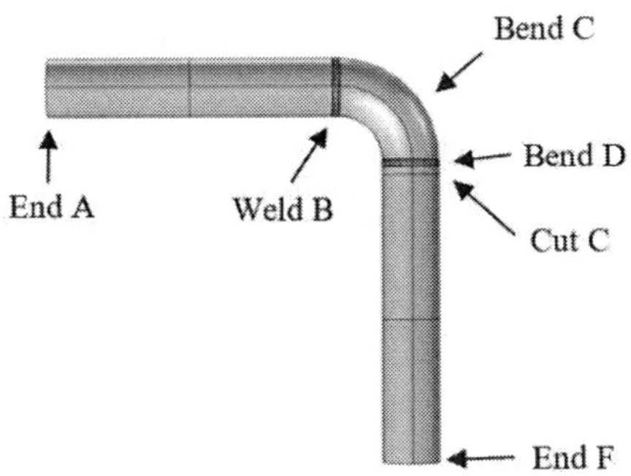

Figure 1. Geometrical setup of the finite element simulation in COMSOL.

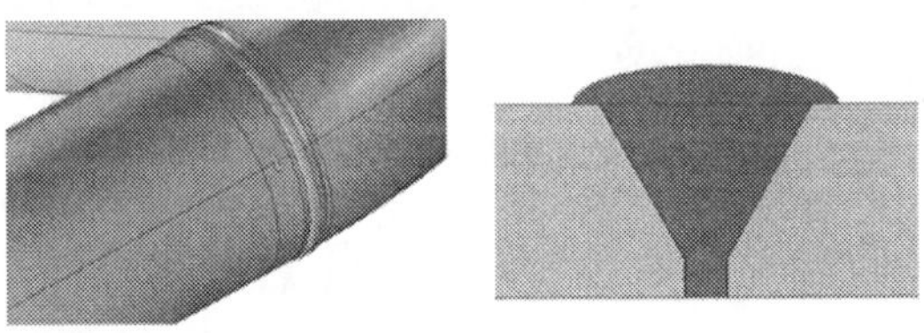

(a) (b)

Figure 2. (a) A closer view of the weld D and cut E. The cut shown has a circumferential extent of 10°. (b) The geometry of the weld used in the simulation. The geometry is used for both weld B and weld D.

Guided wave is excited at the outer edge of the end A and travels across weld B, bend C, weld D and meets defect E before reflecting from end F. The monitoring probes are at end A, coinciding the excitation points. This is to reflect the actual guided wave setup where the transducers ring usually acts as both the excitation source and receiving sensor. Excitation of the torsional T(0,1) and longitudinal L(0,2) with different orientation is shown in Figure 3. The excitation pulse is a 5-cycle Hanning window at 70 kHz. Each mode is excited across the whole circumference of end A with no delays between each other elements. 70 kHz is chosen as the investigated modes are fairly non-dispersive at that frequency as shown from the group velocity curves in Figure 4.

Table 1. The Young's modulus, Poisson ratio and density of the materials used.

Material	Young's Modulus (GPa)	Poisson Ratio	Density (kgm^{-3})
Steel	200	0.33	7850
316L	200	0.3	8000

$$[C] = 10^9 \begin{bmatrix} 194 & 139 & 139 & 0 & 0 & 0 \\ 139 & 233 & 100 & 0 & 0 & 0 \\ 139 & 100 & 233 & 0 & 0 & 0 \\ 0 & 0 & 0 & 65 & 0 & 0 \\ 0 & 0 & 0 & 0 & 106 & 0 \\ 0 & 0 & 0 & 0 & 0 & 106 \end{bmatrix} \mathrm{Pa} \quad (1)$$

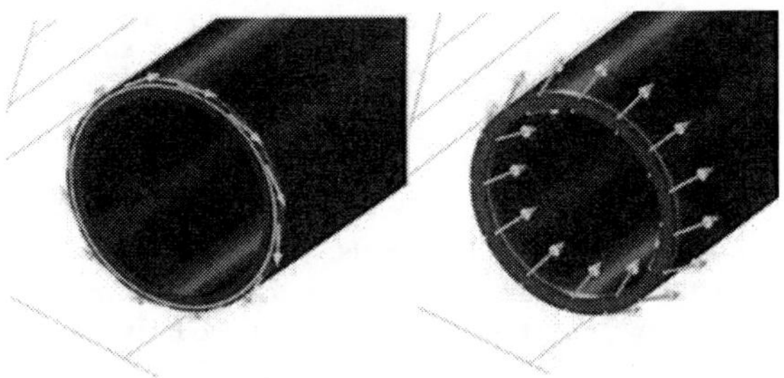

(a) (b)

Figure 3. Excitation of the (a) torsional T(0,1), (b) longitudinal L(0,2) with different orientation. The yellow arrows indicate the direction while the colour map indicates strength.

Grid independency study is conducted and to obtain high accuracy study, the time steps are discretized such that there are 16 nodes in one period of the investigated frequency (i.e. 70 kHz) and the spatial element sizes are discretized such that there are at least 16 nodes in one wavelength corresponding to the interested modes. This results in a time step of 9.52e-7 s and an element size of 3 mm. The wall thickness is also discretized such that there are two elements across the thickness which allows representation of mode shape up to second order guided wave modes.

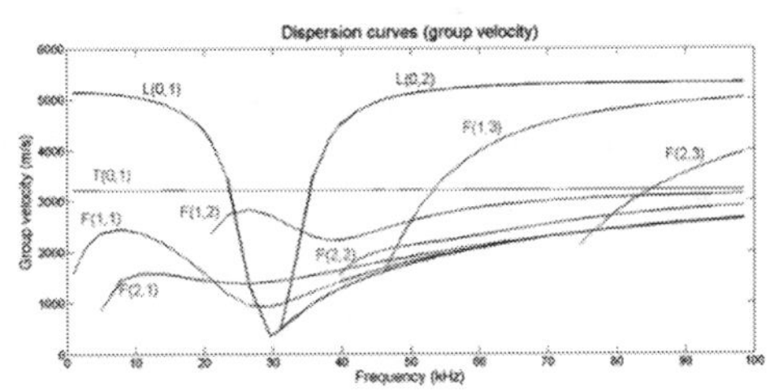

Figure 4. Group velocity dispersion curve of a steel pipe with 30mm external radius and 4mm pipe wall thickness, showing up to second order modes.

3 Defect detection ability

The displacement-time response for T(0,1) guided wave mode at the excitation end for a defect-free pipe is as shown in Fi5. Few wave packets can be identified. The first wave packet is the incident wave packet. The second and third waves, at 220µs and 340µs respectively, are the reflection due to the two welds of the bend. A small wave packet at around 450µs is likely a weak higher order reflection from one of the welds while the large wave from 520µs is the reflection from the end of the pipe. The displacement-time response for L(0,2) mode is similar with just a different travelling time due to its higher group velocity. It is thus omitted for brevity.

The displacement-time response for T(0,1) wave mode at the excitation end for a pipe with a circumferential notch that extends 30° circumferentially is as shown in Fig6. As can be seen, the response looks almost identical to the one without defect. This is because even though there is reflection due to the defect, it is superposed onto the reflection of second weld, consequently making it impossible to detect. Reflection coefficients of the pipe when excited with T(0,1) for various circumferential extents are plotted in Fig7. Defect extent that is less than or at 30° circumferentially is virtually impossible to be detected by T(0,1) mode given its similar reflection coefficient with a defect-free pipe. Monitoring reflection of higher order mode is also not suitable, as even though it has a small increase, the difference in reflection is too small to call for a defect.
Reflection coefficients of the pipe when excited with L(0,2) for various circumferential extents are shown in Figure 8. One biggest difference between the response of L(0,2) and T(0,1) mode is that L(0,2) mode is able to differentiate a defect-free pipe from a pipe that contains a small defect based on reflection of the fundamental mode. Consequently, if the reflection without defect is known, defect with 10° circumferential extent can be detected by observing the fundamental zero-th order reflection. A second order mode, likely to be F(2,2) based on its travelling time, is also present strongly for pipes containing small defects. Both of these modes can be monitored together to improve the confidence to call for a defect. For bigger defect from 30° onwards, in addition to the dominant fundamental zero-th order reflection, first order mode, likely a F(1,3) based on travelling time, can be monitored as well for defect detection. First order mode reflection can be used to characterize the size of the defect for moderately larger cut from 120° onwards.

Two interesting remarks can be made from studying the reflection coefficients of T(0,1) and L(0,2) modes in a pipe with welded bend. Firstly, being a non-dispersive mode, T(0,1) mode has much lesser mode conversion to higher order modes than L(0,2) modes. While the higher order modes can serve as additional information to detect defects, they can also affect and distort the signals significantly. Interpretation can be much more challenging for a more complex case, e.g. a pipe with more bends, welds and defects, pipe under fluid loading etc. On the other hand, the low mode conversion of T(0,1), despite propagating through a bend, makes it a viable candidate to detect a moderately sized defect and potentially on more challenging and complex cases.

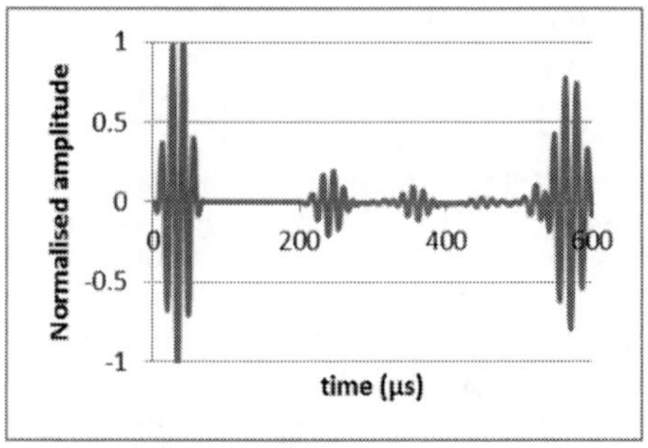

Figure 5. Displacement-time response for T(0,1) of a pipe with welded bend without defect.

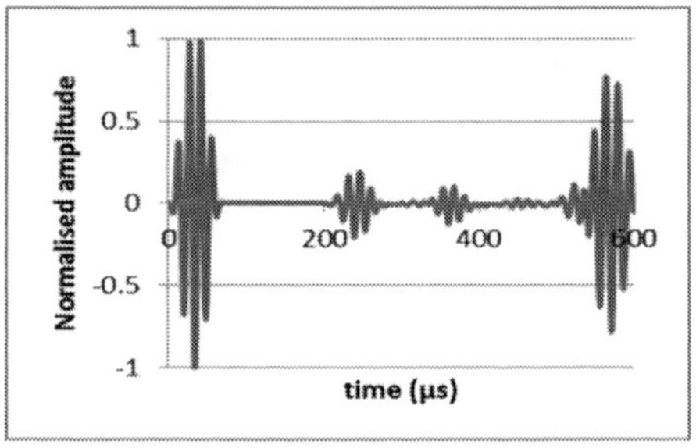

Figure 6. Displacement-time response for T(0,1) of a pipe with welded bend with a circumferential notch that extends 30° circumferentially.

Secondly, when comparing the reflection of L(0,2) mode in a pipe with bend to the reflection in a straight pipe (as can be seen in Fig9), it can be inferred that mode conversion occurs when propagating across and back the bend. While first order mode has its minimum when there is no defect in the case of pipe with bend, its reflection is at its maximum when there is a full circumferential notch. This is in contrast to the result in a straight pipe where first order mode is minimal when there is a full circumferential notch. This indicates that the existence of bend has resulted in the development of a stronger first order mode when the guided wave interacts with the defect.

The existence of second order mode even in a defect free pipe with bend also indicates that mode conversion to second order mode occurs when propagating through and back the bend. However, similar to the reflection of a straight pipe, it exhibits two peaks, first at around 90° and second at around 240°. The first peak coincides with the peak of a straight pipe while the second peak of a straight pipe occurs around 270°. This indicates that in addition to the second order mode developed due to the bend, it is also developed in a predictable fashion when interacting with the defect, i.e. peak at around 90° and 270°.

DOI: 10.1051/matecconf/20167102003

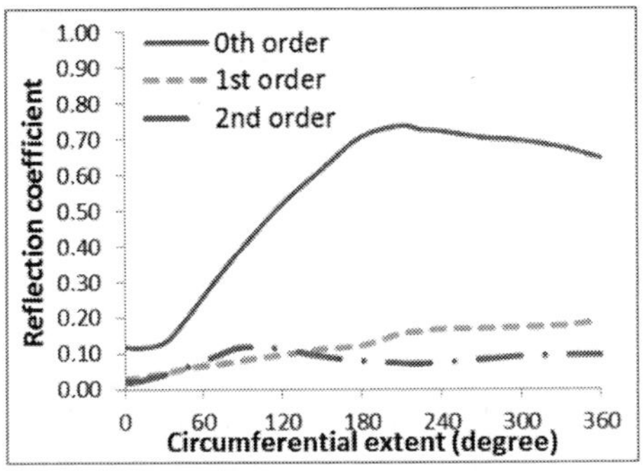

Figure 7. Reflection coefficient for T(0,1) mode against notch of variable circumferential extent.

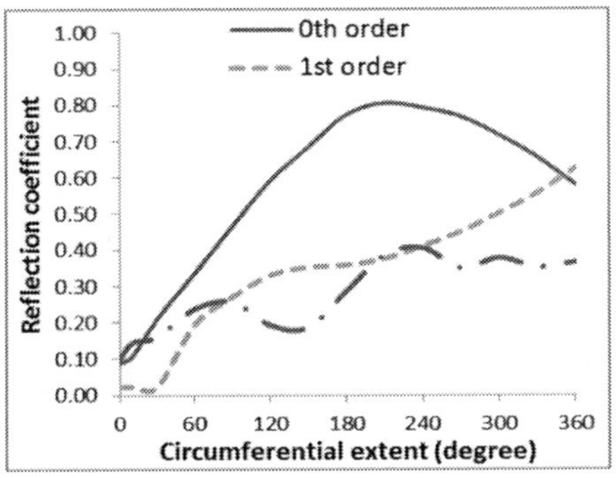

Figure 8. Reflection coefficient for L(0,2) mode against notch of variable circumferential extent.

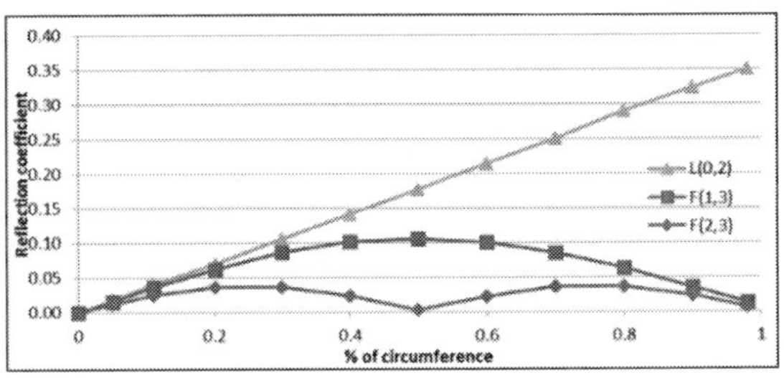

Figure 9. Reflection coefficient against percentage of circumference extent of notches at 50% depth extent for straight pipe.

4 Conclusion

The propagation behaviour of guided ultrasonic waves in a steel pipe with welded bend is studied by finite element simulation. The effectiveness of the longitudinal L(0,2) and torsional T(0,1) guided waves in detecting circumferential cut near the weld is investigated. Detection of small circumferential cut (up to 60° circumferential extent) can be achieved with longitudinal L(0,2) mode. Detection of moderate to large circumferential cut can be achieved by torsional T(0,1) or longitudinal L(0,2) modes, with T(0,1) mode preferred due to its less mode conversion to higher order modes. Reflection of a defect-free pipe and a pipe with a small 10° circumferential defect can be detected by L(0,2) mode. However, strong mode conversion of the L(0,2) mode may affect its performance on more complex cases. On the other hand, the non-dispersive T(0,1) mode cannot detect defect that extends less than or at 30° circumferentially.

References

1. Demma, A, Cawley, P and Lowe, M. The reflection of the fundamental torsional mode from cracks and notches in pipes. Journal of Acoustical Society of America, **2** (2003).

2. Ma, J, Lowe, M and Simonetti, F. Feasibility study of sludge and blockage detection inside pipes using guided torsional waves. Measurement Science and Technology, **14** (2007).

3. Carandente, R, Ma, J and Cawley, P. The scattering of the fundamental torsional mode from axi-symmetric defects with varying depth profile in pipes. Journal of Acoustical Society of America, **127** (2001).

4. Cawley, P and Lovstad, A. The reflection of fundamental torsional guided wave from multiple circular holes in pipes. NDT&E International, **46** (2011).

5. Zhang, L, Luo, W and Rose, J. Ultrasonic guided wave focusing beyond welds in a pipeline. Review of Quantitative Nondestructive Evaluation, **25** (2006).

6. Moreau, L, Velichko, A and Wilcox, P. Accurate finite element modeling of guided wave scattering from irregular defects. NDT&E International, **46** (2012).

7. Carandente, R and Cawley, P. The effect of complex defect profiles on the reflection of the fundamental torsional mode in pipes. NDT&E International, **46** (2012).

8. Aristegui, C, Lowe, MJ and Cawley, P. Guided waves in fluid-filled pipes surrounded by different fluids. Ultrasonic, **39** (2001).

9. Baik, K, Jiang, J and Leighton, T. Acoustic attenuation, phase and group velocities in liquid-filled pipes: Theory, experiment, and examples of water and mercury. Journal of Acoustical Society of America, **128** (2010).

10. Jin, S, et al. Study on ultrasonic guided waves in fluid-filled pipes surrounded by water. Calgary (2006).

11. Kwun, H, et al. Torsional guided-wave attenuation in coal-tar-enamel-coated, buried piping. NDT&E International, **37** (2004).

12. Ma, J, Simonetti, F and Lowe, MJ. Scattering of the fundamental torsional mode by an axisymmetric layer inside a pipe. Journal of Acoustical Society of America, **120** (2006).

13. Kirby, R, Zahari, Z and Mudge, P. On the scattering of longitudinal elastic waves from axisymmetric defects in coated pipes. Journal of Sound and Vibration, **332** (2012).

14. Hayashi, T., Kawashima, K., Sun, Z., and Rose, J.L., Guided wave propagation mechanics across a pipe elbow, Journal of Pressure Vessel Technology, **127** (2005).

15. Sanderson, R., Billson, D., Hutchins, D., and Mudge, P., The investigation of guided wave propagation around a pipe bend using an analytical modelling approach, J. Acoust. Soc. Am., **133** (2013).

16. Verma, B., Balasubramaniam, K., and Rajagopal, P., Studies on the influence of axial bends on ultrasonic guided waves in hollow cylidners (pipes), AIP Conf. Proc., **1511** (2013).

17. Demma, A., Cawley, P., and Pavlakovic, B., The effect of bends on the propagation of guided waves in pipes, Journal of Pressure Vessel Technology, **127** (2005).

18. Yoshida, H., Nishino, K., Cho, H., and Takemoto, M."Propagation phenomena of wideband guided waves in a bended pipe", Ultrasonics, **44** (2006).

19. Predoi, M., and Petre, C., Guided waves scattering by discontinuities near pipe bends, J. Acoust. Soc. Am., **133** (2013).

20. Yamamoto, T., Furukawa, T., and Nishino, H., Efficient defect detections of elbow pipes using propagation characteristics of guided waves, AIP Conf. Proc., **1581** (2014)

21. Miki, M., Nagashima, Y., Endou, M., Kodaira, K., Koike, M., and Odakura, M., Evaluation for defect detection over bending zone in piping by guided wave, AIP Conf. Proc., **1211** (2010).

22. Rose, J.L., Zhang, L., Avioli, M., and Mudge, P., A natural focusing low frequency guided wave experiment for the detection of defects beyond elbows, Journal of Pressure Vessel Technology, **127** (2005).

23. L Zhang, W Luo and J Rose, Ultrasonic guided wave focusing beyond welds in a pipeline, Review of Quantitative Nondestructive Evaluation, **25** (2006).

Heat transfer optimization of SCO_2 porous flow based on Brinkman model

David T.W. Lin[1,a], Jui Ching Hsieh[2], Bo Yen Shih[1]

[1]*Institute of Mechatronic System Engineering, National University of Tainan, No. 33, Sec. 2, Su-Lin St., Tainan, Taiwan, R.O.C*
[2]*Green Energy and Environment Research Laboratories, Industrial Technology Industrial Technology Research Institute, No. 195, Sec. 4, Chung Hsing Rd, Hsinchu 31040, Taiwan, R.O.C.*

Abstract. The purpose of this study is to obtain the optimal operating condition in order to find the maximum supercritical CO_2 heat extraction in the enhanced geothermal system (EGS). In this study, the heat transfer model conjugated with the Brinkman model is used to evaluate the thermal behavior in the reservoir of the EGS. This numerical model is validated by experiment. Optimization is processed based on the Nelder-Mead approach. The optimal operating conditions are proposed with different pressure, porosity. This study will build the optimal platform of heat source of geothermal power plant.

1 Introduction

With the progress of science and technology rapidly, the energy growing is emphasized gradually. Among the growing energy, the renewable energy such as the geothermal energy continues to be upgraded. For this reason, many countries begin to support the research of geothermal energy and hope to obtain the better low-cost power generation.

Nowadays, the deep geothermal power generation technology common with enhanced geothermal system and closed-loop heat collection system (CEEG).The principle of EGS is to inject the fluid with high pressure for generating the multiple artificial fractures under heat reservoir. Then, the working fluid flows through the reservoir to absorb the heat and flows out to ground through production well. Water is first applied on the EGS as the working fluid for its high latent heat, specific enthalpy and cheap cost. However, a number of disadvantages have to overcome, the main ones are viscosity coefficient and density. These will bring the pipeline fouling.

CO_2-EGS is first proposed by Brown in 2000 [1], which leads the research of EGS proliferated recently. It brings many benefits, such as low-temperature and high-density of carbon dioxide injection to production occurred phenomena of thermal siphon. It results in the lower pumping power.

Fouillac et al. [2], Ueda et al. [3] and Wan et al. [4], have also proposed CO_2-EGS for generating electricity, further and even more importantly. It will result in the CO_2 storage and prevent the pipeline fouling. There is a lot of previous research to study the possibility and potential of CO_2, such as Pruess et al. [5] and Zhang et al. [6]. In addition, Zhang et al. [6] compare carbon dioxide and water to evaluate the efficiency. Li et al. assess the renewable energy in the world [7], that will confirm the importance of geothermal energy. Xu et al. conduct a series of studies on the interaction of fluid and rock [8], and Jiang et al. discuss the influence of buoyancy of supercritical carbon dioxide in a porous media tube [9], and Liao et al. find the buoyancy effect significant influence flow direction [10]. Even a lot of researches attend the CO_2-EGS but few experiment of reservoir based on the optimal heat extraction. The purpose of this study proposes an integrated method which combined experiment, optimal scheme to obtain the maximum heat extraction of CO_2-EGS in the reservoir.

2 Experimental Section

The schematic diagram of the experimental system is shown in figure 1. The experimental system consisted of a CO_2 cylinder, high pressure pump, preheated water bath, cooling water bath, heating unit, test section, computer, and data logger. The test section is composed of stainless tube, and ten calibrated T-type thermocouples and pressure transducers are inserted into test section to record experimental data. The test section had inside and outside diameters of 29.9 and 55.0mm, and 235.0mm of length. The test section is filled with silica-base particles. To compare effects of particles, the empty of test section is also examined.

The particle mean mass diameters 1.54 and 2.03mm, respectively. The porosity of porous tube was approximately fixed at 0.2 and 0.39. To simulate reservoir

[a] David T.W. Lin: david@mail.nutn.edu.tw

DOI: 10.1051/matecconf/20167102004

temperature, the test section wall is heated as 200℃. The heated power is about 53W.

The system pressure is set as 7.5, 8.5, 9.5, 10.5, 11.5 and 12.5MPa; the flow rate is set as 20, 40, 60, and 80 ml/min. The overall uncertainty of the heat transfer coefficient of the supercritical CO_2 is evaluated using by the standard approach of Kline and McClintock [11]. The analysis shows that the measured uncertainties of temperature, flow rate, pressure, extraction power and heat transfer coefficient are estimated to be less than ±0.1 °C, ± 0.5 %, ± 2.3 %, ±2.37 % and ±17.6%, respectively.

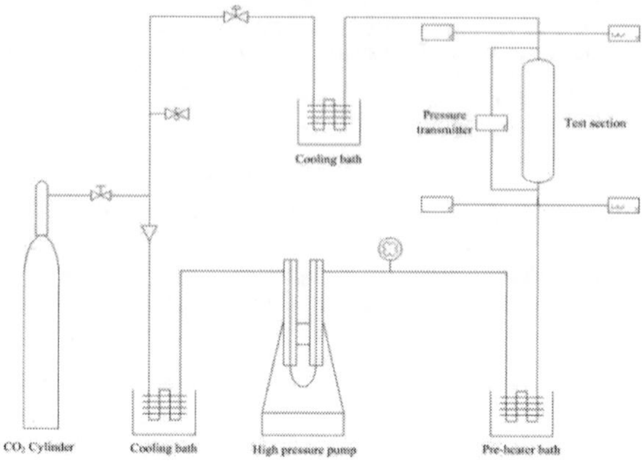

Figure 1. The schematic diagram of experimental system.

3 Simulation

In this model, a 3D free and porous media flow model combined with heat transfer model is established by finite element method – COMSOL multiphysics package. The material of tube is stainless steel. The properties of supercritical carbon dioxide are interpolated function based on the National Institute of Standards and Technology (NIST) standard reference database 69. The schematic diagram of this model is shown in figure 2.

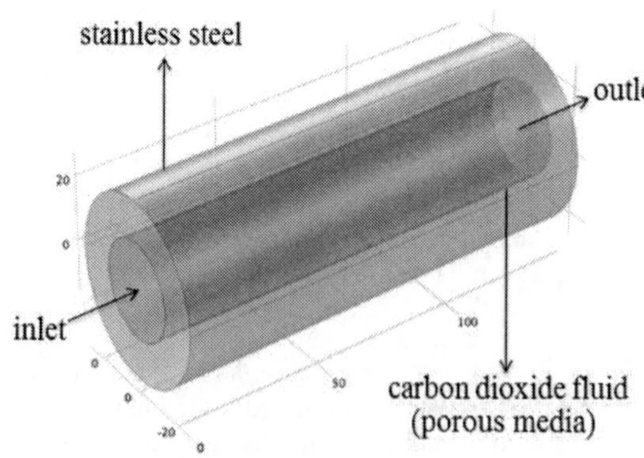

Figure 2. The schematic diagram of test tube.

The non-porous medium in the Navier-Stokes equation is described as below:

Continuity equation is:

$$\frac{\partial\rho}{\partial t}+\nabla\cdot(\rho u)=0 \qquad (1)$$

Momentum equation is:

$$\rho\frac{\partial u}{\partial t}+\rho(u\cdot\nabla)u$$
$$=\nabla\cdot\left[-pI+\mu(\nabla u+(\nabla u)^T)-\frac{2}{3}\mu(\nabla\cdot u)I\right]+F \qquad (2)$$

where μ is viscosity; u is velocity; ρ is fluid density; p is pressure and F is forced term.

The porous medium of the Brinkman equation is described as below:

Continuity equation is:

$$\frac{\partial}{\partial t}(\varepsilon\rho)+\nabla\cdot(\rho u)=Q_{br} \qquad (3)$$

Momentum equation is:

$$=-\nabla p+\nabla\cdot\left[\frac{1}{\varepsilon}\left\{\mu(\nabla u+(\nabla u^T))-\frac{2}{3}\mu(\nabla\cdot u)I\right\}\right]$$
$$-\left(\frac{\mu}{\kappa}+\frac{Q_{br}}{\varepsilon^2}\right)u+F \qquad (4)$$

where ε is porosity; κispermeability and Q_{br} is mass force.

The heat transfer of porous medium of the Fourier's law is described as below:

$$\frac{\partial}{\partial t}\left[(1-\varepsilon)\rho_p C_{p,p}T_p\right]-(1-\varepsilon)\nabla\cdot\left(k_p\nabla T_p\right)=0 \qquad (5)$$

Energy balance equation is:

$$\frac{\partial}{\partial t}\left[\varepsilon\rho_f C_{p,f}T_f\right]+\nabla\cdot\left(\rho_f C_{p,f}DT_f\right)-\varepsilon\nabla\cdot\left(k_f\nabla T_f\right)=0 \qquad (6)$$

where T is temperature; C_p is specific heat; k is thermal conductivity and D is Darcy flow; subscript f means fluid and p is media.

Figure 3 shows the comparison between the heat extraction of experiment and simulation. Through figure 3, the simulation is validated by the experiment. The error of simulation is limited within 2.3%. Therefore, we will process the optimization under this validated model.

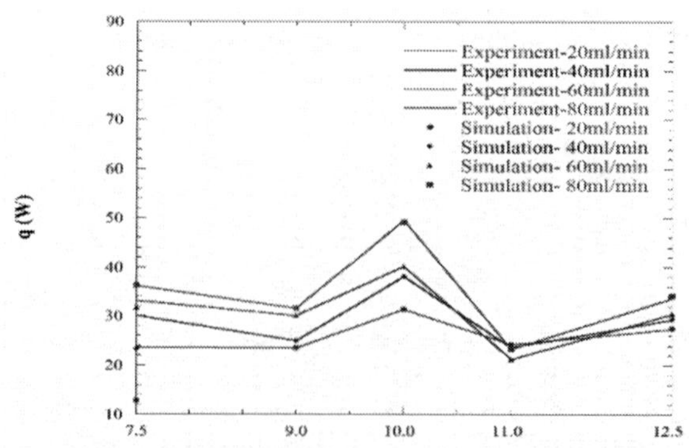

Figure 3. The comparison between experiment and simulation (porosity is 0.2).

The Nelder-Mead non-gradient method is used as the optimal method [11].

here, the constraints of design variables (pressure and flow rates) are shown as:

DOI: 10.1051/matecconf/20167102004

$$7.5 \leq P \leq 12.5\ ;\ 20 \leq U \leq 80 \qquad (7)$$

where P = pressure (MPa); U = flow rate (ml/min).

4 Results and Discussion

This study processes the optimization of heat extraction under the varied pressure and flow rate. Based on the validated model, three kinds of test tube is optimized.

Case 1: empty tube, porosity is 0.

Case 2: particle size 1.54mm, porosity is 0.2.

Case 3: particle size 2.03mm, porosity is 0.39.

First, we can find the best heat extraction is 26.86W occurred at 9.5MPa and 20ml/min as shown figure 4. After optimization, we can see more clearly the changes around. The optimal results show that the result is happened at 9.4MPa and 28.007ml/min, the corresponding heat extraction is 27.966W as shown figure 5.

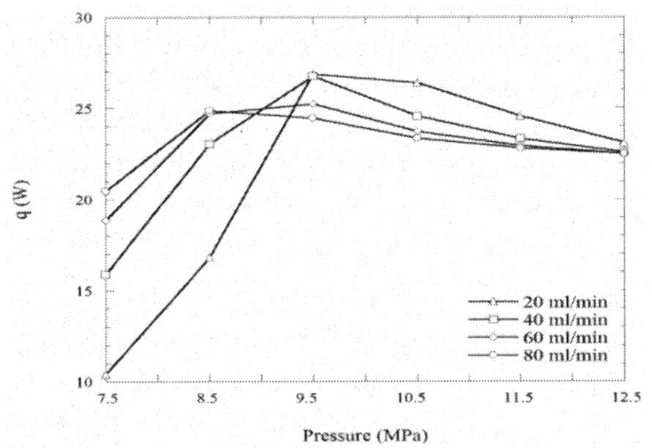

Figure 4. The initial simulation results of case 1.

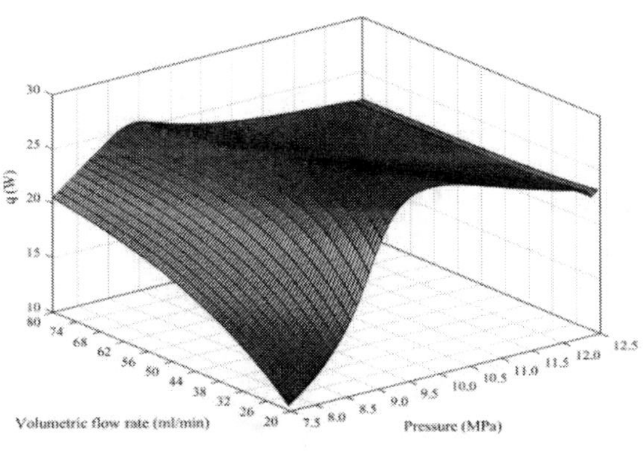

Figure 5. The optimal results of case 1.

In the porous medium as case 2 and case 3, the similar results are obtained. As shown in figures 6 and 7, the optimal results of case 2 is obtained. We can find the best heat extraction is 28.166W, occurred at 9.4MPa and 27.418ml/min.

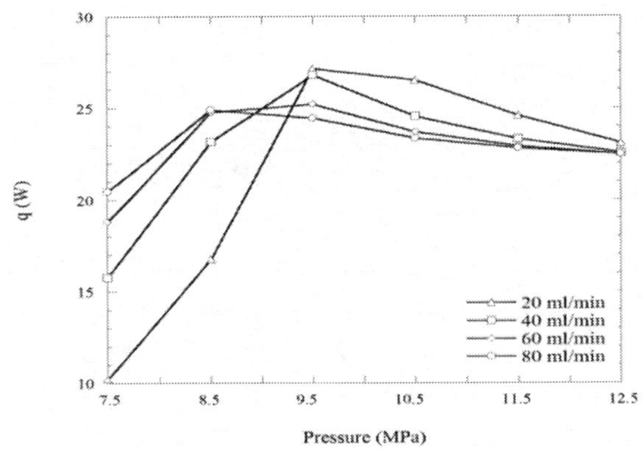

Figure 6. The initial simulation results of case 2.

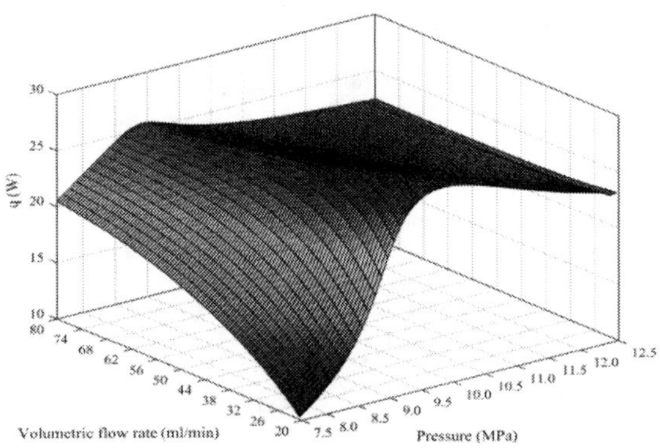

Figure 7. The optimal results of case 2.

In addition, the optimal heat extraction of case 3 is 28.168W, occurred at 9.4MPa and 27.409ml/min. The results are shown in figures 8 and 9.

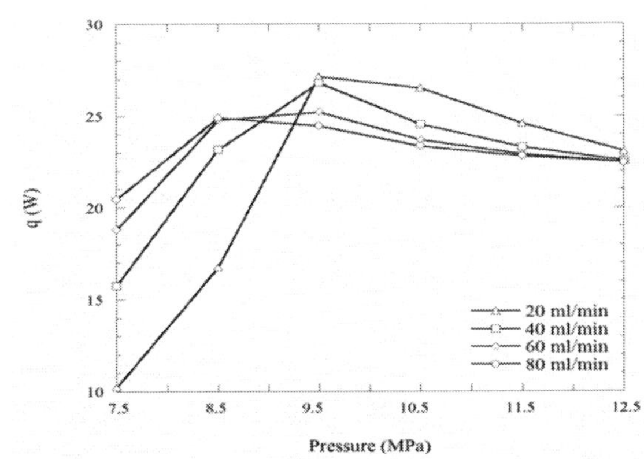

Figure 8. The initial simulation results of case 3.

DOI: 10.1051/matecconf/20167102004

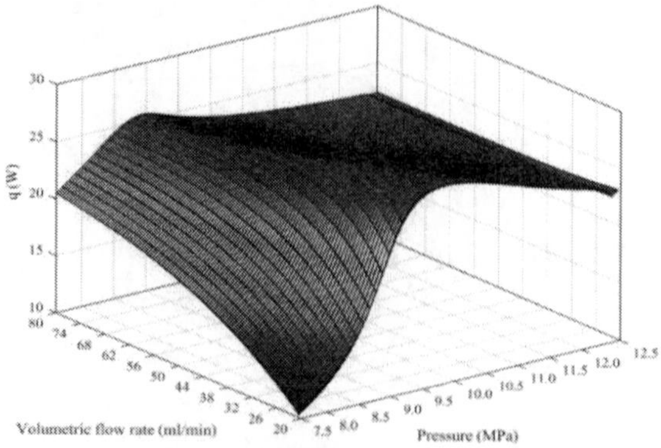

Figure 9. The optimal results of case 3.

The detailed results are listed in Table 1.

Table 1. Optimal results of the objective function

		q(W)	*P*(MPa)	*U*(ml/min)
Case 1-Empty	Initial	26.860	9.5	20.000
	optimum	27.966	9.4	28.007
Case 2-1.54mm	Initial	27.138	9.5	20.000
	optimum	28.166	9.4	27.418
Case 3-2.03mm	Initial	27.142	9.5	20.000
	optimum	28.168	9.4	27.409

5 Conclusions

This study is to find the optimal operating condition of the heat extraction of supercritical CO_2 in the geothermal reservoir. The results show that the optimal heat extraction can be obtained from the optimal process. Through the optimization, it shows the suitable flow rate is necessary to obtain the highest capability of heat extraction. It can be observed that the best heat extraction is close 9.4MPa. It results from the characteristics of supercritical CO_2. In addition, while as can be seen from the result, the influence of porosity on the heat extraction is minor. This study can improving the heat extraction and reduces the cost of actual test of enhanced geothermal system.

Acknowledgments

The financial support provided to this study by the Ministry of Science and Technology of the Republic of China under Contract No. MOST 104-3113-M-024 -001 is gratefully acknowledged.

References and Note

1. D. Brown, *Proc. Workshop on Geothermal Reservoir Eng.* **25**, 233-238 (2000).
2. C. Fouillac, B. Sanjuan, S. Gentier, C. L. Isabelle, *Annu. Conf. on Carbon Capture and Sequestration* **3**, 3-6 (2004).
3. A. Ueda, K. Kato, T. Ohsumi, T. Yajima, H. Ito, H. Kaieda, R. Meycalf, H. Takase, *Geochem. J.* **39**, 417-425 (2005).
4. Y. Wan, T. Xu, K. Pruess, *Proc. Workshop on Geothermal Reservoir Eng.* **36**, SGP-TR-191 (2011).
5. K. Pruess, *Energy Convers. Manage.* **49**, 1446-1454 (2008).
6. F. Z. Zhang, P. X. Jiang, R. N. Xu, *Appl Therm Eng.* **61**, 236-244 (2013).
7. L. Kewen, Y. B. Hui, W. L. Chang, F. Z. Dan, Y. Yanan, *Renew. Sust. Energ. Rev.* **42**, 1464-1474 (2015).
8. T. Xu, S. Eric, N. Spycher, K. Pruess, *Comput. Geosci.* **32**, 145-165 (2006).
9. P. X. Jiang, Y. J. Xu, R. F. Shi, S. He, J. D. Jackson, *Appl Therm Eng.* **24**, 1255-1270 (2004).
10. S. M. Liao, T. S. Zhao, *Int. J. Heat Mass Transfer* **45**, 5025-5034 (2002).
11. S. J. Kline, F. A. McClintock, *Mech Eng* **73**, 3-8 (1953).
12. H. M. John, D. F. Kurtis, *Numerical methods using matlab Ed. 4*, Chap8.

MATEC Web of Conferences 71, 02005 (2016)
ICCPE 2015 DOI: 10.1051/matecconf/20167102005

Modeling of Intermesh Schedule for Strong Heavy Plate Roller Straightener and Finite Element Method Modification

Xu Meng [1,2], Sun Dengyue [1,a], Xu Shimin[1] , Chen Peng[1] and Sun Hao[1]

[1] National Cold Rolled Strip Equipment and Process Engineering Technology Research Center,Yan Shan University ,066004 Hebei ,China
[2] Capital Spaceflight Machinery Company,100076 Beijing,China

Abstract. Regarding straightening unit as the research object, considered the different of roller spacing, a mathematical model of intermesh schedule suitable to the 2800 seven-roller strong heavy plate straightening machine was deduced by the geometric method. According to the mathematical model, the intermesh schedules of several specifications of plate were calculated, and the finite element model of straightening process was established in the finite element analysis software Abaqus. By analysing, it was found that those plates after straightening cannot meet the requirement of flatness, due to the existence of the work hardening. The bending deflection of the last straightening unit was modified and the new calculation formula of intermesh schedule was obtained. The values of the modified coefficients were determined by the finite element method. The intermesh schedules of other specifications of plate were calculated by the modified calculation formula and then verified it in Abaqus. The simulation results showed that those plates after straightening meet the requirement of flatness. So, the research results provide a theoretical basis for development of a new plate straightening machine and formulation of intermesh schedule.

1 General instructions

At present , the straightening intermesh schedule of strip and medium thin plate is studied more [1], but the heavy plate more than 50mm is usually straightened by the less efficient pressure straightening machine. And the algorithm of intermesh schedule for some thick plate roller straightening machine ignore the different of roller spacing, and it is inaccuracy. To obtain the straightening intermesh schedule meeting the requirements the repeated adjusting with experience is needed [2].

In order to improve the speed of formulating intermesh schedule and the straightening work efficiency , for the 2800 seven-roller strong heavy plate straightening machine designed by Yanshan University, a mathematical model of intermesh schedule is deduced. Introduced modified coefficient X ,through the simulation of the finite element software Abaqus, the effect of plate thickness and yield strength on straightening result is studied, and the value curves of thickness effect coefficient and strength effect coefficient are fitted , then the value of modified coefficient X is determined and verified by the finite element method (FEM) . The modified calculation formula Improve the accuracy of formulate intermesh schedule and reduce the number of adjustment.

2 Deduction of intermesh schedule mathematical model

2.1 Brief Introduction of 2800 Seven-Roller Strong Heavy Plate Straightening Machine

The distribution of upper three and under four is adopted by the 2800 seven roller plate straightening machine, as shown in Figure 1. The upper rollers can be adjusted vertically or inclinedly as a whole, and the import and export guide rollers which can be adjusted independently are set in the under roller system, and the intermediate two rollers are fixed. The straightening process is reversible, and it's each work roller is vertically supported by three back rollers. Work roller diameter D_w=480 mm, back roller diameter D_b=500 mm, roller spacing of under edge guide rollers t_1=600 mm, roller spacing of middle rollers t_2=520 mm; the maximum total straightening force ΣP=30000 kN, the maximum straightening power N=1000 kW; straightening range: the thickness is 30-75 mm,the maximum width is 2650 mm,the maximum yield strength is 700 MPa.

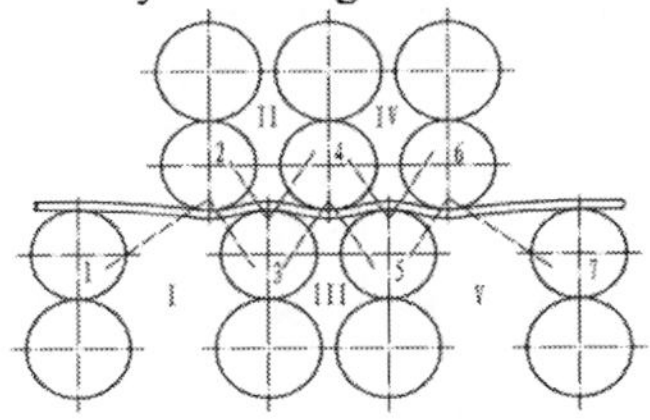

Figure 1. Distribution of roll system and straightening unit

[a] Sun Dengyue: sdy@ysu.edu.cn

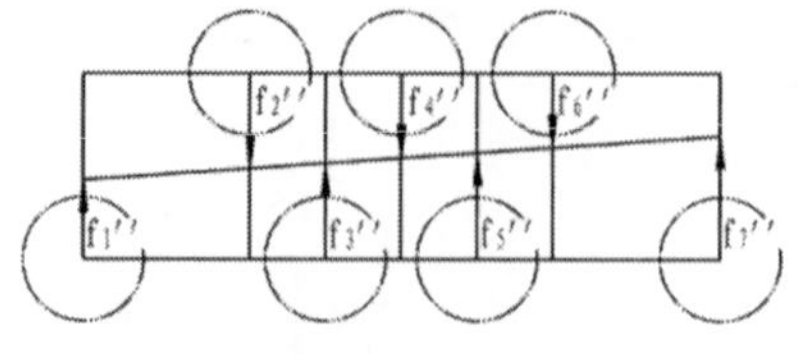

Figure 2. Diagram of the roll system adjustment

2.2 The Modeling Process

There are 5 straightening units in the roll system, whose distribution is shown in Figure 1. Due to the adopting of large deformation straightening scheme, the amount of intermesh is large at import terminal, and the plate is not so easy to enter the roll system. So the intermesh amount of the first straightening unit is decreased, just eliminating the inconsistency of plate curvature in a certain extent, so the second straightening unit is the first straightening unit of large deformation straightening scheme in the real sense.

Assuming the 2#,4# and 6# rollers were fixed, and the 3# and 5# rollers could move vertically. Let the reverse bending curvature of 3# roller is equal to the elastic back curvature $1/\rho_3$ which is needed to straighten the original curvature $-1/r_0$, according to the iterative formula [3]:

$$\frac{1}{\rho_3}=\left[3-\left(\frac{\frac{1}{\rho_w}}{\frac{1}{r_0}+\frac{1}{\rho_3}}\right)^2\right]\frac{1-\eta}{2\rho_w}+\eta\left(\frac{1}{r_0}+\frac{1}{\rho_3}\right) \tag{1}$$

Where: η——the material hardening coefficient;

$1/\rho_w$——the yield curvature, whose value is $2\sigma_s/(Eh)$, the σ_s is yield strength of material and the E is elastic modulus of material

The elastic back curvature $1/\rho_3$ is gotten through iteration.

According to the "virtual fulcrum" theory [5], the zero bending moment point of intermediate roll approximate to the midpoint of the adjacent rollers. Considering the large deformation scheme, the anti bending deflection of 3# roller in the second straightening unit is:

$$f_3''=\xi_3 f_3'''=\xi_3\frac{t_2^2}{48\rho_3} \tag{2}$$

Where: ξ_3 —— the coefficient considering large deformation degree

Let the anti bending deflection of 5# roller equal to the maximum value of elastic bending deflection:

$$f_5''=\frac{t_2^2}{48\rho_w} \tag{3}$$

For the first and fifth straightening unit, due to the diffidence of roller spacing, the zero bending moment point is no longer at the midpoint of straightening unit. The anti bending deflection of 2# and 6# roller respectively is:

$$f_2''=\frac{t_2}{24}(2t_1-t_2)\frac{1}{\rho_2} \tag{4}$$

$$f_6''=\frac{t_2}{24}(2t_1-t_2)\frac{1}{\rho_6} \tag{5}$$

Let the reverse bending curvature of 2# roller equal to the the elastic back curvature of 3# roller $1/\rho_3$, and let the reverse bending curvature of 6# roller equal to the the yield curvature $1/\rho_w$.

According to geometrical relationship the adjustment amount of each roller is deduced, ignoring the displacement of the upper rollers in the horizontal direction because of the overall inclination adjustment of upper roller system.

For the movable roller, its adjustment amount is equal to the sum of its reverse bending deflection and a half of the sum of reverse bending deflections of its adjacent rollers.

For the fixed rollers, there is no need to get the adjustment amount, but the reverse bending deflection should be calculated, whose value is the average value of its adjacent two rollers.

Assuming the upper rollers were fixed, so the adjustment amount of 3# and 5# roller respectively is:

$$\Delta h_3=f_3''+\frac{1}{2}(f_2''+f_4'') \tag{6}$$

$$\Delta h_5=f_5''+\frac{1}{2}(f_4''+f_6'') \tag{7}$$

According to geometrical relationship of relative motion shown in Figure 2, adjustment amount of 4# roller is equal to the average value of adjustment amount of 3# roller and 5# roller.That is:

$$\Delta h_4=\frac{1}{2}(\Delta h_3+\Delta h_5) \tag{8}$$

And because the upper roller system can be adjusted overall in practice, the reverse bending deflection of 4# roller is also calculated:

$$\Delta h_4=f_4''+\frac{1}{2}(f_3''+f_5'') \tag{9}$$

So reverse bending deflection of 4# roller is obtained according to the equation (6)-(9):

$$f_4''=\frac{1}{2}(f_2''+f_6'') \tag{10}$$

According to geometrical relationship of relative motion once again:

$$\Delta h_2=\frac{3}{2}\Delta h_3-\frac{1}{2}\Delta h_5 \tag{11}$$

$$\Delta h_6=\frac{3}{2}\Delta h_5-\frac{1}{2}\Delta h_3 \tag{12}$$

According to the reverse bending deflections and adjustment amounts of 2# and 6# roller and the geometrical relationship of relative motion, the adjustment amounts of 1# and 7# roller are calculated:

$$\Delta h_1=\frac{2t_1}{t_2}(f_2''-\Delta h_2) \tag{13}$$

$$\Delta h_7=\frac{2t_1}{t_2}(f_6''-\Delta h_6) \tag{14}$$

Finaly, through arranging these equations above, the adjustment amounts of 1#, 2#, 4#, 6#, and 7# roller are obtained:

$$\left.\begin{aligned}
\Delta h_1 &= \frac{t_1 t_2}{48}\left(\frac{2\sigma_s}{hE} - 3\xi_3\frac{1}{\rho_3}\right)\\
\Delta h_2 &= \frac{t_2}{8}\left(\frac{2t_1 - t_2}{3}\frac{1}{\rho_3} + \xi_3\frac{t_2}{4}\frac{1}{\rho_3} - \frac{t_2\sigma_s}{6hE}\right)\\
\Delta h_4 &= \frac{t_2}{48}\left[\frac{t_2}{2}\left(\xi_3\frac{1}{\rho_3} + \frac{2\sigma_s}{hE}\right) + (2t_1 - t_2)\left(\frac{1}{\rho_3} + \frac{2\sigma_s}{hE}\right)\right]\\
\Delta h_6 &= \frac{t_2}{8}\left(\frac{2t_1 - t_2}{3}\frac{2\sigma_s}{hE} + \frac{t_2}{4}\frac{2\sigma_s}{hE} - \xi_3\frac{t_2}{12}\frac{1}{\rho_3}\right)\\
\Delta h_7 &= \frac{t_1 t_2}{48}\left(\xi_3\frac{1}{\rho_3} - \frac{6\sigma_s}{hE}\right)
\end{aligned}\right\} \tag{15}$$

3 Finite element simulation

3.1 Establishment of Finite Element Model

Using the formula (15), the straightening intermesh schedules of the specifications of plate as shown in Table 1 are calculated, and the finite element models are established in the Abaqus.

Take the 400 MPa/50 mm specification plate as the example to descript the finite element modeling process.

Table 1. Specifications of plate need to be straightening Simulated

Specifications	**Spec1**	**Spec2**	**Spec3**	**Spec4**	**Spec5**	**Spec6**
Yield strength/MPa	400	400	400	300	500	700
Thickness/mm	30	50	70	50	50	50

To improve the computational efficiency, the 1/2 model is adopted, and the work roller is analytical rigid, the whole model is shown in Figure 3. The unevenness of plate is generated randomly; the coordinate distribution along the length direction is shown in Figure 4. The maximum difference of plate in high direction (Y direction) is about 38 mm, and the plate unevenness is 7.3 mm•m^{-1}. The elastic modulus of steel is 21000 MPa, the poison's ratio is 0.3, and the material stress-strain curve is shown in Figure 5.The plate model is divided by C3D8R (eight nodes linear grid), whose size is 50 mm*25 mm*5 mm, 57200 elements in total.

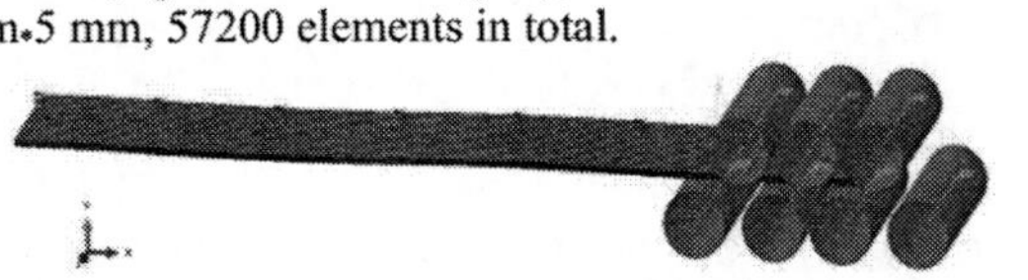

Figure 3. The whole model

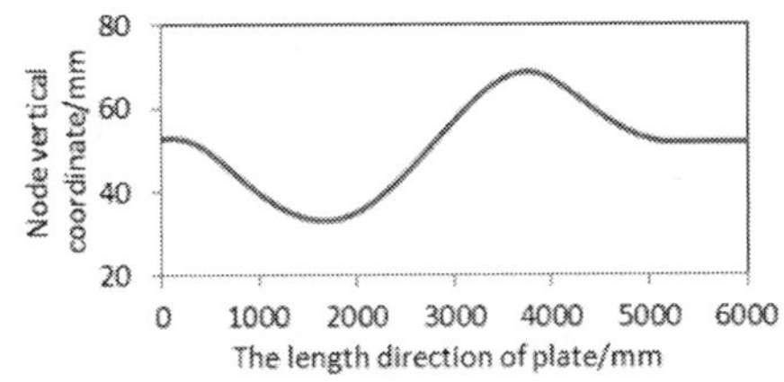

Figure 4. Coordinate distribution of plate intermediate nodes along the length direction before straightening

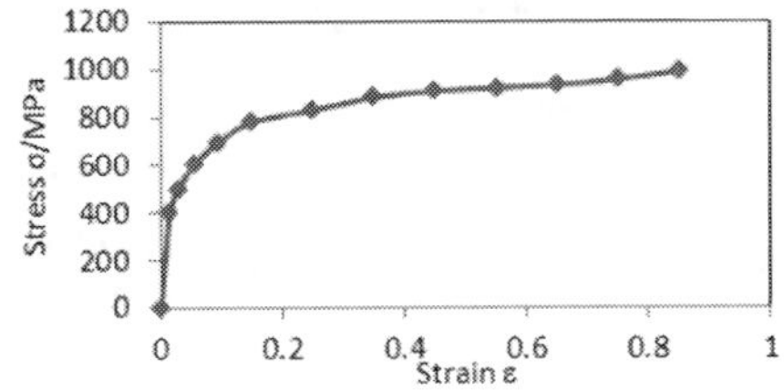

Figure 5. The material stress-strain curve

3.2 Analysis of Results

Through simulation, the stress distribution of 400 MPa/50 mm plate in the roller system is shown in Figure 6. The maximum stress value is 639 MPa, and most of the stress value exceeds the yield strength of 400 MPa. Therefore, plastic deformation of plate in the straightening simulation process occurs. In addition, due to the work hardening, with reverse bending deformation, although the intermesh amount of last straightening unit is smaller ,the stress value is increased.

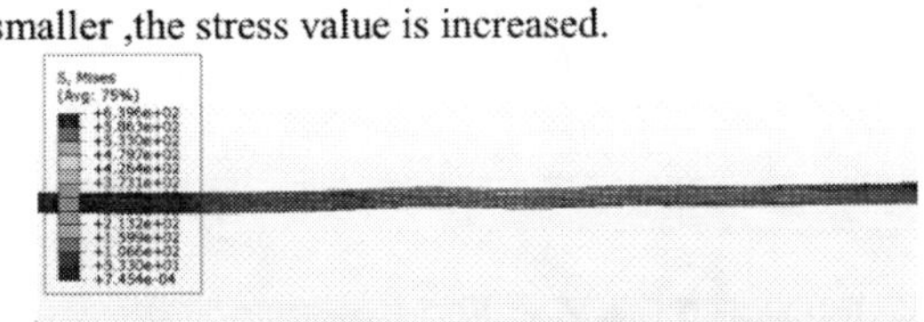

Figure 6. Mises stress distribution of the plate in the roll system

The node coordinates in the middle of the plate after straightening are extracted and rotated according to the coordinate rotation formula (16), and the starting point of the curve is translated to the origin of the coordinate. So the curve is shown in Figure 7.

$$\left.\begin{aligned}
x &= x_1\cos\theta + y_1\sin\theta\\
y &= -x_1\sin\theta + y_1\cos\theta
\end{aligned}\right\} \tag{16}$$

Where: (x,y): coordinate after rotation, (x_1,y_1): coordinate before rotation, θ :the angle between original graph and rotation graph.

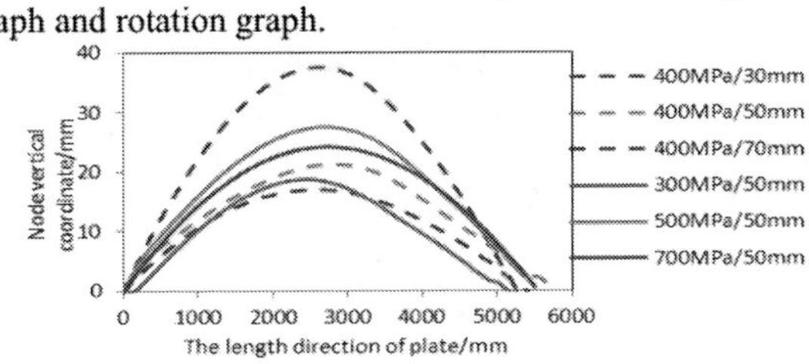

Figure 7. Coordinate value curve of plate nodes after straightening

DOI: 10.1051/matecconf/20167102005

As shown in the figure, the unevenness of the plates after straightening get some improvement, but cannot meet the flatness requirements of high quality plate (less than 1 $mm.m^{-1}$). In addition it can be found from the figure that along the length of the plate the two bends turn into one, and the residual curvature tends to be consistent. Through analysis and inference, the reason for this phenomenon is that after the plastic deformation in front of the straightening unit the steel plate produces work hardening and the yield strength of the material increases, while the yield curvature increases, but the intermesh amount in the simulation is calculated by the original yield strength, so the anti bending deflection of the last one straightening unit is too small, namely the intermesh of 6# roller is not enough . In this case, the modified coefficient X is introduced to improve this phenomenon. The anti bending deflection of 6# roller is:

$$f_6'' = X\frac{t_2}{24}(2t_1 - t_2)\frac{1}{\rho_6} \tag{17}$$

After modifying, the calculation formula of intermesh schedule is:

$$\left.\begin{aligned}
\Delta h_1 &= \frac{t_1 t_2}{48}\left(\frac{2\sigma_s}{hE} - 3\xi_3\frac{1}{\rho_3}\right)\\
\Delta h_2 &= \frac{t_2}{8}\left(\frac{2t_1 - t_2}{3}\frac{1}{\rho_3} + \xi_3\frac{t_2}{4}\frac{1}{\rho_3} - \frac{t_2\sigma_s}{6hE}\right)\\
\Delta h_4 &= \frac{t_2}{48}\left[\frac{t_2}{2}\left(\xi_3\frac{1}{\rho_3} + \frac{2\sigma_s}{hE}\right) + (2t_1 - t_2)\left(\frac{1}{\rho_3} + X\frac{2\sigma_s}{hE}\right)\right]\\
\Delta h_6 &= \frac{t_2}{8}\left(X\frac{2t_1 - t_2}{3}\frac{2\sigma_s}{hE} + \frac{t_2}{4}\frac{2\sigma_s}{hE} - \xi_3\frac{t_2}{12}\frac{1}{\rho_3}\right)\\
\Delta h_7 &= \frac{t_1 t_2}{48}\left(\xi_3\frac{1}{\rho_3} - \frac{6\sigma_s}{hE}\right)
\end{aligned}\right\} \tag{18}$$

3.3 Determine the Modified Coefficient

Adjusting the modified coefficient, the intermesh schedules corresponding different modified coefficients for each specifications of plates are calculated. Through simulating respectively, then extracting and processing the result data, the modified coefficients corresponding straightening effect (the unevenness is less than 1 $mm{\bullet}m^{-1}$), as shown in Figure 8., meeting the flatness requirement is determined, as shown in the Table 2.

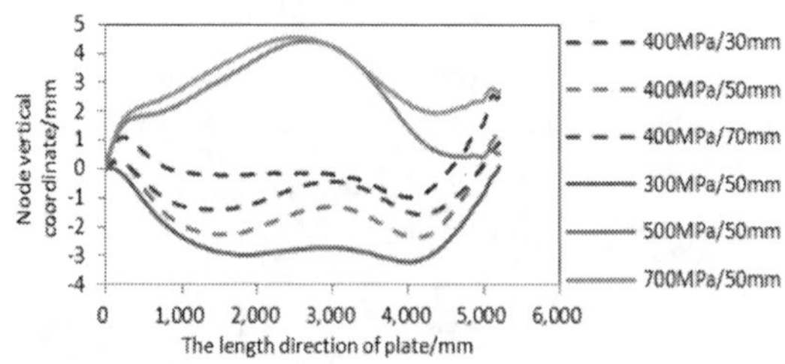

Figure 8. Coordinate value curves of 400MPa with different thickness plate after straightening

Table 2. Modified coefficients corresponding to different specification of plates

Specifications	**Spec 1**	**Spec 2**	**Spec 3**	**Spec 4**	**Spec 5**	**Spec 6**
Yield strength/MPa	400	400	400	300	500	700
Thickness/mm	30	50	70	50	50	50
modified coefficient	1.13	1.45	1.90	1.46	1.42	1.43

From the table, it can be found that the thickness of the plate has a great influence on the modified coefficient in the formula, and the value of the modified coefficient increases with the increase of the thickness. Through analysis, the reason is that the work hardening of plate after anti bending deformations in the first four straightening units increases with the increasing of thickness, and the insufficient degree of intermesh amount of 6# roller also increases, so the modified coefficient is larger. On the other hand, the yield strength of the plate has a small influence on the modified coefficient. This is because that the insufficient of intermesh amount of 6# roller is related to the degree of variation of yield strength of the material before and after work hardening , but is not directly related to the size of the yield strength before work hardening.

In order to apply modified coefficient to practice more conveniently for the plates with different thickness and different yield strength, the modified coefficient X is summarized as the product of thickness effect coefficient X_h and strength effect coefficient X_σ, namely $X=X_hX_\sigma$。 Because the influence of plate yields strength on the modified coefficient is small, on the basis of the modified coefficient of 400 MPa/50 mm plate, each 50 mm thickness plate corresponding to the correction coefficient in the Table 2. is divided by 1.45 that is modified coefficient of 400 MPa/50 mm plate , and the values of strength effect coefficient are obtained ,as shown in Table 3. And the value of the plate thickness effect coefficient is the modified coefficient of different thickness plate with 400MPa yield strength , that is the modified coefficients corresponding to Spec 1—Spec 3 in the Table 2.

Table 3. The values of strength effect coefficient

Yield strength /MPa	300	400	500	600	700
Strength effect coefficient	1.034	1.000	0.979	0.986	0.986

The thickness effect coefficient and the strength effect coefficient are anchored point in the respective coordinate system and the value curves are fitted then, as shown in Figure 9 and Figure 10. The two curves can be used to read value for a certain specification plate in practice, and the modified coefficient X is calculated, then the values are taken into the formula (18), and the intermesh schedule is obtained finally.

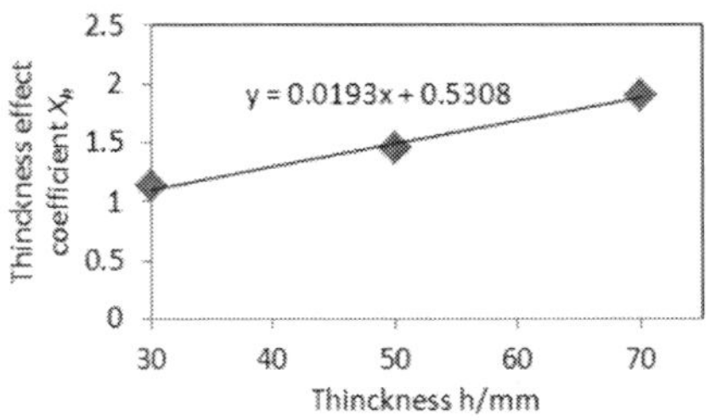

Figure 9. The value curve of thickness effect coefficient

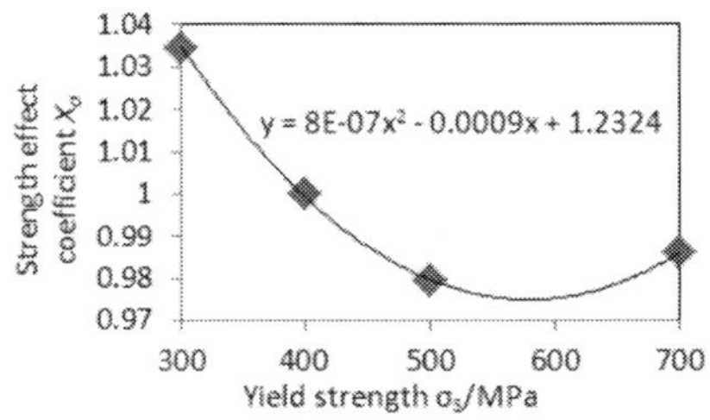

Figure 10. The value curve of strength effect coefficient

4 Verification

In order to verify whether the straightening quality of the calculation formula of intermesh schedule after modifing is improved , the straightening intermesh schedules of plates of 400 MPa/40 mm,400 MPa/60 mm and 600 MPa/50 mm specifications are calculated by the modified formula , then the finite element models are established and simulated . The result dates are processed to observe whether the plates after straightening meet the flatness requirement.

According to the thickness and yield strength of the plates need to be verified, thickness effect coefficient and strength effect coefficient are obtained from the figure 9 and 10. The modified coefficient is the product of thickness effect coefficient and strength effect coefficient, and then the modified coefficient is brought into the formula (18), the intermesh amount of each roller is calculated. The finite element models of straightening process are established and simulated, and then plate node coordinates of simulation results are extracted and processed according to the above method. The unevennesses of the plates verified after straightening are obtained and shown in the table 4. From the table, it is found that each unevenness of the plates after straightening is less than 1 mm.m^{-1}, and meets the flatness requirement of high quality plate. Therefore, the modified calculation formula of straightening intermesh schedule meets the accuracy requirements, and can be used in formulating the straightening ingtermesh schedule for the newly developed 2800 seven-roller strong heavy plate straightening machine in the practical production.

Table 4. The values of modified coefficient of the plates need to be verified and the unevenness after straightening

Plate specifications	**400MPa /40 mm**	**400MPa /60 mm**	**600MPa /50 mm**
Thickness effect coefficient Xh	1.31	1.68	1.45
Strength effect coefficient Xσ	1	1	0.98
Modified coefficient X	1.31	1.68	1.42
Unevenness/ mm•m-1	0.65	0.77	0.56

5 Conclusions

Regarding straightening unit as the research object, considered the different of roller spacing, a mathe-matical model of intermesh schedule suitable to the 2800 seven-roller strong heavy plate straightening machine is deduced by the geometric method, and a new calculationg method of intermesh schedule is offered for other heavy plate straightening machine.

According to mathematical model, the intermesh schedule is calculated, and the finite element model of straightening process is established in the FEA software Abaqus . Through simulation, it is found that the residual curvature decreases and tends to be consistent, but has not yet reached the flatness re-quirements . By analyzing, the reason is that because of the existence of work harding, the anti bending deflection of the last straightening unit is too small.

Introducing modified coefficient *X*, a new calculation formula of straightening intermesh schedule is obtained, and the value of the modified coefficient is obtained by the finite element method.

Through the verification of other specification plates, the modified calculation formula of straightening intermesh schedule meets the accuracy requirements, and together with the value curves of the effect coefficient, can provide theoretical basis for the development of the new heavy plate straightening machine and the formulation of straightening intermesh schedule.

References

1. W. Zhigang, W. Hui, L. Yourong . J.W.U.S.T., *Method for Calculating the Intermesh of Straightener and Its Influence on Straightening Quality*. **32(4)**, 347-350 (2009).
2. W. Xiaogang, H. Qingxue, M. Qin, et al. J.P.E., *Research of Deficient Deformation Level ing Model in Heavy Plate*. **15(5)**, 191-194 (2008).
3. L. Jiachuang. C.M.P., *Straightening Theory and Curling Theory* . 1-18 (2011).
4. L. Jiachuang. M.E., *Draft Calculation and Setting for Roller Leveller of Strip and Plate*.**182**, 1-4 (2010).
5. H. Yuhua, X. Youzhong, Y. Huilin. H.M., *New Exploration of Straightening Deformation Theory on the Roller-strainghtener*. **6**, 28-30 (1999).
6. L. Yilun. J.C.S.U., *A Math Model on Bending Correction and Its Solution*. **19(3)**:316-324(1988).
7. B. Xiangzhong. S.P.,*Mechanics of Materials*. 159-161 (2007).
8. C. Pu. M.I.P., *Straightening Principle and Straightening Machine*. 95-98 (2002).

9. G. Ben, Z. Yong, Q. Weizhuang, et al. J.M.E., *Stress Evolvement and Its Influence on Bending Behavior during Roller Leveling Process*. **48(2)**, 81-86 (2012).
10. W. Xiaogang, H. Qingxue, H. Ying. C.M.E.,*Correction and Application of Numerical Model on Plate Roller Levelling*. **03**: 335-338 (2012).
11. M.Fang. J.A.,*Complements of Tableaux and Straightening Bideterminants*. **322**,1170–1186(2009).
12. L.A.De Bkjar, P.F.R.A. M.M.,*Integrating Mathematical CAD and Artificial Simulation of Plastic Rotation in Heat-straightening Processes*. Appl.,**17**, 34-40 (1993).
13. Z. Cunlong, W. Guodong ,L. Xianghua ,et al. J.P.E., *Work Hardening Modulus and Reverse Bend Curvature in Plate Hot Roller Leveling*. **14(4)**, 133-135 (2007).
14. G. Benyue, W. Zhigang, L. Yourong, et al. J.P.E., *Finite Element Analysis for the Residual Stress of Plate During Straightening Process*., **17(6)**, 116-120 (2010).
15. Z. Huajian, D. Lei. E.I.P., *ABAQUS Basics and Cases* Master. 28-163 (2012).

MATEC Web of Conferences 71, 03001 (2016) DOI: 10.1051/matecconf/20167103001
ICCPE 2015

A Study of Effects of Open Ceremony on e-learning Account Sustainability

L.H. Kuo, H.J. Yang & L.M. Chen[a]

National Kaohsiung Normal University, Taiwan, R.O.C.

Abstract. The purpose of this study was to identify whether an open ceremony affects on-line learning platform account keeping behavior. This is a longitudinal non-experimental study. Total 9959 user accounts on a e-learning server since 2009 were reviewed. The server provides on-line learning courses to the publics. There are two types of courses, one with open ceremony for learners in classroom and the other is not. A survival analysis was performed to assess the effectiveness of the ceremony for user account sustainability. It was found that after five years, the survival rate of ceremony group is 0.53 and non-ceremony group is 0.01. Based upon the statistically significant effect, it was concluded that open ceremony had shown effects on user account sustainability.

1 Introduction

E-learning, Open-Courses, & Platforms
In this information age, information system provides a flexible communication solution. It also plays an important way for supporting personal learning by establishing well organized educational environment.

E-learning had become a foundation of life-long learning. No matter when, or where, one can always rely on e-learning to exploring knowledge according one's will. E-learning provides a new concept on learning.

Open-course is another break though on the other side of e-learning. Open-courses are well organized learning activities and freely open to the publics. Based upon the concept of e-learning on the learner side, open-courses on the content side, and internet connected platform, a fully functional education environment is established. A conceptual model is illustrated in Figure 1.

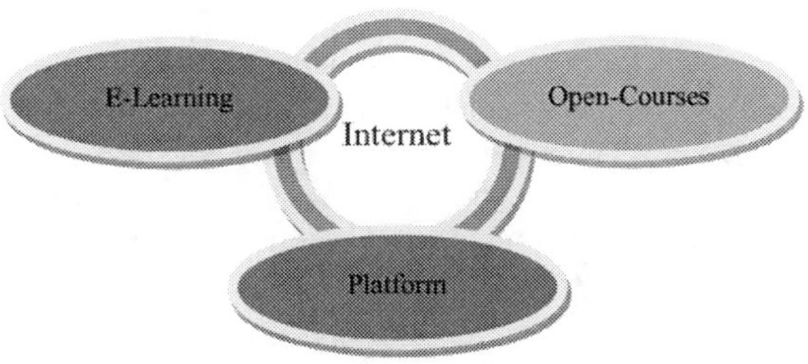

Figure 1. A fully functional education environment

1.1 Problem Statement

For managing an information system, there is always a need to understand user's behavior on system service. In an open learning environment, how an information service provider can keep their users on using service is an important problem.

The purpose of this study was to identify whether an open ceremony affects on-line learning platform account keeping behavior.

2 Literature review

2.1 Massive Open Online Courses

A MOOC is an online educational platform that provides the dissemination of content to large numbers of users who have access to the Internet. The fundamental design of a MOOC affords users the ability to watch videos, access assignments, utilize resources, and connect with fellow users and instructors. Some MOOCs require users to complete tests and quizzes that can be automatically graded, while others rely on peer reviewed feedback from other current users.

Most MOOCs are free but do not provide transferable credit to other institutions. Some institutions do provide users the ability to obtain a "certificate of completion" by completing a final exam or standardized assignments, but this credential is not transferable as college-level credit. Recently, the American Council on Education accredited five MOOCs for which users can earn transferable credit. These courses require that students pay a fee and

[a]Corresponding author: Li-Min Chen <chenlm7@gmail.com>

successfully complete assignments and exams before being awarded credit.

Several companies, organizations, and institutions support MOOCs and opensource education to users. As of January 2013, Coursera, Udacity, and edX were the three leading organizations to provide online platforms for MOOCs(Coursera, 2013).

3 Methodology

3.1 Survival / Failure Analysis

Survival/failure analysis is a family of techniques dealing with the time it takes for something to happen: a cure, a failure, an employee leaving, a relapse, a death, and so on.

One interesting feature of the analysis is that survival time (the DV) often is unknown for a number of cases at the conclusion of the study. Some of the cases are still in the study but have not yet failed: some employees have not yet left, some components are still functioning, some patients are still apparently well, or some patients are still living. For other cases, the outcome is simply unknown because they have withdrawn from the study or are for some reason lost to follow-up. Cases whose DV values—survival time— are unknown for whatever reasons are referred to as censored.

The term survival analysis is based on medical applications in which time to relapse or death is studied between groups who have received different medical treatments. Is survival longer for a group receiving ceremony rather than another?

3.2 Life tables

Within the family of survival- analysis techniques, different procedures are used depending on the nature of the data and the kinds of questions that are of greatest interest.

Life tables describe the survival (or failure) times for cases, and often are accompanied by a graphical representation of the survival rate as a function of time, called a survivor function. Survivor functions are frequently plotted side-by-side for two or more groups (e.g., treated and untreated patients) and statistical tests are used to test the differences between groups in survival time.

3.3 Proportions Surviving at Various Time

The primary goal of one type of survival analysis is to describe the proportion of cases surviving at various times, within a single group or separately for different groups. The analysis extends to statistical tests of group differences. The primary goal of the other type of survival analysis is to assess the relationship between survival time and a set of covariates (predictors), with treatment considered one of the covariates, to determine whether treatment differences are present after statistically controlling for the other covariates. We may ask that what the survival rate at various points in time is.

3.4 Group Differences in Survival

This study also asks that If there are different groups, are their survival rates different? Several tests are available to evaluate group differences. If statistically significant group differences are found, separate life tables and survivor functions are developed for each group.

3.5 Limitations to survival Analysis

One problem with survival analysis is the nature of the outcome variable, time itself. Events must occur before survival or failure time can be analyzed: Components must fail, users must leave, and patients must succumb to the illness. However, the purpose of treatment often is to delay this occurrence or prevent it altogether. The more successful the treatment, then, the less able the researcher is to collect data in a timely fashion.

Survival analysis is subject to the usual cautions about causal inference.

3.6 Fundamental Equations for Survival Analysis

Life tables are built around time intervals; in this study, the intervals are of width 3 months. The survivor function, P, is the cumulative proportion of cases surviving to the beginning of the i + 1st interval, estimated as:

$$P_{i+1} = p_i P_i$$

where:

$$p_i = 1 - q_i$$

$$q_i = \frac{d_i}{r_i}$$

and

where d_i = number responding (dropping out) in the interval and

$$r_i = n_i - \frac{1}{2} c_i$$

where n_i= number entering the interval

c_i= number censored in the interval (lost to follow up for reasons other than dropping out).

The proportion of cases surviving to the (i+1)st interval is the proportion who survived to the start of ith interval times the probability of surviving to the end of the ith interval.

4 Findings

4.1 Descriptive Statistics

In Table 1., the months of account sustainability was listed. The minimum is four months and the maximum is fifty months. The average account sustains 34.9 months. The standard deviations are 15.271 months. The total accounts are 9959.

Table 1. N, Minimum, Maximum, Mean, & Std. Deviation of Account Sustainability (Months)

Descriptive Statistics					
	N	Minimum	Maximum	Mean	Std. Deviation
MONTHS	9959	4	50	34.90	15.271
Valid N (listwise)	9959				

In Table 3, the frequency of account status was listed in two types under still using and not using. There are 4052, 40.7%, active accounts and 5907, 59.3%, discontinued accounts.

Table 2. Life table of account sustainability

First-order Controls		Interval Start Time	Number Entering Interval	Number Withdrawing during Interval	Number Exposed to Risk	Number of Terminal Events	Proportion Terminating	Proportion Surviving	Cumulative Proportion Surviving at End of Interval	Std. Error of Cumulative Proportion Surviving at End of Interval	Probability Density	Std. Error of Probability Density	Hazard Rate	Std. Error of Hazard Rate
CEREMONY	NON_CEREMONY	0	2402	0	2402.000	0	.00	1.00	1.00	.00	.000	.000	.00	.00
		3	2402	0	2402.000	30	.01	.99	.99	.00	.004	.001	.00	.00
		6	2372	0	2372.000	169	.07	.93	.92	.01	.023	.002	.02	.00
		9	2203	0	2203.000	8	.00	1.00	.91	.01	.001	.000	.00	.00
		12	2195	0	2195.000	1072	.49	.51	.47	.01	.149	.003	.22	.01
		15	1123	0	1123.000	20	.02	.98	.46	.01	.003	.001	.01	.00
		18	1103	0	1103.000	25	.02	.98	.45	.01	.003	.001	.01	.00
		21	1078	0	1078.000	23	.02	.98	.44	.01	.003	.001	.01	.00
		24	1055	0	1055.000	437	.41	.59	.26	.01	.061	.003	.17	.01
		27	618	0	618.000	18	.03	.97	.25	.01	.002	.001	.01	.00
		30	600	0	600.000	20	.03	.97	.24	.01	.003	.001	.01	.00
		33	580	0	580.000	16	.03	.97	.23	.01	.002	.001	.01	.00
		36	564	0	564.000	408	.72	.28	.06	.01	.057	.003	.38	.02
		39	156	0	156.000	123	.79	.21	.01	.00	.017	.001	.43	.03
		42	33	0	33.000	0	.00	1.00	.01	.00	.000	.000	.00	.00
		45	33	0	33.000	0	.00	1.00	.01	.00	.000	.000	.00	.00
		48	33	33	16.500	0	.00	1.00	.01	.00	.000	.000	.00	.00
	CEREMONY	0	7557	0	7557.000	0	.00	1.00	1.00	.00	.000	.000	.00	.00
		3	7557	0	7557.000	21	.00	1.00	1.00	.00	.001	.000	.00	.00
		6	7536	0	7536.000	112	.01	.99	.98	.00	.005	.000	.00	.00
		9	7424	0	7424.000	5	.00	1.00	.98	.00	.000	.000	.00	.00
		12	7419	0	7419.000	715	.10	.90	.89	.00	.032	.001	.03	.00
		15	6704	0	6704.000	14	.00	1.00	.89	.00	.001	.000	.00	.00
		18	6690	0	6690.000	16	.00	1.00	.88	.00	.001	.000	.00	.00
		21	6674	0	6674.000	16	.00	1.00	.88	.00	.001	.000	.00	.00
		24	6658	0	6658.000	880	.13	.87	.76	.00	.039	.001	.05	.00
		27	5778	0	5778.000	12	.00	1.00	.76	.00	.001	.000	.00	.00
		30	5766	0	5766.000	13	.00	1.00	.76	.00	.001	.000	.00	.00
		33	5753	0	5753.000	11	.00	1.00	.76	.00	.000	.000	.00	.00
		36	5742	0	5742.000	1527	.27	.73	.56	.01	.067	.002	.10	.00
		39	4215	0	4215.000	107	.03	.97	.54	.01	.005	.000	.01	.00
		42	4108	0	4108.000	43	.01	.99	.54	.01	.002	.000	.00	.00
		45	4065	0	4065.000	18	.00	1.00	.54	.01	.001	.000	.00	.00
		48	4047	4019	2037.500	28	.01	.99	.53	.01	.002	.000	.00	.00

Table 3. Frequency, & Percent of accounts status

N_USING					
		Frequency	Percent	Valid Percent	Cumulative Percent
Valid	STILL USING	4052	40.7	40.7	40.7
	NOT USING	5907	59.3	59.3	100.0
	Total	9959	100.0	100.0	

In Table 3, the frequency information of ceremony was illustrated. Those accounts with ceremony experience are 7559, 75.9%. Those who without ceremony experience are 2402, 24.1%.

Table 4. Frequency, & Percent of Ceremonial Experience

CEREMONY					
		Frequency	Percent	Valid Percent	Cumulative Percent
Valid	NON_CEREMONY	2402	24.1	24.1	24.1
	CEREMONY	7557	75.9	75.9	100.0
	Total	9959	100.0	100.0	

4.2 Life table of account sustainability

In Table 4., the intervals are width three months. The first-order controls are non-ceremony and ceremony. Items listed in Table 4. are number entering interval, number withdrawing during interval, number exposed to risk, number of terminal events, proportion terminating, proportion surviving, cumulative proportion surviving at end of interval, std. error of cumulative proportion surviving at end of interval, probability density, std. error of probability density, hazard rate, and std. error of hazard rate.

For the group without ceremony experience, there are 2,402 accounts in the beginning. In the first 15 months, the number drop less then half of the original, the number for the fifteenth months is 1,123.

For the group with ceremony experience, there are 7,557 accounts in the beginning. Till the end of the fiftieth month, the value is still higher than half of the beginning. The value is 4,047.

4.3 Group Comparison

In Table 5, the median survival time of non-ceremonial experience are 14.78 months and the median survival time of ceremonial experience are 48 months.

Table 5. Median survival time

Median Survival Time		
First-order Controls		Med Time
CEREMONY	NON CEREMONY	14.78
	CEREMONY	48.00

In Table 6, the significant level is lesser than .05, the groups are significantly different by the log-rank test.

Table 6. Comparisons for factor ceremony status

Overall Comparisons [a]		
Wilcoxon (Gehan) Statistic	df	Sig.
2860.997		.000

a. Comparisons are exact.

Figure 1, the survival function for cumulating survival rate by months was drawn. The dotted line was drawn according the ceremonial experience group.

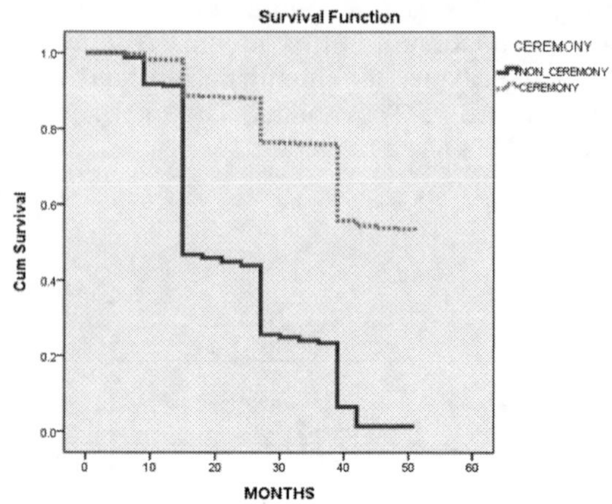

Figure 2. Survival Function

The solid line is for the non-ceremonial experience group.

In Table 7, the mean account sustainability was listed. The value of non-ceremony group is 20.53 months. The value of ceremony group is 39.47 months.

Table 7. N, Mean, Std. Deviation of Account Sustainability for both non-ceremony & ceremony groups

Descriptives								
MONTHS								
					95% Confidence Interval for Mean			
	N	Mean	Std. Deviation	Std. Error	Lower Bound	Upper Bound	Minimum	Maximum
NON CEREMONY	2402	20.53	11.140	.227	20.08	20.97	4	50
CEREMONY	7557	39.47	13.465	.155	39.17	39.78	4	50
Total	9959	34.90	15.271	.153	34.60	35.20	4	50

Table 8. One-way ANOVA for Comparing Account Sustainability of Groups

ANOVA					
MONTHS					
	Sum of Squares	df	Mean Square	F	Sig.
Between Groups	654339.151	1	654339.151	3906.391	.000
Within Groups	1667845.036	9957	167.505		
Total	2322184.188	9958			

In Table 8, the result of comparing was significant. There exists significant difference between groups.

5 Conclusions

The purpose of this study was to identify whether an open ceremony affects on-line learning platform account keeping behavior. This is a longitudinal non-experimental study. Total 9959 user accounts on a e-learning server since 2009 were reviewed. The server provides on-line learning courses to the publics.

There are two types of courses, one with open ceremony for learners in classroom and the other is not. The mean value of non-ceremony group is 20.53 months. The mean value of ceremony group is 39.47 months. Based upon one-way ANOVA test, there is significant difference between ceremony group and non-ceremony.

In the other words, the mean value of ceremony group is significantly higher than the mean value of non-ceremony group.

A survival analysis was performed to assess the effectiveness of the ceremony for user account sustainability. It was found that after five years, the survival rate of ceremony group is 0.53 and non-ceremony group is 0.01. Based upon the statistically significant effect, it was concluded that open ceremony had shown effects on user account sustainability.

This study was conducted from two view points. The first one is seeing accounts sustainability as a whole. The second one is based upon the time line. From both viewpoints, the conclusion is the same as "ceremony had shown significant effects on account sustainability."

Reference

1. COURSERA. 2013. Coursera website [Online]. Available: https://www.coursera.org [Accessed June 8 2015].

Exploring Effects of Different Roles on Using an E-learning Service

H.J. Yang & L.H. Kuo

National Kaohsiung Normal University, Taiwan, R.O.C.

Abstract. The purpose of this study was to explore effects of different roles on learning satisfaction. For better hosting a e-learning service, there is a need to understand user's satisfaction of the learning service. Data was collected from an on-line learning system. Based upon course learning activities, each course was assign to a certain role for the learner. There are six types of learning role from all course. 1) Reporter, 2) Listener, 3) Researcher, 4) Reporting Listener, 5) Reporting Researcher, 6) Report Listen Researcher; One-way ANOVA procedure was applied to exploring whether the role difference affects users satisfaction. According to the significant level less than .05, it was concluded that learning satisfaction of different roles are significant different. On the other word, learning roles affect users' learning satisfaction.

1 Introduction

1.1 E-learning Platform

In this information age, information system provides a flexible communication solution. It also plays an important way for supporting personal learning by establishing well organized educational environment.

E-learning is becoming an increasingly large part of higher education (Anderson, 2014, Duck and Parente, 2014, Kim, 2011). Over 7.1 million college and university students took at least one online course by the end of the fall 2012 semester in the United States (Allen and Seaman, 2014). More than 71% of US colleges and universities offered online courses in 2012(Allen and Seaman, 2013) and one-third of higher education students took at least one online course in 2012 (Allen and Seaman, 2014).

1.2 Problem Statement

The purpose of this study was to explore effects of different roles on learning satisfaction. For better hosting a e-learning service, there is a need to understand user's satisfaction of the learning service.

Data was collected from an on-line learning system. Based upon course learning activities, each course was assign to a certain role for the learner. There are six types of learning role from all course. 1) Reporter, 2) Listener, 3) Researcher, 4) Reporting Listener, 5) Reporting Researcher, 6) Report Listen Researcher

2 E-learning satisfaction

E-learning has been described as technology-based learning (Ali and Leeds, 2009, Kim, 2011), web-based learning (Bakia et al., 2012), network- and computer-based learning (Angelina et al., 2007), or "instructional environments supported by the Internet" (Bakia et al., 2012). Although each researcher uses different terms to describe the phenomenon of e-learning, the common element in all of the research is that learners need to be familiar with using computer technology and the Internet to take online courses.

E-learning environments in higher education in the United States have been

expanding rapidly (Allen and Seaman, 2014). Online enrollment as a percentage of the total enrollments in U.S. universities was less than 10% in 2002, whereas it reached more than 33.5% in 2012(Allen and Seaman, 2014). The benefit of e-learning is that distance learners can study anytime anywhere at their own pace without the limitations of time and space, a factor which led to the rapid growth of e-learning. However, e-learning is still confronted with a number challenges, such as student readiness for taking an online course and lower retention rates when compared to the traditional face-to-face course.

3 Methodology

3.1 Data Collection

For better hosting a e-learning service, there is a need to understand user's satisfaction of the learning service. Data was collected from an on-line learning system. Based upon course learning activities, each course was

[a] Corresponding author: admi@nknucc.nknu.edu.tw

assign to a certain role for the learner. There are six types of learning role from all course. 1) Reporter, 2) Listener, 3) Researcher, 4) Reporting Listener, 5) Reporting Researcher, 6) Report Listen Researcher.

Learning Satisfaction is represented as Likert Scale from 1 to 5 as not satisfied to fully satisfied and 3 is the middle value.

3.2 One-way ANOVA

One can use the One-Way ANOVA procedure to test the hypothesis that the means of two or more groups are not significantly different.

One-Way ANOVA also offers:

- Group-level statistics for the dependent variable
- A test of variance equality
- A plot of group means
- Range tests, pairwise multiple comparisons, and contrasts, to describe the nature of the group differences

Descriptive statistics are shown in followings:

$$\text{Sample size} = W_q$$
$$\text{Mean} = \overline{T}_q$$
$$\text{Standard deviation} = S_q$$
$$\text{Standard error} = S_q / \sqrt{W_q}$$

95% confidence interval for the mean:

$$\overline{T}_q \pm t_{W_q - 1} S_q / \sqrt{W_q}$$

Where t_{W_q-1} is the upper 2.5:% critical value for the t distribution with Wq-1 degrees of freedom.

In a ANOVA table, there are three source of variation.

Between (BSS), df= k'-1

$$\sum_{l=1}^{k} T_l^2 / n_l - G^2 / W$$

Within (WSS), df=W-k'

$$\sum_{l=1}^{k} SS_l$$
$$(s_p^2 D \text{ for matrix input})$$

Total (TSS), df= W-1

$$BSS + WSS$$

Mean squares are calculated by dividing each sum of squares by its degree of freedom. The F ratio for testing equality of group means is

$$F = \frac{\text{Mean Square Between}}{\text{Mean Square Within}} = \frac{BSSM}{WSSM}$$

The significance level is obtained from the F distribution with numerator and denominator degrees of freedom.

Establishment of homogenous subsets could be defined by following procedure. If the sample sized in all groups are equal, or the harmonic mean for all groups has been selected, or the multiple comparison procedure is SNK or DUNCAN, homogeneous subsets are established as follows:

The means are sorted into ascending order from

$$\overline{T}_{(1)} \text{ to } \overline{T}_{(k')}$$

Values of i and q such that

$$\left| \overline{T}_{(q)} - \overline{T}_{(i)} \right| \le R_{q-i+1} M_{q,i} (*)$$

Are systematically searched for and

$$\left\{ \overline{T}_{(i)}, \ldots, \overline{T}_{(q)} \right\}$$

Is considered a homogeneous subset. The search procedure is as follows:

At each step t, the value of i is incremented by 1 (the starting value is 1), and q=k'. The value of q is then decremented by one until (*) is true. Call this value qt. If qt > qt-1 and (*) is true,

$$\left\{ \overline{T}_{(i)}, \ldots, \overline{T}_{q_t} \right\}$$

is considered homogeneous. Otherwise I is incremented and the next step is done. The procedure terminates when i=k or qt = k.

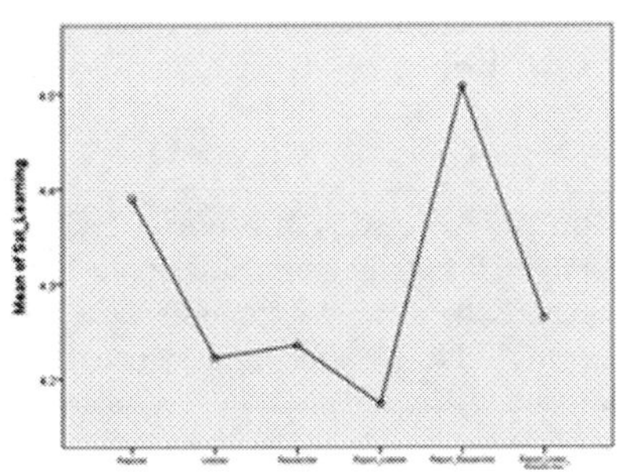

Figure 1. Satisfaction Means plots by each role

4 Findings

In this session, the findings would be reported in descriptive statistics, and One-way ANOVA.

4.1 Descriptive Statistics

In Table 1., the descriptive statistics of role were listed. The overall subjects are 837 with 13.5 % reporter, 14% listener, 20.3% researcher, 14.5 report listener, 22.8 report researcher, and 14.9 report listen researcher.

Table 1. frequency and percent of Roles

Roles					
		Frequency	Percent	Valid Percent	Cumulative Percent
Valid	Reporter	113	13.5	13.5	13.5
	Listener	117	14.0	14.0	27.5
	Researcher	170	20.3	20.3	47.8
	Report Listener	121	14.5	14.5	62.2
	Report Researcher	191	22.8	22.8	85.1
	Report_Listen_Researcher	125	14.9	14.9	100.0
	Total	837	100.0	100.0	

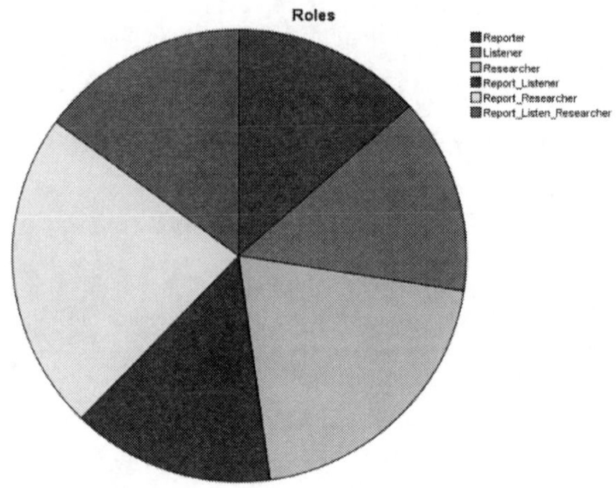

Figure 2. Learning Role distribution

Table 2. Leaning Satisfaction by Roles

	N	*Mean*	*Std. Deviation*	*Std. Error*	*95% Confidence Interval for Mean*		*Minimum*	*Maximum*
					Lower Bound	*Upper Bound*		
Repoter	*113*	*4.39*	*.647*	*.061*	*4.27*	*4.51*	*2*	*5*
Listener	*117*	*4.22*	*.696*	*.064*	*4.09*	*4.35*	*1*	*5*
Researcher	*170*	*4.24*	*.655*	*.050*	*4.14*	*4.33*	*2*	*5*
Report_Listener	*121*	*4.17*	*.715*	*.065*	*4.04*	*4.30*	*1*	*5*
Report_Researcher	*191*	*4.51*	*.680*	*.049*	*4.41*	*4.60*	*1*	*5*
Report_Listen_Researcher	*125*	*4.26*	*.824*	*.074*	*4.12*	*4.41*	*2*	*5*
Total	*837*	*4.31*	*.710*	*.025*	*4.26*	*4.36*	*1*	*5*

Satisfaction is represented as Likert Scale from 1 to 5 as not satisfied to fully satisfied and 3 is the middle value. The learning satisfaction mean for the role of Report Researcher is the highest, 4.51. The standard deviation is .68.

1.1 One-Sample T-Test & One-way ANOVA

For exploring overall satisfaction, one-sample T-Test was applied to compare the mean with middle value, 3, of the Likert Scale. In Table 3, the significance value less than .05. The value of satisfaction is significantly higher thane middle value 3.

Table 3. One-sample T-Test of Satisfaction

	Test Value = 3					
					95% Confidence Interval of the Difference	
	t	df	Sig. (2-tailed)	Mean Difference	Lower	Upper
Sat_Learning	53.442	836	.000	1.312	1.26	1.36

In Table 4, the significant level is less than .05. The test result supports that there exists satisfaction difference among role groups.

Table 4. ANOVA table of learning satisfaction

	Sum of Squares	df	Mean Square	F	Sig.
Between Groups	12.554	5	2.511	5.101	.000
Within Groups	409.059	831	.492		
Total	421.613	836			

In Teable 6., Duncan test suggested three subsets. Both Reporter and Report Researcher roles is the highest satisfaction set.

Table 5. Homogeneous Subsets by Duncan test

Duncan[a,b]				
		Subset for alpha = 0.05		
Roles	N	1	2	3
Report_Listener	121	4.17		
Listener	117	4.22	4.22	
Researcher	170	4.24	4.24	
Report_Listen_Researcher	125	4.26	4.26	
Reporter	113		4.39	4.39
Report_Researcher	191			4.51
Sig.		.343	.075	.167

Table 6. Post Hoc Tests using Scheffe for multiple comparisons

Multiple Comparisons
Dependent Variable: Sat_Learning

	(I) Roles	(J) Roles	Mean Difference (I-J)	Std. Error	Sig.	95% Confidence Interval Lower Bound	95% Confidence Interval Upper Bound
Scheffe	Reporter	Listener	.167	.093	.660	-.14	.48
		Researcher	.154	.085	.658	-.13	.44
		Report_Listener	.216	.092	.356	-.09	.52
		Report_Researcher	-.118	.083	.846	-.40	.16
		Report_Listen_Researcher	.125	.091	.863	-.18	.43
	Listener	Reporter	-.167	.093	.660	-.48	.14
		Researcher	-.013	.084	1.000	-.29	.27
		Report_Listener	.049	.091	.998	-.25	.35
		Report_Researcher	-.286*	.082	.035	-.56	-.01
		Report_Listen_Researcher	-.042	.090	.999	-.34	.26
	Researcher	Reporter	-.154	.085	.658	-.44	.13
		Listener	.013	.084	1.000	-.27	.29
		Report_Listener	.062	.083	.990	-.22	.34
		Report_Researcher	-.273*	.074	.019	-.52	-.03
		Report_Listen_Researcher	-.029	.083	1.000	-.30	.25
	Report_Listener	Reporter	-.216	.092	.356	-.52	.09
		Listener	-.049	.091	.998	-.35	.25
		Researcher	-.062	.083	.990	-.34	.22
		Report_Researcher	-.334*	.082	.005	-.61	-.06
		Report_Listen_Researcher	-.090	.089	.961	-.39	.21
	Report_Researcher	Reporter	.118	.083	.846	-.16	.40
		Listener	.286*	.082	.035	.01	.56
		Researcher	.273*	.074	.019	.03	.52
		Report_Listener	.334*	.082	.005	.06	.61
		Report_Listen_Researcher	.244	.081	.105	-.03	.51
	Report_Listen_Researcher	Reporter	-.125	.091	.863	-.43	.18
		Listener	.042	.090	.999	-.26	.34
		Researcher	.029	.083	1.000	-.25	.30
		Report_Listener	.090	.089	.961	-.21	.39
		Report_Researcher	-.244	.081	.105	-.51	.03

*. The mean difference is significant at the 0.05 level.

In Table 6, those comparisons with significant level less than .05 are:

- Listener to Report Researcher at .035
- Researcher to Report Researcher at .019
- Report Listener to Report Researcher at .005

5 Conclusions

The purpose of this study was to explore effects of different roles on learning satisfaction. For better hosting a e-learning service, there is a need to understand user's satisfaction of the learning service. Data was collected from an on-line learning system. Based upon course learning activities, each course was assign to a certain role for the learner.

There are six types of learning role from all course. 1) Reporter, 2) Listener, 3) Researcher, 4) Reporting Listener, 5) Reporting Researcher, 6) Report Listen Researcher; One-way ANOVA procedure was applied to exploring whether the role difference affects users satisfaction. According to the significant level less than .05, it was concluded that learning satisfaction of different roles are significant different. On the other word, learning roles affect users' learning satisfaction.

References

1. ALI, R. & LEEDS, E. 2009. The Impact of Face-to-Face Orientation on Online Retention: A Pilot Study. Online Journal of Distance Learning Administration, 12.
2. ALLEN, I. E. & SEAMAN, J. 2013. Changing Course: Ten Years of Tracking Online
3. Education in the United States [Online]. Babson Survey Research Group and Quahog Research Group, LLC. . Available: http://www.onlinelearningsurvey.com/reports/changingcourse.pdf [Accessed June 10 2015].
4. ALLEN, I. E. & SEAMAN, J. 2014. Grade Change: Tracking Online Education in the United States [Online]. Babson Survey Research Group and Quahog Research Group, LLC. Available: http://www.onlinelearningsurvey.com/reports/gradechange.pdf [Accessed June 10 2015].
5. ANDERSON, P. K. 2014. University education as relational: The challenge of online studies. Contemporary PNG Studies: DWU Research Journal, 19, 14.
6. ANGELINA, L. M., WILLIAMS, F. K. & NATVIG, D. 2007. Strategies to Engage Online Students and Reduce Attrition. Journal of educators online, 4, 12.
7. BAKIA, M., SHEAR, L., TOYAMA, Y. & LASSETER, A. 2012. Understanding the Implications of Online Learning for Educational Productivity [Online]. Washington, DC: U.S. Department of Education, Office of Educational Technology. Available: http://www.sri.com/sites/default/files/publications/implications-online-learning.pdf [Accessed June 10 2015].
8. BERNARD, R. M., BRAUER, A., ABRAMI, P. C. & SURKES, M. 2004. The development of a questionnaire for predicting online learning achievement. Distance Education, 25, 17.
9. DUCK, J. M. & PARENTE, D. H. 2014. How Do We Get To Tomorrow? The Path to Online Learning. Developments in Business Simulation and Experiential Learning, 35, 2.
10. KIM, J. 2011. Developing an instrument to measure social presence in distance higher education. British Journal of Educational Technology, 42, 15.
11. LEE, Y. & CHOI, J. 2013. A structural equation model of predictors of online learning retention. The Internet and Higher Education, 16, 7.

MATEC Web of Conferences 71, 03003 (2016)
ICCPE 2015
DOI: 10.1051/matecconf/20167103003

Identify User's Satisfaction from Platform Using Behavior

L.H. Kuo[1], W.H. Chen[2,a], & H.J. Yang[1]

[1]National Kaohsiung Normal University, Taiwan, R.O.C.
[2]Taipei College of Maritime Technology, Taiwan, R.O.C.

Abstract.. The purpose of this study was to verify a model of user's satisfaction of an e-learning environment based upon platform using behaviors. This is a non-experimental study. The data was collected from system management logs and users satisfaction survey after learning service. Total 314 users were invited in this study. First, theory model was identified. Second, the satisfaction survey results were prepared. Third, the behavior data of each survey subject were prepared. A CFA procedure were conducted to verify whether the data fits into the model. The model fit is positive with $\chi 2$ = 2.06, p-value=.151, df= 1, RMSEA=.058. The proposed two-factor theory model with simple structure fit the data. The love e-learning and satisfaction of e-learning factors are significantly supporting the hypothesis of a relationship between the factors. The findings suggested that identifying satisfaction from behavior is possible.

1 Introduction

1.1 E-learning Platform

E-learning is becoming an increasingly large part of higher education (Anderson, 2014, Duck and Parente, 2014, Kim, 2011).

Over 7.1 million college and university students took at least one online course by the end of the fall 2012 semester in the United States (Allen and Seaman, 2014). More than 71% of US colleges and universities offered online courses in 2012(Allen and Seaman, 2013) and one-third of higher education students took at least one online course in 2012 (Allen and Seaman, 2014).

1.2 Problem Statement

For managing an information system, there is always a need to understand user's behavior on system service. In an open learning environment, how can an information service provider keeping their users on using service is an important problem.

The purpose of this study was to identify whether an open ceremony affects on-line learning platform account keeping behavior.

2 Literature review

2.1 Massive Open Online Courses

According to the U.S. Department of Education Distance Learning Report (Bakia et al., 2012), the benefits of e-learning are: a) to broaden access to the educational resources, b) to personalize learning, c) to provide flexibility in time and location for students, and d) to reduce school-based facilities' costs. However, the benefits of e-learning also bring some challenges into the field of education.

First, the retention rates in e-learning courses are 10-25% less than those for traditional face-to-face classes(Ali and Leeds, 2009, Angelina et al., 2007, Lee and Choi, 2013) in higher education. In other words, over one half of distance students may dropout of their education as a result of online courses (Duck and Parente, 2014). Second, students who take online courses for the first time tend to feel lonely and socially isolated not only because previously, tend to focus more on computer or Internet skills, technology accessibility, and general learner characteristics such as attitude toward online education or personal learning preferences (Bernard et al., 2004).

E-learning has been described as technology-based learning (Ali and Leeds, 2009, Kim, 2011), web-based learning (Bakia et al., 2012), network- and computer-based learning (Angelina et al., 2007), or "instructional environments supported by the Internet" (Bakia et al., 2012). Although each researcher uses different terms to describe the phenomenon of e-learning, the common element in all of the research is that learners need to be

[a] Corresponding author: aska1205@gmail.com

familiar with using computer technology and the Internet to take online courses.

3 Methodology

3.1 Model Hypotheses

Structural equation modeling (SEM) is a collection of statistical techniques that allow a set of relationships between one or more IVs, either continuous or discrete, and one or more DVs, either continuous or discrete, to be examined. Both IVs and DVs can be either factors or measured variables. Structural equation modeling is also referred to as causal modeling, causal analysis, simultaneous equation modeling, analysis of covariance structures, path analysis, or confirmatory factor analysis (CFA). The latter two are actually special types of SEM.

One interesting feature of the analysis is that survival time (the DV) often is unknown for a number of cases at the conclusion of the study. Some of the cases are still in the study but have not yet failed: some employees have not yet left, some components are still functioning, some patients are still apparently well, or some patients are still living. For other cases, the outcome is simply unknown because they have withdrawn from the study or are for some reason lost to follow-up. Cases whose DV values— survival time— are unknown for whatever reason are referred to as censored.

SEM allows questions to be answered that involve multiple regression analyses of factors. Several conventions are used in developing SEM diagrams. Measured variables, also called observed variables, indicators, or manifest variables, are represented by squares or rectangles.

Factors have two or more indicators and are also called latent variables, constructs, or unobserved variables. Factors are represented by circles or ovals in path diagrams. Relationships between variables are indicated by lines; lack of a line connecting variables implies that no direct relationship has been hypothesized. Lines have either one or two arrows. A line with one arrow represents a hypothesized direct relationship between two variables, and the variable with the arrow pointing to it is the DV. A line with an arrow at both ends indicates an unanalyzed relationship, simply a covariance between the two variables with no implied direction of effect.

Parameters (path coefficients, variances, and covariances of IVs) are estimated to create an estimated population covariance matrix. If the model is good, the parameter estimates will produce an estimated matrix that is close to the sample covariance matrix. "Closeness" is evaluated primarily with the chi- square test statistic and fit indices.

3.2 Covariance Algebra

Within the family of survival- analysis techniques, different procedures are used depending on the nature of the data and the kinds of questions that are of greatest interest.

The idea behind SEM is that the hypothesized model has a set of underlying parameters which correspond to (1) the regression coefficients, and (2) the variances and covariances of the independent variables in the model (Bentler, 1995). These parameters are estimated from the sample data to be a "best guess" about population values. The estimated parameters are then combined by means of covariance algebra to produce an estimated population covariance matrix. This estimated population covariance matrix is compared with the sample covariance matrix and, ideally, the difference is very small and not statistically significant.

Covariance algebra is a helpful tool in calculating variances and covariances in SEM models; however, matrix methods are generally employed because covariance algebra becomes extremely tedious as models become increasingly complex. Covariance algebra is useful to demonstrate how parameter estimates are combined to produce an estimated population covariance matrix for a small example.

The three basic rules in covariance algebra appear below where c is a constant and Xi is a random variable:

1. $COV(c,X1)=0$
2. $COV(cX1, X2)=cCOV(X1,X2)$
3. $COV(X1 + X2, X3) = COV(X1, X3) + COV(X2,X3)$

In SEM, as in multiple regression, we assume that the residuals do not correlate with each other or with other variables in the models. In this model, both degree of motivation (Y1) and exam score (Y2) are DVs. Recall that a DV in SEM is any variable with a single- headed arrow pointing toward it. Treatment group (X1) with no single-headed arrows pointing to it is an IV. To specify the model, a separate equation is written for each DV.

3.3 Model Hypotheses

The data set contains four continuous measured variables:

Hours, the number of a participant has learning on-line.

Course, the number of a participant has taken from the e-learning service.

Plat.Sa, a Likert scale measure of degree of the satisfaction of the platform.

Cou.Sa, a Likert scale measure of degree of the satisfaction of the course.

The hypothesized model for these data is diagrammed in Figure 1. Latent variables are represented with oval and measured variables are represented with squares. A line with an arrow indicates a hypothesized direct relationship between the variables.

 DOI: 10.1051/matecconf/20167103003

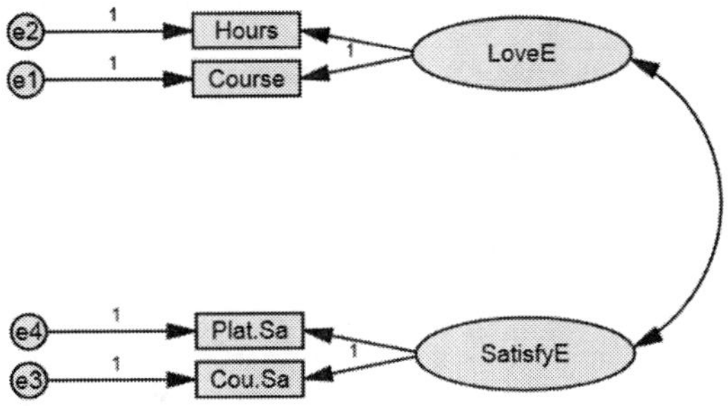

Figure 1. Hypothesized model for the study

4 Findings

In this session, the findings would be reported in descriptive statistics, CFA model, estimates of CFA model, and model fits summary.

For the model fits summary, three sets of model fit index are reported. Those are CMIN, RMSEA, and RMR & GFI.

4.1 Descriptive Statistics

In Table 1., the descriptive statistics of variables were listed. The mean value of course is 9.22. The mean value of hours is 19.45.

The mean course satisfaction is 4.49. The mean platform satisfaction is 4.51. The sample number is 314.

Table 1. N, Minimum, Maximum, Mean, & Std. Deviation of Variables

Descriptive Statistics					
	N	Minimum	Maximum	Mean	Std. Deviation
course_num	314	6	10	9.22	.869
hours	314	12	22	19.45	2.853
CourseSa	314	3	5	4.49	.599
PlatformSa	314	3	5	4.51	.636
Valid N (listwise)	314				

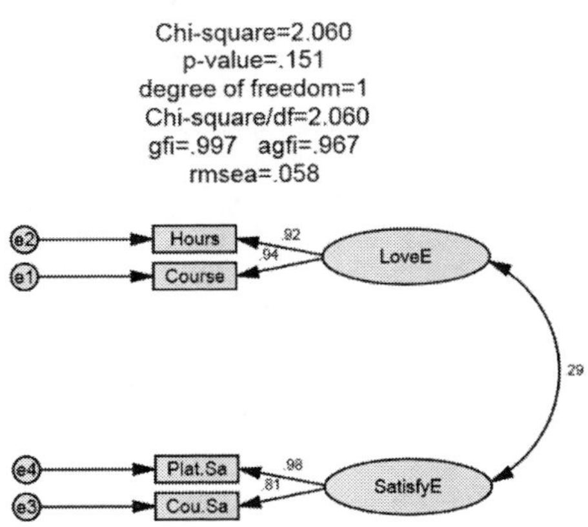

Figure 2. CFA Model of this study

All variables in the model are listed here, classified as either observed or unobserved, and as either endogenous or exogenous. A summary table shows the number of variables in each category, as well as the total number of variables in the model.

In Table 2, parameters of the proposed model were listed. This table shows the numbers of model parameters that fall into various categories.

The columns of the table are:

- **Weights**: regression weights
- **Covariances**: self explanatory
- **Variances**: self explanatory
- **Means**: self explanatory
- **Intercepts**: self explanatory

The rows of the table are:

- **Fixed**: parameters whose values are fixed at a constant value
- **Labeled**: parameters that are labeled
- **Unlabeled**: parameters that are neither fixed nor labeled. Such parameters are free to take on any value. (Labeled parameters can also be free -- a parameter that has been associated with a unique label is free to take on any value.)

Table 2. Parameter Summary

	Weights	Covariances	Variances	Means	Intercepts	Total
Fixed	6	0	0	0	0	6
Labeled	0	0	0	0	0	0
Unlabeled	2	1	6	0	0	9
Total	8	1	6	0	0	15

4.2 CFA model

The first hypothesis, that the model fits the data, has been evaluated and supported. The model fit is positive with $\chi 2$ = 2.06, p-value=.151, df= 1, RMSEA=.058. The proposed two-factor theory model with simple structure fit the data.

4.3 Estimates of CFA Model

Estimates of the following model parameters:

- Regression weights
- Variances of exogenous variables
- Covariances among exogenous variables
- Means of exogenous variables
- Intercepts for predicting endogenous variables

Estimates of the following quantities also appear here when requested.

- Squared multiple correlations
- Correlations among the exogenous variables
- Standardized regression weights

In Table 3, the regression weights were listed and all indicators are significant.

Table 3. Regression weights

			Estimate	S.E.	C.R.	P	Label
course_num	<---	LoveE	1				
hours	<---	LoveE	3.212	0.346	9.28 2	***	par_1
CourseSa	<---	SatisfyE	1				
PlatformSa	<---	SatisfyE	1.29	0.198	6.51 9	***	par_2

In Table 4, the standardized regression weights were listed.

Table 4. Standardized Regression Weights

			Estimate
course_num	<---	LoveE	0.945
hours	<---	LoveE	0.924
CourseSa	<---	SatisfyE	0.807
PlatformSa	<---	SatisfyE	0.982

In Table 5, the covariance between two factors was listed.

Table 5. Covariances of Model

			Estimate	S.E.	C.R.	P	Label
LoveE	<-->	SatisfyE	0.114	0.0 3	3.82 6	***	par_ 3

4.4 Model Fit Summary

4.4.1 CMIN

In Table 6, the model is fit since the p value > .05. The CMIN is 2.060. The CMON/DF is 2.06. The model is acceptable fit.

Table 6. Model Fit of CMIN

Model	NPAR	CMIN	DF	P	CMIN/DF
Default model	9	2.060	1	.15 1	2.060
Saturated mod-el	10	.000	0		
Independence model	4	785.138	6	.00 0	130.856

4.4.2 RMSEA

In Table 7, the RMSEA is .058. The value indicates close approximate fit.

Table 7. Model Fit of RMSEA

Model	RMSEA	LO 90	HI 90	PCLOSE
Default model	.058	.000	.174	.301
Independence model	.644	.606	.683	.000

4.4.3 RMR & GFI

In Table 8, the RMR is 0.008. The range of the RMR is calculated based upon the scales of each indicator. The SRMR of 0 indicates perfect fit. The GFI is 0.997. The GFI value suggests a good fit model.

Table 8. Model Fit of RMR & GFI

Model	RMR	GFI	AGFI	PGFI
Default model	0.008	0.997	0.967	0.1
Saturated model	0	1		
Independence model	0.716	0.552	0.253	0.331

5 Conclusions

The purpose of this study was to verify a model of user's satisfaction of an e-learning environment based upon platform using behaviors. This is a non-experimental study.

The data was collected from system management logs and users satisfaction survey after learning service. Total 314 users were invited in this study. First, theory model was identified. Second, the satisfaction survey results were prepared. Third, the behavior data of each survey subject were prepared. A CFA procedure were conducted to verify whether the data fits into the model.

The model fit is positive with $\chi 2$ = 2.06, p-value=.151, df= 1, RMSEA=.058. The proposed two-factor theory model with simple structure fit the data. The love e-learning and satisfaction of e-learning factors are significantly supporting the hypothesis of a relationship between the factors. The findings suggested that identifying satisfaction from behavior is possible.

References

1. ALI, R. & LEEDS, E. 2009. The Impact of Face-to-Face Orientation on Online Retention: A Pilot Study. Online Journal of Distance Learning Administration, 12.
2. ALLEN, I. E. & SEAMAN, J. 2013. Changing Course: Ten Years of Tracking Online
3. Education in the United States [Online]. Babson Survey Research Group and Quahog Research Group, LLC. . Available: http://www.onlinelearningsurvey.com/reports/changingcourse.pdf [Accessed June 10 2015].
4. ALLEN, I. E. & SEAMAN, J. 2014. Grade Change: Tracking Online Education in the United States [Online]. Babson Survey Research Group and Quahog Research Group, LLC. Available: http://www.onlinelearningsurvey.com/reports/gradechange.pdf [Accessed June 10 2015].
5. ANDERSON, P. K. 2014. University education as relational: The challenge of online studies. Contemporary PNG Studies: DWU Research Journal, 19, 14.

6. ANGELINA, L. M., WILLIAMS, F. K. & NATVIG, D. 2007. Strategies to Engage Online Students and Reduce Attrition. Journal of educators online, 4, 12.
7. BAKIA, M., SHEAR, L., TOYAMA, Y. & LASSETER, A. 2012. Understanding the Implications of Online Learning for Educational Productivity [Online]. Washington, DC: U.S. Department of Education, Office of Educational Technology. Available: http://www.sri.com/sites/default/files/publications/implications-online-learning.pdf [Accessed June 10 2015].
8. BERNARD, R. M., BRAUER, A., ABRAMI, P. C. & SURKES, M. 2004. The development of a questionnaire for predicting online learning achievement. Distance Education, 25, 17.
9. DUCK, J. M. & PARENTE, D. H. 2014. How Do We Get To Tomorrow? The Path to Online Learning. Developments in Business Simulation and Experiential Learning, 35, 2.
10. KIM, J. 2011. Developing an instrument to measure social presence in distance higher education. British Journal of Educational Technology, 42, 15.
11. LEE, Y. & CHOI, J. 2013. A structural equation model of predictors of online learning retention. The Internet and Higher Education, 16, 7.

Homophily Effect of Academic Achievement on Adolescent Friendship Network Evolution

Chyi-In Wu[1], and Hsieh-Hua Yang[2,*]

[1]Institute of Sociology, Academia Sinica, 11529 Taipei, Taiwan
[2]Department of Health Care Administration, Oriental Institute of Technology, 22061 New Taipei City, Taiwan

Abstract. The objective is testing the homophily effect of academic achievement on adolescent friendship network evolution. There were 93 boys and 82 girls of high school students participating in this study. Panel data was collected 10 waves during 3 semesters from Sep. 2008 to Jan. 2010. The students were asked to nominate up to 16 intimate classmates. Academic achievement of computer science, mathematics, Chinese, and English was measured in the beginning of 3 semesters. The model was constructed by the program SIENA. The results demonstrated the homophily effect of perceived academic achievement was significant for computer science at the first semester, for mathematics and English at the second semester, and for Chinese at the third semester.

1 Introduction

The importance of peer influence has been confirmed by many researchers. The process of influence is either by homophily principle or assimilation principle. Homophily principle was proposed by [1] to explain the network autocorrelation, which stands for the argument that it is easier or more rewarding for an actor to interact with a similar other than with a dissimilar other. An alternative explanation of the same phenomenon is the assimilation principle according to which network actors adapt their own individual characteristics to match those of their social neighborhood [2, 3].

Friendship plays a distinct role in adolescent development. It provides social support, promoting social skills, and also feeling of belongingness. It represents unique relationships that differ from other relationships. Friendship stability may have impact on adolescent adjustment or reinforcing behaviour patterns. Friendship network formation is an evolution process. It is likely to adapt to an individual's changing needs and goals. Some friendships are stable and last for a long time, some end temporarily or permanently after several days, and some others are newly-formed. Many theories are applied to explain the mechanism of social network evolution and behaviors. Social comparison theory [4] holds that people have a drive to evaluate his abilities by comparison with the abilities of others. And if this evaluation is possible with others who are close enough, then there should be some attraction to groups where others are relatively close with respect to abilities. Social learning theory [5] proposes that individuals' behaviors may arise and be reinforced through observing and imitating others, including peers. Therefore, individual interaction with peers, social support from peers and their understanding of situations are important factors which have effect on individual learning outcome.

Many researchers studied the effect of social context on educational outcomes. Daw, Guo, and Harris [6] use a regression model to test the shared and nonshared environment exert important influences on the academic achievement. Levy-Tossman, Kaplan, & Assor [7] find the connection between academic motivation and social relationships. Frank et al. [8] agree on the importance of peer influences and find that girls are highly responsive to the social norms in their local positions. Hasan and Bagde [9] found that peer effects are a consequential determinant of academic achievement.

Friendship network formation is an evolution process. During the evolution process, strangers are converted to acquaintances and acquaintances to friends. van Duijn et al. [10] developed a theory to explain how changes in the network structure. The theory posits that proximity plays a dominant role in the formation of networks, visible similarity plays a role in the beginning of the friendship development process, and invisible similarity will have effect at later stages of the friendship development. A friendship is more likely to be continued if each individual perceives the other as similar in a variety of ways. We argue that the similarity of academic achievement has effect on friendship network evolution. And the research questions are whether the similarity of academic performance has effect on friendship network evolution, and at what stages which subject is important.

In Taiwan's education system, the junior high school students' basic competence test (BCT) has 5 components: English, mathematics, Chinese, natural sciences, and social sciences. The senior high school students' subject

* Corresponding author: yansnow@gmail.com

competency test (SCT) covers English, Chinese, mathematics, the natural sciences and humanities. Both BCT and SCT have 3 subjects, English, Chinese, and mathematics. In this study, these 3 subjects and computer science were evaluated as academic performance.

2 Method

2.1 Participants

Data were from the project "Luminous Shine and Dark Shadow: The Duality of Late Adolescents to Early Adults' Friendship Networks" sponsored by the Academia Sinica in Taiwan. In this study, the participants were 175 high school students of 93 boys and 82 girls.

2.2 Measurement

Panel data was collected 10 waves during 3 semesters from Sep. 2008 to Jan. 2010. Sociometric data were collected by having each student nominate up to 16 intimate classmates.

Academic performance of computer science, mathematics, Chinese, and English was measured in the beginning of 3 semesters, namely at wave 1, 5, and 8. Academic achievement was measured by asking the student if his/her performance was better than others.

2.3 Data analysis

The program SIENA (Simulation Investigation for Empirical Network Analysis) was applied to estimate the models for the evolution of social networks according to the dynamic actor-oriented model of Snijders [11, 12]. For the estimation, sociometric data was transformed into adjacency matrices. The academic performance was coded as 1 if it was worse than others, 2 if it was the same as others, and 3 if it was better than others. Boy was coded as 1 and girl as 2.

The actor-oriented model [13] assumes that the network $X(t)$ is evolving in continuous time, although being observed only at the discrete moments t_m, $m=1,\ldots, M$. M is the number of repeated observations. This actor orientation means that, for each change in the network, the perspective is taken of the actor whose tie is changing. It is assumed the actor i controls the set of outgoing tie variables ($Xi1,.., Xig$), collected in the ith row of the adjacency matrix. The network changes only by one tie at a time. The moment when actor i changes one of his ties can depend on the network structure and on attributes represented by observed covariates. The "moment when" is stochastically determined in the model by the rate function, and "the particular change to make" by the objective function and the gratification function.

In the study, the evolution model components include density, reciprocity, transitivity, gender similarity, and similarities of perceived academic performance. Density effect is defined as the number of outgoing ties. Reciprocity effect is defined the number of reciprocated ties. A transitive triplet effect is defined by the number of transitive patterns in i's relations of actors (j,h) (an ordered pairs of actors to both of whom i is tied, while also j is tied to h) for which $i \rightarrow j \rightarrow h$ and also $i \rightarrow h$.

The program can estimate all possible effects simultaneously, and is suitable to test our hypotheses. The SIENA program is included in the StOCNET system, and can be downloaded from http://stat.gamma.rug.nl/stocnet/ [14].

3 Results

3.1 Description of participants

The analyses were run on the 175 students that were present at all ten measurement points. The participants' perceived academic achievement of computer science, mathematics, Chinese, and English was analysed for 3 observations. The result, as table 1, 2, and 3, showed that the perceived performance of computer science had no gender difference, while the perceived performance of mathematics and English had gender difference at all observations. And for the subject of Chinese, there was gender difference at the 2nd and 3rd semesters.

For mathematics, there were more boys who perceived their performance were better than others. For Chinese, there were more girls who perceived their performance were better than others. For English, there were more girls who perceived their performance were better than others.

Table 1. Academic achievement at 1st semester

Variable	Boys		Girls		Chi-square	p
	N	%	N	%		
Computer science					0.25	0.883
1	28	30.1	24	29.3		
2	41	44.1	39	47.6		
3	24	25.8	19	23.2		
Mathematics					17.12	<.001
1	35	37.6	48	58.5		
2	21	22.6	24	29.3		
3	37	39.8	10	12.2		
Chinese					5.98	0.050
1	35	37.6	17	20.7		
2	27	29.0	31	37.8		
3	31	33.3	34	41.5		
English					6.32	0.043
1	42	45.2	22	26.8		
2	24	25.8	28	34.1		
3	27	29.0	32	39.0		

 DOI: 10.1051/matecconf/20167103004

Table 2. Academic achievement at 2nd semester

Variable	Boys N	%	Girls N	%	Chi-square	p
Computer science					4.08	0.130
1	28	30.1	19	23.2		
2	38	40.9	46	56.1		
3	27	29.0	17	20.7		
Mathematics					6.88	0.032
1	40	43.0	50	61.0		
2	23	24.7	18	22.0		
3	30	32.3	14	17.1		
Chinese					6.07	0.048
1	31	33.3	14	17.1		
2	32	34.4	34	41.5		
3	30	32.3	34	41.5		
English					14.13	0.001
1	40	43.0	18	22.0		
2	33	35.5	26	31.7		
3	20	21.5	38	46.3		

Table 3. Academic achievement at 3rd semester

Variable	Boys N	%	Girls N	%	Chi-square	p
Computer science					3.51	0.173
1	31	33.3	23	28.0		
2	36	38.7	43	52.4		
3	26	28.0	16	19.5		
Mathematics					19.36	<0.001
1	33	35.5	53	64.6		
2	20	21.5	17	20.7		
3	40	43.0	12	14.6		
Chinese					11.66	0.003
1	31	33.3	12	14.6		
2	33	35.5	26	31.7		
3	29	31.2	44	53.7		
English					12.84	0.002
1	39	41.9	27	32.9		
2	33	35.5	16	19.5		
3	21	22.6	39	47.6		

3.2 Model of friendship network evolution and academic achievement

The evolution models for 3 semesters are shown as table 4-6. The rate parameter describes the average number of changes in network ties between measurement points. The highest rate parameter is in the beginning of each semester. And it is higher between 2 semesters than during the semesters.

The endogenous network effects are all statistically significant. The density, reciprocity and transitive triplets have effects during all the stages. The negative density effect represents the hindrance of dissolved relationships. The reciprocity effect indicates a preference for reciprocating relationships. The transitivity effect indicates a preference for being friends with friends' friends.

The effect of gender similarity was significant during all waves. The positive parameter implies that actors prefer ties to others with same gender.

Table 4. Evolution model of wave 1 to 4

Variable	w1-2 Estimate(se)	w2-3 Estimate(se)	w3-4 Estimate(se)
Rate parameter	7.06(0.24)	5.37(0.23)	5.52(0.23)
Network effects			
Density	-2.022(0.06)*	-1.88(0.06)*	-2.19(0.07)*
Reciprocity	0.93(0.09)*	1.20(0.10)*	1.34(0.10)*
Transitive triplets	0.11(0.01)*	0.13(0.01)*	0.11(0.01)*
Gender similarity	0.37(0.08)*	0.40(0.09)*	0.42(0.09)*
Academic achievement similarity			
Computer science	0.03(0.10)	0.29(0.12)*	-0.01(0.11)
Mathematics	0.10(0.09)	-0.02(0.10)	-0.03(0.11)
Chinese	-0.08(0.09)	-0.08(0.10)	0.19(0.11)
English	-0.18(0.09)	-0.191(0.10)	0.02(0.11)

*p<.05

Table 5. Evolution model of wave 4 to 7

Variable	w4-5 Estimate(se)	w5-6 Estimate(se)	w6-7 Estimate(se)
Rate parameter	5.90(0.26)	5.70(0.28)	6.19(0.29)
Network effects			
Density	-2.03(0.07)*	-2.10(0.06)*	-1.89(0.06)*
Reciprocity	1.08(0.09)*	1.37(0.10)*	1.13(0.10)*
Transitive triplets	0.16(0.01)*	0.13(0.01)*	0.14(0.01)*
Gender similarity	0.61(0.09)*	0.37(0.09)*	0.50(0.08)*
Academic achievement similarity			
Computer science	0.040(0.11)	-0.06(0.12)	0.08(0.11)
Mathematics	0.34(0.10)*	-0.11(0.11)	-0.14(0.09)
Chinese	-0.07(0.10)	-0.20(0.11)	-0.03(0.11)
English	0.24(0.10)*	-0.06(0.12)	0.11(0.10)

*p<.05

Table 6. Evolution model of wave 7 to 10

Variable	w7-8 Estimate(se)	w8-9 Estimate(se)	w9-10 Estimate(se)
Rate parameter	6.50(0.30)	6.60(0.34)	4.99(0.24)
Network effects			
Density	-2.17(0.07)*	-1.96(0.06)*	-1.89(0.07)*
Reciprocity	1.51(0.10)*	1.34(0.09)*	1.13(0.10)*
Transitive triplets	0.14(0.01)*	0.13(0.01)*	0.13(0.01)*
Gender similarity	0.28(0.09)*	0.476(0.09)*	0.59(0.09)*
Academic achievement similarity			
Computer science	0.14(0.12)	0.13(0.11)	0.21(0.12)
Mathematics	0.06(0.11)	-0.16(0.10)	0.13(0.10)
Chinese	-0.11(0.11)	-0.11(0.10)	0.25(0.12)*
English	0.06 (0.11)	0.17(0.10)	0.09(0.10)

*p<.05

During wave 1 to wave 10, the homophily effect of perceived academic achievement was significant for computer science at w2-3 stage, for mathematics and English at w4-5, and for Chinese at w9-10. The positive parameter indicates the tendency to be friends with the same perceived performance of academic achievement.

4 Discussion and conclusion

Consistent with previous studies, the similarity of academic achievement has effect on friendship network evolution. Longitudinal network data can yield important insights into social processes. Ten waves' observations may offer rich materials for understanding adolescents' friendship network development and its relationship with academic performance.

4.1 Homophily effect of academic achievement

Our findings revealed that the similarity effect of perceived achievement of computer science was significant during the first semester, the effects of mathematics and English were significant during the second semester, and the effect of Chinese was significant during the last semester.

Whether the achievement of academic is visible or invisible depends on social context and social norms. If most of them have good performance, the achievement would be visible. Otherwise, it would be invisible. About 70% of the participants were perceived themselves not worse than others for the achievements of computer science. The perceived achievement of computer science might be visible and has homophily effect in the first semester. The other reason may be due to the learning context. The computer science must learn by practicing on the computer. All the students learn in a computer classroom together. The learning process is visible. They can perceive others' achievement immediately and the perceived performance of computer science had homophily effect in the first semester.

During w4-5, the similarity effects of perceived performance are significant for the subjects of mathematics and English. W4 is the end of the first semester, and w5 is the beginning of the second semester. After a semester, the performances of these 2 subjects became visible and exert homophily effect.

Social comparison theory [4] suggests the drive to make comparisons with others who are more similar. The measurement of perceived performance is the result of comparison with others. Since people prefer to compare with others who are more similar, the comparison result of "worse than others" is not their favorable outcome. The unfavorable result of comparison will produce tendencies to change one's evaluation of the ability or lead to action for reducing the discrepancy and so as to move closer to others in the groups. Namely, the drive to have a favorable comparison result may trigger their behaviors by observing and imitating proposed by social learning theory. The students who perceived themselves better than others in Chinese and English have increasing tendency. It may be explained by this mechanism. The students will imitate others to modify their behavior for improving their performance.

The perceived performance of Chinese had effect in the last stage. Chinese is the native language of the participants. The comparison with others in this subject seems ridiculous or meaningless. This subject may be invisible or be ignored. It might be the reason why the homophily effect existed in the last stage.

Based on the hypothesis of van Duijn et al. [10], in early stage, visible similarity plays important role to initiate friendship, and in later stages, dyadic relations will be strengthened by invisible similarities. Since the achievement of computer science can be easily observed, the similarity effect of computer science exerts in early stage. The performances of other 3 subjects need more time to be observed and become visible and exert effect at later stages.

4.2 Evolution of friendship networks and gender effect

The rate parameter is up and down, then decreasing at the end of observation. It represents the friendship network is dynamic but not steady. The effects of network are the same as the results of van Duijn et al. [10]. Network density, transitive triplets, and reciprocity are important for friendship evolution during all stages. These three factors are the most important effects for the structural part of the model for network dynamics [14]. The density effect and the reciprocity effect must always be included in the dynamic model. There are four network closure effects. And it will be sufficient to express the tendency to network closure by including at least one factor. The transitive triplet effect is selected to be included in the model. It is the representation of network closure by the number of transitive triplets. For this effect the contribution of the tie $i \rightarrow j$ is proportional to the total number of transitive triplets that it forms which can be transitive triplets of the type $\{i \rightarrow j \rightarrow h; i \rightarrow h\}$ as well as $\{i \rightarrow h \rightarrow j; i \rightarrow j\}$.

In our findings, positive parameter of gender similarity indicated the tendency of preference to make same-sex friends. Gender is a visible variable and exerts effect in the beginning of friendship formation. After initiation of friendship, gender keeps its effect during all the waves. Obviously, gender is a powerful organizer of peer relationships throughout development. Adolescents report feeling closer to their same-sex friends than to their other-sex friends [15, 16]. Thus, same-sex friendships are an important resource for the development of social health during adolescence. However, heterosex friendships represent unique opportunities both for healthy development and adverse development [17]. Other-sex friendships may aid youth in learning how to interact productively with individuals with interests, experiences, and backgrounds that differ from their own.

4.3 Gender difference of academic performance

Our findings are consistent with most of the previous studies. The results revealed gender difference of perceived performance for mathematics, Chinese and English, but not computer science.

Falaye [18] indicated that academic achievement gap pertaining to gender exists based on subject. Most studies found a positive gender gap in language art; and a traditional negative gap in STEM (science, technology, engineering, and mathematics). O'Reilly and McNamera [19] investigated the performance of high school students and found significant gender differences with males scoring higher on measures of passage comprehension and science knowledge and females scoring higher on measures of reading skills, strategy knowledge, and those pertaining to course grades. Preckel et al. [20] indicated that male students can obtain higher grades in mathematics whereas female students are better in oral skills. Gong, Ding, and Tsang [21] estimated gender gaps in Chinese and math among primary and lower-secondary school students and found small positive gaps favoring girls in Chinese and negative gaps in math for all grades. Jansen, Schroeders, and Ludtke [22] analyzed data and found that female students possess a lower self-concept in chemistry and physics.

The gender difference may come from psychological factors. Ghazvini, and Khajehpour [23] found that the girls take greater responsibility for academic failures and it is associated with the girls' obtaining better results in the subject of Literature. The same research also reveal that boys use concentration, information processing and selecting main ideas strategies more, and getting better marks in mathematics. Furthermore, Sinclair, Carlsson & Bjorklund [24] suggested social networks are of importance for gender differences in field of study and the differences is rooted in the tendency for adolescents to form same-gender friendships. Female adolescents who adjust their choice in line with friends will be more likely to choose a female-dominated field of study.

4.4 Conclusion

In conclusion, social network analysis is uniquely suited for measuring and understanding the behavior of peers because it provides a formal means for "mapping" friendships and measuring properties of those friendships [25]. Friendship network is initiated by chance; in the early stages the proximity and visible similarity variables determined dyadic relations. In later stages, dyadic relations will be strengthened by invisible similarities. Our findings of homophily effects for different subjects in different stages offered some evidence to support this argument.

References

1. J.M. McPherson, L. Smith-Lovin, J.M. Cook, Birds of a feather: homophily in social networks. Annu. Rev. Sociol. **27**, 415-444 (2001)
2. N. Friedkin, Norm formation in social influence networks. Soc. Networks. **23**, 167-189 (2001)
3. E.R. Oetting, J.F. Donnermeyer, Primary socialization theory: the etiology of drug use and deviance. Subst. Use. Misuse. **33**, 995-1026 (1998)
4. L. Festinger, A theory of social comparison processes. Human Relations **7**, 117-140 (1954)
5. A. Bandura, Social Learning Theory. Prentice Hall, Englewood Cliffs, N.J. (1977)
6. J. Daw, G. Guo, K.M. Harris, Nurture net of nature: Re-evaluating the role of shared environments in academic achievement and verbal intelligence. Social Science Research **52**, 422-439 (2015)
7. I. Levy-Tossman, A. Kaplan, A. Assor, Academic goal orientations, multiple goal profiles, and friendship intimacy among early adolescents **32**, 231-252 (2007)
8. K.A. Frank, C. Muller, K.S. Schiller, C. Riegle-Crumb, A.S. Mueller, R. Crosnoe, J. Pearon, The social dynamics of mathematics coursetaking in high school, AJS **113(6)**, 1645-1696 (2008)
9. S. Hasan, S. Bagde, The mechanics of social capital and academic performance in an Indian college, American Sociological Review **78(6)**, 1009-1032 (2013)
10. M.A.J. van Duijn, E.P.H. Zeggelink, M. Huisman, F.N. Stokman, F.W. Wasseur, Evolution of sociology freshmen into a friendship network, Journal of Mathematiscal Sociology **27**, 153-191 (2003)
11. T.A.B. Snijders, Stochastic Actor-oriented Models for Network Change, Journal of Mathematical Sociology **21**, 149-172 (1996)
12. T.A.B. Snijders, The Statistical Evaluation of Social Network Dynamics. In M.E. Sobel & M.P. Becker (eds), Sociological Methodology, Boston: Basil Blackwell (2001)
13. R. Veenstra, C. Steglich, Actor-based model for network and behavior dynamics: A tool to examine selection and influence processes. Chapter 34 (pp. 598-618) in B. Laursen, T. D. Little, and N. A. Card (Eds.), Handbook of Developmental Research Methods. New York: Guilford Press (2012)
14. T.A.B. Snijders, C. Steglich, M. Schweinberger, M. Huisman, Manual for SIENA version 3, University of Groningen: ICS. Oxford: University of Oxfor, Department of Statistics. http://stat.gamma.rug.nl/stocnet (2007)
15. W. Bukowski, L. Sippola, B. Hoza, Same and other: Interdependency between participation in same- and other-sex friendships, Journal of Youth and Adolescence **28**, 439-459 (1999)
16. B. Lundy, T. Field, C. McBride, T. Field, S. Largie, Same-sex and opposite-sex best friend interactions among high school juniors and seniors, Adolescence **33**, 279-289 (1998)
17. L.K. Sippola, Getting to know the "other": the characteristics and developmental significance of other-sex relationships in adolescence, Journal of Youth and Adolescence **28**, 407-418 (1999)
18. F.V. Falaye, Numerical ability, course of study and gender differences in students' achievement in practical geography, Research in Education, **76**, 33-42 (2006)
19. T. O'Reilly, D.S. McNamera, The impact of science knowledge, reading skill, and reading strategy knowledge on more traditional "High-Stakes" measures of high school students' science achievement, American Educational Research Journal **44**, 161-196 (2007)
20. F. Preckel, , T. Goetz, , R. Pekrun, & M. Kleine, Gender differences in gifted and average-ability students: Comparing girls' and boy's achievement, self-concept, interest, and motivation in Mathematics, The Gifted Child Quarterly **52**, 146-160 (2008)

21. X. Gong, Y. Ding, M.C. Tsang, Gender differences of academic performance in compulsory education in rural Southwestern China, International Journal of Educational development **39**, 203-214 (2014)
22. M. Jansen, U. Schroeders, O. Ludtke, Academic self-concept in science: Multidimensionality, relations to achievement measures, and gender differences, Learning and Individual Differences **30**, 11-21 (2014)
23. S.D. Ghazvini, M. Khajehpour, Gender differences in factors affecting academic performance of high school students, Procedia Social and Behavioral Sciences **15**, 1040-1045 (2011)
24. S. Sinclair, R. Carlsson, F. Bjorklund, The role of friends in career compromise: same-gender friendship intensifies gender differences in educational choice, Journal of Vocational Behavior **84**, 109-118 (2014)
25. S.T. Ennett, K.E. Bauman, Adolescent social networks: school, demographic, and longitudinal consideration, Journal of Adolescent Research **11(2)**, 194-215 (1996)

Exploring subjective well-being in social network sites? - Supplementary fit and social support perspective

Ying Chieh Liu[1], Hung Yi Chen[1a], and Bo Yen Yao[1]

[1]*Chaoyang University of Technology, 168, Jifeng E. Rd., Wufeng District, Taichung, 41349, Taiwan*

Abstract. This study formed a framework theorized on supplementary fit and social support to explore users' perceptions of Subject Well-Being in Social Network Sites. Online surveys were conducted in Facebook and Google+. Empirical results showed that value-based fit and personal-based fit promoted the obtaining of emotional and informational support but not instrumental support, and emotional and informational support had significant impacts on life satisfaction and happiness while instrumental support had no significant impacts on life satisfaction and happiness. Our model pinpointed novel ideas to present how individuals obtain social supports through Social Network Sites and how these social supports influence happiness and life satisfaction.

1 Introduction

Recent research has proposed many factors affecting SWB (Subjective Well-Being) and empirically examined them, such as socializing, self-esteem, loneliness, self-disclosure, social integration, social bonding, social bridging [1, 2]. However, little research has studied SWB in the SNS (Social Network Sites) in the viewpoint of social support. The traditional application of social support focuses on the linkages between health treatment and supports provided by people around patients ex., [3, 4]. But our study focuses on ordinary people who interact with their friends in SNS instead of patients with specific health issue. Additionally, little research has explored the antecedents influencing the obtaining of social support. We applied supplementary fit to fill in this gap. Supplementary fit refers to the similar or matching characteristics between a person and other group members [5]. It occurs when "a person supplements, embellishes, or possesses characteristics which are similar to other individuals in the environment" [6]. From this point of view, we assume that individuals perceive social supports when supplementary fit occurs between individuals and the other members in SNS. Thus, the purpose of this study is to: (a) Identify how supplementary fit affects people's perception of social support in the context of SNS; (b) Explore how people's perception of social support influences SWB in the context of SNS.

2 Literature review and hypotheses development

In the context of SNS, individuals interact through website platform to obtain emotional or practical support to satisfy their needs and improve their ability to adapt the environment. People around the world can be in individuals' social support sys-tem, as long as they are using the same SNS. We synthesize social support as emotional, instrumental and informational support. Emotional support refers to demonstrations of love and caring, esteem and value, encouragement, and sympathy. Instrumental support consists of offering or supplying behavioral or material assistance with practical tasks or problems, such as goods and services. Informational support is the provision of facts or advice that may help a person solve problems.

According to the similarity theory, people are more attracted to those who are similar to them in aspects like demographic background, personality trait, values and socio-economic background. In a relationship consisting of like-minded individuals, sharing the same viewpoints allows them to feel that they are accepted by others, and thereby their own confidence increases and the relationship reinforces. Supplementary fit, which is applied in this study, can be categorized into value-based fit and personality-based fit [7]. The former one emphasizes on the match of value and the latter one focuses on math of personality traits between individuals and other group members. In a context of SNS, value-based fit regards to value congruence between individuals and the SNS. Individuals use values to make choices of the extent they participate in the SNS activities. The SNS members use values to develop a norm to guide members' beliefs and behaviors. The match between the individual's values and the SNS values can facilitate the interactions between individual and other members, and

[a] Corresponding author: hychen39@gmail.com

DOI: 10.1051/matecconf/20167103005

further improve harmony of the SNS and intentions providing supports for each other. Thus, we assume that value-based fit promotes more exchange of social support and the three hypotheses are described below.

H1a: Value-based fit promotes the obtaining of emotional support for individuals.
H1b:Value-based fit promote the obtaining of instrumental support for individuals.
H1c: Value-based fit promote the obtaining of in-formational support for individuals.

Personality, in this study, instead of specifying to demographic variables (such as race, age, gender) but a stable characteristics and is proximal to behavior. In a context of SNS, personality-based fit implies to the psychological and behavioral compatibility between individuals and other members in SNS. With personality-based fit occurs, individuals feel more comfortable and are willing to help others. Thus, we assume that personality-based fit promotes more exchange of social support which includes emotional, instrumental and information support. The three hypotheses are described below.

H2a: Personality -based fit can promote the obtaining of emotional support for individuals.
H2b: Personality -based fit can promote the obtaining of instrumental support for individuals.
H2c: Personality -based fit can promote the obtaining of informational support for individuals.

We classified SWB as twofold: life satisfaction and happiness. Life satisfaction is more reflective of the broad and continuing circumstances of an individual's life, while happiness is more volatile. We assume that three kind of social supports (emotional, instrumental and informational support) can improve the outlook of life, letting people more optimistic, lightens the stress of life and further increase life satisfaction and happiness. Thus, the hypotheses are proposed below.

H3a: Emotional support is positively related to life satisfaction.
H3b: Emotional support is positively related to happiness.
H4a: Instrumental support is positively related to life satisfaction.
H4b: Instrumental support is positively related to happiness.
H5a: Informational support is positively related to life satisfaction.

H5b: Informational support is positively related to happiness.

3 Research methodology

To examine the proposed hypotheses, we conducted surveys on Facebook and Google+. My3q (www.my3q.com) was used as our questionnaire distribution platform. The duration lasted for one month and 450 validated questionnaires were collected. Most of the subjects had bachelor degree or higher (87%). Females (53%) were higher than males (47%). The ages ranged from 20 to 30 accounted for 74%. Subjects reported that 78% of Facebook or Google+ friends come from people they know in real life.

The overall fit of our structural model was quite satisfactory by CFA ($\chi 2$/df.=2.42, RMSEA = 0.58, CFI = 0.97, IFI=0.95, GFI = 0.91, AGFI=0.88). SEM (Structural Equation Modeling) then was applied to estimate the path diagram of the framework in the method of maximum likelihood. Figure 1 shows the path diagram. As seen in Figure 2, there are four insignificant paths (value-based fit, personality-based fit • instrumental support, Instrumental support • Life satisfaction, Happiness). The hypothesis test was engaged according to the path diagram. Thus, the Hypothesis 1a, 1c, 2a, 2c, 3a, 3b, 5a and 5b were supported while the Hypothesis 1b, 2b, 4a and 4b were not.

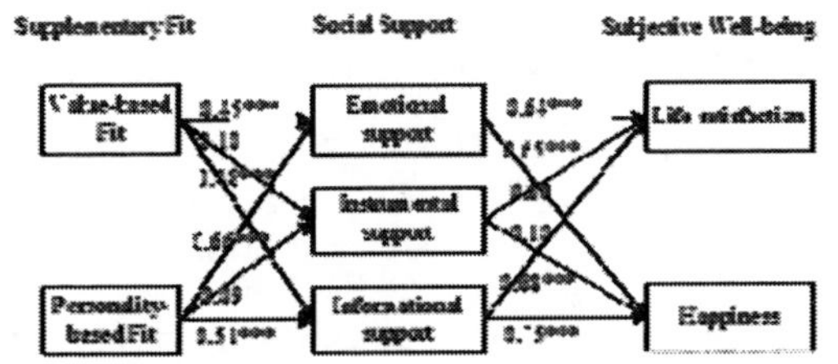

Figure 1. The result of structural equation model

4 Discussions

The hypothesis 1a and 1c were supported but not 1b, which proves that value-based fit promotes emotional and informational support but not instrumental support. That is to say, individuals perceive more emotional and informational support when the match between the individual's values and other SNS members' values occurs. Also, the hypothesis 2a and 2c were supported but not 2b, which proves that personality-based fit promotes emotional and informational support but not instrumental support. That is to say, individuals perceive more emotional and in-formational support when the match between the individual's personality traits and other SNS members' personality traits occurs. These results give us two implications. First, emotional and information support are provided when value-based fit and personality-based fit occur. Second, instrumental sup-port seems to be difficult to appear even value-based fit and personality-based fit exists.

By the significant support of the hypothesis 3a, 3b, 5a and 5b, both emotional support and informational support had significant impacts on life satisfaction and happiness, which is correspondent to the past studies. It can be inferred that emotional and in-formational support can be a buffer to people's stress and a solution resource to solve the issues in people's lives. For example, when users post their job frustration and express bad emotions, their friends may comfort them by showing sympathy or even providing some tips to solve the issues. This would improve users' degree of happiness and life satisfaction. These results also indicate that SNS has the same functions as traditional interpersonal relationship network in term of providing emotion and information support to help people be happier and satisfy with their lives.

However, the insignificant support of hypothesis 3a and 3b indicated that instrumental support had no significant impacts on life satisfaction and happiness. The reason could come from the essence of SNS. Instrumental support refers to specific material and practical aids provided by others. In a virtual environment of SNS, online friends can hardly pro-vide immediate and practical aids due to they may not check the messages anytime or they are geo-graphically far from the help-seeker, which leads that users perceive little instrumental support from their online friends. To improve this, SNS managers can develop a "call for help" push function to enable users to transmit their messages immediately to the assigned friends through various communication tools, such as text message, phone call, Line, WhatsApp, WeChat…etc, or even combine GPS (Global Positioning System) to search friends nearby to require their assistance.

To summarize the discussion above, emotional and information support are provided but not instrumental support when value-based fit and personality-based fit occurs in the context of SNS. Individuals gain emotional and information support and improve life satisfaction and happiness when they share their inhibition, pressure, depression and excitement with online friends through SNS. In terms of emotional and information support, SNS can play the same role as interpersonal relationship in real world. Additionally, subjects reported that 78% of online friends in SNS come from people they know in real life, which means SNS may help users improve intimacy with their friends. That is to say, SNS does not weaken substantial relationships in re-al world; on the contrary, it enhances existing relationships through emotional and information sup-port, and in turn improves users' satisfaction and happiness.

5 Conclusions

This study made several contributions. First, we applied value-based fit and personal-based fit as antecedent variables to explain why social support can be obtained. Additionally, the empirical test showed that emotional and information support are available when value-based fit and personality-based fit occur. Second, different from the past studies regarding social support as a single-dimensional factor, this study regarded it as a multi-dimensional factor and conceptualized it as emotional, instrumental and in-formational support, which identifies the supports that people are able to retrieve from SNS. Also, SWB was conceptualized as momentary happiness and long-term life satisfaction. Our model pinpointed novel ideas to present how individuals obtain social supports through SNS and how the three sort of social supports influence users' happiness and life satisfaction. Third, supplementary fit and social sup-port theory were proved well explained people's interactive relationships in SNS.

References

1. V. Apaolaza, P. Hartmann, E. Medina, J.M. Barrutia, C. Echebarria, Comput. Hum. Behav. **29**, 4, 1282–1289 (2013)
2. K. T. Lee, M. J. Noh, D. M. Koo. Cyberpsychol Behav. Soc. Netw, **16**, 6, 413-418 (2013)
3. C. H. Kroenke, M. L. Kwan, A. I. Neugut, I. J. Er-gas, J. D. Wright, B. J. Caan. Breast. Cancer. Res. Treat, **139**, 2, 515-527 (2013)
4. F. J. H. Lu, Y. Hsu. J. Athl. Training, **48**, 1, 92-98. (2013)
5. D. M. Cable, J. R. Edwards. J. App. Psych, **89**, 5 822-834 (2004)
6. P. M. Muchinsky, C. J. Monahan. J. Vocat. Behav, **31**, 3, 268–277 (1987)
7. J. Y. Seong, A. L. Kristof-Brown. J. Manag. Psych, **27**, 6, 536-556 (2012)

The Study of Inter-organizational Cooperation in Hospital —A Perspective of Resource-based Theory

Chi-Wen Juan[1], Hung-Chih Lai[2] , Ying -Chieh LIU[3,a] and Ping-Liang Chen[1]

[1]Department of Management, Kuang Tien General Hospital (The Shalu hospital Area), Taiwan
[2]Department of International Business Studies, National Chi Nan University, Taiwan
[3]Associate Professor at Department of Information Management, Chaoyang University of Technology, Taiwan

Abstract. Based on the resource-based theory, this study analyzed cross-organization cooperation and resource value of hospital medical departments in the viewpoint of rare, irreplaceable, and inimitability of a resource base model developed by scholars Grant and Barney. We interviewed 10 respondents in three regional hospitals and analyzed the interview contents according to the model. The important results are twofold: (a) The rare and irreplaceable resources owned by hospitals relate to the competitive advantage of the hospital; (b) Hospital medical collaboration across organizational divisions or self-development and medical divisions in professional competence are positively correlated.

1 Introduction

A concept, "Hospital Medical Division Collaborating", has become an important managerial strategy for hospitals. This study was based on a study by [1] who established a medical research model of resources value and rareness for across-organizations of medical divisions. There are differences between various types of hospitals in the domestic medical organizations. Based on the relation-ships between the use of medical across-organizational cooperation model and the resource-based theory, the research propositions can be developed. They are LVLR 、LVHR 、HVLR and HVHR (they will be explained in the following sections).

2 Literature review

2.1 Across organizational cooperation of medical divisions

Charles and Essinger [2] developed the outsourcing model based on the service agreement and the responsibilities of the third party. Outsourcing is known as companies are entrusted to external organizations for their parts of business processes. Out-sourcing is the appropriate allocation of company's technologies and resources to produce the maximum benefit. Companies focus on its core business and outsource other unimportant activities ([3]).

2.2 Resources based view

Barney [1] pointed out that for achieving lasting competitive advantages, an organization`s resources must have characteristics of value and rareness. Overall, resource-based theory is particularly focus on the resources which have the following characteristics: value, rareness, inimitability and irreplaceable. Enterprises are able to create sustainable competitive advantage by resources with these characteristics.

2.3 Value and rareness integration model

We adopted the value and rareness model in re-source-based view to discuss across-organizational cooperation of medical divisions. The definitions in value and rareness of this study are as follows. First, value means a company's existing resources which create performance for companies. Second, rareness means currently only a few competitors have the re-sources in the market, but the owner amounts of the resources in the market are less than the demands. Resources have both characteristics of value and rareness or have either are important for across-organizational cooperation (Figure 1).

[a] Corresponding author: allanliu@cyut.edu.tw

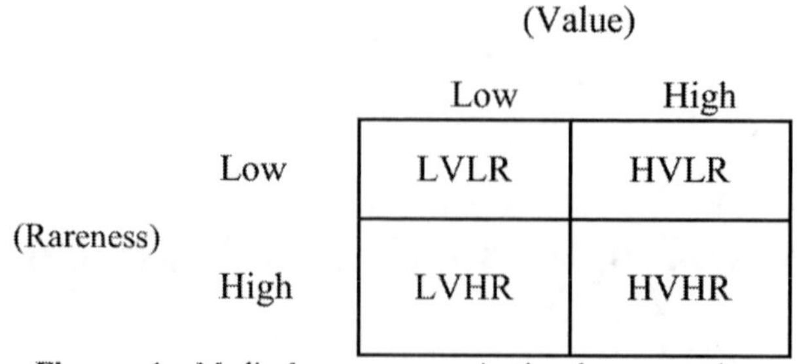

Figure 1. Medical across-organizational cooperation model.

HVHR: resources are with high value and high rareness. These are the hospital's core resources which have a high degree of sustaining competitive ad-vantage and are rare in the market.

LVHR: resources are with low value and high rareness. These contribute less to hospital, and are rare in the market.

HVLR: resources are with high value and low rareness. These contribute to the maintenance of competitive advantages but are not rare in the market.

LVLR: resources are with low value and low rareness. These contribute less to maintain competitive advantages for hospitals and are common resources in the market.

3 Research methodology

To understand the impacts of resources on the cooperation of across organizations, we collected information by in-depth interviews which consisted of three parts: pre-visit interview, interview and data analysis.

The interviewees were the heads or directors of the regional hospital or the competent outsourcing medical divisions. Because they can provide a wealth of health related problems while they are operating di-visions, and the problems with across-organizational cooperation. We have interviewed 10 respondents in three hospitals. Table 1-3 show their information.

Table 1. A hospital respondents and time.

Hospital lieu of that	A General	A branch	A branch	A branch
Respondents	Director Ho	Lee Nursing length	Premier Lee	Director Su
Established	1913	1996	2006	2006
Hospital location	Shalu	Dajia	Tunghsiao	Tunghsiao
Hospital grade	Regional Teaching	Regional Teaching	Regional Hospital	Regional Hospital
Hospital Ownership	Corporation aggregate	Corporation aggregate	Corporation aggregate	Corporation aggregate
Preparation of beds	657	473	50	50
Interview Time	90 minutes	85 minutes	100 minutes	95 minutes

Table 2. B hospital respondents and time

Hospital, lieu of that	B Branch	B Branch	B Branch	B Branch
Respondents	Wang Director	Minister Li	Chen Commissioner	Director Wang
Established	1980	1980	1980	1980
Hospital location	Nantou	Nantou	Nantou	Nantou
Hospital grade	Regional Hospital	Regional Hospital	Regional Hospital	Regional Hospital
Hospital Ownership	Corporation aggregate	Corporation aggregate	Corporation aggregate	Corporation aggregate
Preparation of beds	214	214	214	214
Interview Time	70 minutes	100 minutes	60 minutes	90 minutes

Table 3. C hospital respondents and time

Hospital, lieu of that	C General Hospital	C General Hospital
Respondents	Chen, vice president	Director Chen
Established	N.A	N.A
Hospital location	Taichung	Taichung
Hospital grade	Regional Teaching	Regional Teaching
Hospital Ownership	Corporation aggregate	Corporation aggregate
Preparation of beds	653	653
Interview Time	120 minutes	80 minutes

4 Analysis results

4.1 HVHR

When hospitals with high value and rareness, they would like to adopt their own development in-stead of cooperate with others.

"We have established over 100 years as public health in the central coastal line Tunghsiao, Dajia and Shalu area. We have grown up by more than one thousand beds and have reached three thou-sand beds so far. We have a total of five different branches. (An orthopedic surgeons)"

Proposition 1: Hospital divisions which have high value and rareness are more likely to adopt their own development.

4.2 LVHR

These resources contribute less to hospitals' competitive advantages when hospitals have resources with low-value and high rareness.

"Our hospital's dialysis department is basically the same as other hospitals in terms of equipment, technology or mode of operation. The difference is the convenience for the people in this region. We are the only renal dialysis department in the region. Each patient is charged 5,000NT when they do dialysis treatment. If they do too much, the fixed costs would increase and the profit would be reduced. Our economic size is not as large as Chang Gung Memorial Hospital, and our human re-sources are not very enough, often physician man-power faults occur. (A branch of Premier Lee)"

Proposition 2: hospital medical divisions which have high rareness and low value are more likely to take across organizational cooperation.

4.3 HVLR

When hospitals have resources with high value and low rareness, these resources are able to create competitive advantages for hospitals. Thus, they prefer to develop by their own.

"Our hospital does not invest on or develop specialized equipment only for dialysis operation, be-cause dialysis equipment is almost the same be-tween different hospitals. Dialysis treatment is a standard operation. Each hospital can set up and operate the equipment appropriately. The threshold is not high, so we can develop by ourselves. (A Branch Director Wang) "

Proposition 3: hospital medical divisions which have high value and low rareness are more likely to take their own development.

4.4 LVLR

When hospitals have resources with low value and low rareness, these resources fail to create competitive advantages for hospitals.

"If the laboratory area hospitals are not the most unique and with good reputation, unless they are special hospitals like psychiatric hospitals, they would proceed across-organizational cooperation, like Taipei Wan Fang Hospital's laboratory. The hospitals would depend on inter-organizational co-operation to achieve higher performance than those do not pursue across-organizational co-operation. The efficiency improvement of the cooperation weighs the cost reduction. Inter-organizational cooperation is often more efficient than a single hospital. Such as the purchase procedures of hospitals are often a period of assessment. Through inter-organizational cooperation, the purchases procedures can be completed within a few days, as our equipment or multi-mode operation are about the same. (A Branch Director Li) "

Proposition 4: hospital medical divisions which have low value and low rareness are more likely to take across-organizational cooperation.

5 Conclusions

This study concluded the followings:

(A) Medical division value and other core competencies are positively related to the value of hospitals and medical divisions when the hospitals own more core competence.

Core competence is the set of medical subjects including skills and knowledge, not products and capabilities. A flexible and long-term mature platform is able to adapt to the challenges of a changing environment. Core competence is a combination of cross-business sector capacity in enterprises, but also various business sectors have the skills, knowledge, interact with each other and the result of integration. These results need to constantly stick together to business, in-vestment and commitments. So when the medical divisions of the hospital have resources with more value, it would enhance the competitive-ness of the hospital. Interview data showed that medical division with the higher value of the hospital when the hospital owns more core competence.

(B) The irreplaceable and rare resources owned by hospitals relate to the competitive advantage of the hospital.

Other hospitals have the resources with same value are not competitors. In factor markets, demands are more than the supply of the resources, and therefore valuable resources can produce more competitive advantages. Furthermore, competitive advantages can be converted into customer values, which is one of the key functions of resources. Hospitals must consider the value of resources, but also must consider the scarcity of resources. Emphasizing the role of market-based assets, the scarce of resources is one of the elements to maintain competitive advantages. Hospital's scarce resources used in across organizations come from somewhere, because there are a lot of ways including many domestic and international relevance of the research. Through in-depth interviews, this study found that hospital's medical divisions have more resources with higher values, the more competitive advantages the medical department have.

(C) Hospital medical collaboration across-organizational divisions or self-development and medical divisions in professional competence are significantly positive correlated.

There are many domestic and foreign scholars pointed out: the more the higher degree of in-house expertise, the higher tendency to self-development of internal system (insourcing) when in-sufficient capacity is more. Its purpose is nothing more than a cost-effective, complementary technology, flexibility of response and the development of core business issues. Through in-depth interviews, this study found that hospital medical collaboration across organizational di-visions or self-development and medical divisions in professional competence are significantly positive correlated

(D) Health department and other areas of the hospital where the value is related to the medical di-visions with the higher value of the region when hospitals have more abilities.

In addition to considering the profitability of the decisions whether cooperating with other divisions, hospitals should pay more attention in the medical branches of their local roles, especially the regional hospitals. They cannot simply be measured by the cost-effectiveness. The future operation strategy should be evaluated seriously. Through in-depth interviews, this study found that the same amount of scholars and foreign re-search paper, all noted the hospital therapy department of the region, the higher the value of the more part of the core capacity of the hospital.

(E) That medical divisions across organizations quickly acquire technology does not exist within hospitals.

In recent years, although the quality of care for people's attention, and when the domestic and foreign medical department and other equipment and technology are more advanced, hospitals currently have insufficient intentions to adapt them. It may due to hospitals consider the cost and capabilities of developed technologies. Through in-depth interviews, this study found that the same number of scholars and foreign re-search paper reported that outsourcing is a quick way to get into the advanced technology.

References

1. J. Barney, J, J. Manag, **17**, 1, 99-120 (1991)
2. L. G. Charles., J. Essinger. *Inside out-sourcing: The insider's guide to managing strategic sourcing* (Naperville, Brealey Pub, 2000)
3. J. B. Quinn., F.G. Hilmer, Sloan. Manage. Rev, **Summer**, 43-55 (1994)

DOI: 10.1051/matecconf/20167103007

Investigation on the Relationship between Information Communication Technology and Reading Literacy for Northeast Asian Students

Yi-Horng Lai

Oriental Institute of Technology, Department of Healthcare Administration, 22061 New Taipei City, Taiwan

Abstract. The purpose of this study was to investigate the relationship of internet communication technologies (ICT) usage and reading literacy of Northeast Asian Students in PISA 2009. ICT was crucial for all governments in the world towards promoting equity. ICT had been considered a low cost opportunity towards equalizing educational systems. A multilevel modelling (MLM) was applied to identify at which school-level the largest variations in the three indicators exist in this study. With MLM, it could be examine simultaneously the effects of different hierarchical school-level variables; to take account of possible correlations of students' reading literacy in this study within higher levels (Urbanization, Total number of students Computers for education) which may otherwise lead to incorrect standard errors and inefficient estimates; to treat higher levels as related; and to examine inter-area variations at each level. The empirical results include the different between group components was significant. Self-confidence in ICT high level tasks was positive with students' reading literacy, and self-confidence in ICT high level tasks was negative with students' reading literacy. The urbanization of schools' area and total number of students of schools were positive with students' reading literacy. The urbanization of schools' area was negative with the relationship of ICT for school related tasks and students' reading literacy. Total number of students was negative with the relationship of ICT for school related tasks and students' reading literacy, and ICT availability in school and students' reading literacy.

1 General instruction

The purpose of this study was to investigate the relationship of Internet Communication Technologies (ICT) usage and reading literacy of Northeast Asian Students in PISA 2009. Studies on educational efficiency have been on the rise in the past decade. Since, investments in education pays back in the long run, this leads countries to evaluate educational systems with their functions based on student achievement. ICT was crucial for all governments in the world towards promoting equity. ICT had been considered a low cost opportunity towards equalizing educational systems.

These years, scholars were often confronted with the collection of correlated data. This generic term hold a multitude of data structures, such as multivariate observations, repeated measurements, clustered data, longitudinal data, and spatially correlated data. Instances of this type of research can be encountered in virtually every empirical branch of science. Different areas of research will refer to the same or similar concepts with different terminology. Such as multilevel modelling [1] was a frequently encountered term in sociological applications, whereas in classical experimental design research one often re-fers to variance component models [2]. A multilevel modelling can consist of more than two levels and examples also abound in practice. Schooling systems, for instance, present an obvious multilevel structure, with pupils grouped into classrooms, which are nested within schools which themselves may be clustered within education authorities [3, 4].

A multilevel modelling (MLM) was applied to identify at which school-level the largest variations in the three indicators exist in this study. With MLM, it could be examine simultaneously the effects of different hierarchical school-level variables; to take account of possible correlations of outcomes (students' reading literacy in this study) within middle or higher levels which may otherwise lead to in-correct standard errors and inefficient estimates; to treat middle and higher levels as related; and to ex-amine inter-area variations at each level.

The ICT resources for students include home ICT resource and school ICT resource [5, 6, & 7]. ICT connects information, products, ideas, people, individuals as well as communities globally at a low cost. Peoples worry the importance of preparation of young generations with ICT skills in their education policy documents. ICT brings new structures in learning. In PISA survey, students are asked about their familiarity with ICT, mainly with computers.

Previous studies that survey the relationship be-tween ICT usage and problem-solving found that when students have access to computers at home, they had better

[a] Corresponding author: author@e-mail.org

achievement scores [8]. With 297295 fifteen-year-old students in PISA 2009 data, Lee and Wu indicate that ICT availability at home and self-confidence in ICT high level tasks was positive with students' reading literacy [9]. Besides, with enough ICT resource at home, students need more help in self-confidence in ICT high level tasks and ICT for school related tasks for improve reading literacy, now [10]. Although lots of studies point out that computers are helpful in students' reading literacy, Tse, Yuen, Loh, Lam, & Ng recommend that computers can have a beneficial impact on students' reading but only if computer usage is well targeted and does not consist of students playing repetitive games [11].

2 Materials and methodology

The research frame was as Figure 1. School-level variables include urbanization, total number of students, and computers for education. Student-level variables include self-confidence in ICT, high level tasks, ICT for school related tasks, ICT availability at home, and ICT availability in school. The Hypothesis 1 is the student-level variables were positive with reading literacy. The Hypothesis 2 is the school-level variables were positive with reading literacy. The Hypothesis 3 is the student-level variables were positive with reading literacy dependent on school-level variables.

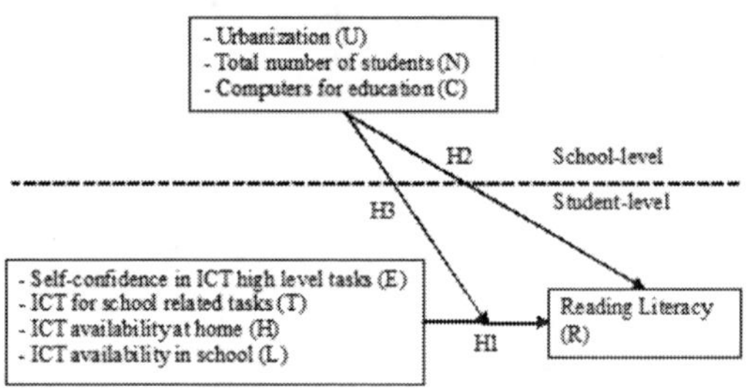

Figure 1. Research framework of this study.

2.1 Research Data

The primary source of data for our analysis is the fourth wave of the Programme for International Stu-dent Assessment (PISA) administered in 2009. PISA is a cross national survey that, every three years since 2000, has assessed 15 year-old students' performance in mathematics, reading and science, as well as cross-curricular problem-solving skills. PISA considers students' knowledge in these areas not in isolation, but in relation to their ability to reflect on their knowledge and experience and apply them to real-world issues. The emphasis is on mastering pro-cesses, understanding concepts and functioning in various contexts within each assessment area [12].

2.2 Data Analysis

In this study, a two multi-level be estimated with random intercepts models. The multilevel modelling (MLM) was applied to identify at which school-level the largest variations in the three indicators exist in this study. The two-level model of reading literacy was as Figure 2. Multilevel models were derived using the computer package Mplus 7.0 and R 3.2.0.

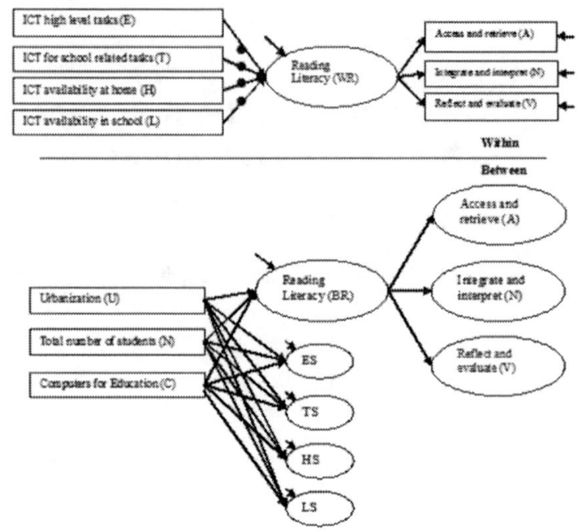

Figure 2. Two-level CFA model of reading literacy.

3 Results

The sample includes 10199 students from 343 schools in PISA 2009 data in Japan and Korea. The sample consisted of 41.92% female and 58.08% male students. 186 Japanese schools and 157 Korean schools, and the data summarize of student-level and school-level was as Table 1. Besides, only Japan and Korea do ICT survey in PISA 2009 data. The Cronbach's α of reading literacy was .96.

Table 1. Data summarize of student-level and school-level.

		Student-level		School-level	
Geographic		**Freq.**	**(%)**	**Freq.**	**(%)**
Gender	Male	4923	41.92	-	-
	Female	5276	58.08	-	-
Country	Japan	5455	53.49	186	54.23
	Korea	4744	46.51	157	45.77
Urbanization	Village	69	.68	2	.58
	Small Town	330	3.24	14	4.08
	Town	1948	19.10	64	18.66
	City	4552	44.62	151	44.03
	Large City	3300	32.36	112	32.65
Total		10199	100.00	343	100.00

For testing the hypotheses in this study (Figure 1), multilevel linear modelling been used for data analysis in this study. This study explored the relationship in the cross-level variables with multilevel linear modelling.

3.1 Null model

This study explored the relationship of students' personal opinion factors and class factors, and detecting the effect of cross-level with multilevel linear modelling. The result

 DOI: 10.1051/matecconf/20167103007

of data analysis was as Table 4. The different between each class with null model were significant. With the result of the null model in this study, it could be found that the different between group components (u0) was significant (u0=3935.25, S.E.<.01).

3.2 Random Coefficient Model

Random coefficient model was the model with all individual level variables. It means a regression model with all individual level variables in individual level and null model in group level. The relationship of dependent variables and independent variables in individual level could be detected with random coefficient model. The result of random coefficient model was as Table 2. It could be found that self-confidence in ICT high level tasks (E) (γ_{10}=5.78, S.E.= .75) and ICT availability at home (H) (γ_{30}=.-2.97, S.E.= .80) were significant, but CT for school related tasks (T) (γ_{20}=-.48, S.E.= .88) and ICT availability in school (L) (γ_{40}=.24, S.E.= .68) were not significant. With the result, self-confidence in ICT high level tasks was positive with students' reading literacy, and self-confidence in ICT high level tasks was negative with students' reading literacy.

3.3 Contextual Model

The percent of the direct effects that can be explained by intercept variance could be detected with contextual model. The result of contextual model was as Table 4. It could be find that urbanization (U) (γ_{01}=10.63, S.E. = 4.29), total number of students (N) (γ_{02}=.01, S.E. =.01), and computers for education (C) (γ_{03}=.03, S.E. =.03) were significant. With the result, the urbanization of schools' area and total number of students of schools were positive with students' reading literacy. Computers for education in schools were negative with students' reading literacy.

3.4 Full Model

The result of full model was as Table 4. It could be find that urbanization (U) was a moderator for ICT for school related tasks (T) with reading literacy (R) (γ_{21}=-2.42, S.E. = .99). Total number of students (N) were moderators for ICT for school related tasks (T) with reading literacy (R) (γ_{22}<-.1, S.E. <.01) and ICT availability in school (L) with reading literacy (R) (γ_{42}<-.1, S.E.<.01). With the result, the urbanization of schools' area was negative with the relationship of ICT for school related tasks and students' reading literacy. Total number of students was negative with the relationship of ICT for school related tasks and students' reading literacy, and ICT availability in school and students' reading literacy.

Table 2. The result of multilevel linear modelling.

		Null Model	Random Coefficient Model	Contextual Model	Full Model
I	γ_{00}_1	536.54* (3.45)	538.13* (3.55)	504.19* (16.90)	507.48* (16.76)
	γ_{00}_2	529.63* (3.39)	531.23* (3.49)	496.95* (17.18)	500.29* (17.04)
	γ_{00}_3	531.01* (3.63)	532.54* (3.71)	494.51* (18.83)	498.05* (18.73)
U	γ_{01}	-	-	10.63* (4.29)	9.68 (4.28)
N	γ_{02}	-	-	.01* (.01)	-.01* (.01)
C	γ_{03}	-	-	-.11* (.03)	-.10* (.03)
E	γ_{10}	-	5.78* (.75)	5.76* (.75)	2.60 (.48)
E	γ_{11}	-	-	-	.60 (.90)
E	γ_{12}	-	-	-	<.01 (<.01)
E	γ_{13}	-	-	-	.01 (.01)
T	γ_{20}	-	-.49 (.88)	-.56 (.88)	8.42* (.01)
T	γ_{21}	-	-	-	-2.42* (.99)
T	γ_{22}	-	-	-	<-.01* (<.01)
T	γ_{23}	-	-	-	.01 (.01)
H	γ_{30}	-	-2.98* (.80)	-2.97* (.80)	-1.84 (.43)
H	γ_{31}	-	-	-	-.18 (.91)
H	γ_{32}	-	-	-	<.01 (<.01)
H	γ_{33}	-	-	-	<.01 (.01)
L	γ_{40}	-	.25 (.68)	.26 (.68)	3.32 (.32)
L	γ_{41}	-	-	-	-.47 (.83)
L	γ_{42}	-	-	-	<-.01* (<.01)
L	γ_{43}	-	-	-	-.01 (.01)
	r	3518.83* (126.38)	3534.07* (390.72)	3115.67* (400.37)	3084.30* (402.55)
	u_0	3935.25* (384.13)	3903.16* (125.95)	3904.37* (126.04)	3846.94* (127.91)
	u_1	-	-	-	1089.14 (12.60)
	u_2	-	-	-	54.91 (15.85)
	u_3	-	-	-	978.87 (14.57)
	u_4	-	-	-	3846.94 (11.59)
	AIC	316790.22	316720.26	316791.06	316799.69
	BIC	316898.67	316857.63	316942.89	317067.20

4 Discussion and conclusion

The data on the Northeast Asian student sample regarding their use of computers on a number of variables and their reading literacy was investigated. The results indicated that there is significant relationship between their use of ICT and reading literacy. The descriptive statistics revealed that a quite high percentage of students had access to computers. The results indicated that there was a high difference in terms of percentages between different schools.

Self-confidence in ICT high level tasks was helpful in students' reading literacy, and this result was the same as Lee and Wu's study [9, 10]. If students want finish high

DOI: 10.1051/matecconf/20167103007

level tasks themselves, they should read and learn more with Internet or other information communication technology. Students would improve reading literacy by ICT with the object of self-confidence high level tasks. But ICT availability (or resource) at home was harmful in students' reading literacy, and this result was the same as Tse, Yuen, Loh, Lam, and Ng's study [11]. For most of students or youth, information communication technology was a tool for online games. Most students that without teachers' advice only want play internet game. With these two student-level result, it could find that teachers play an important role in improving students' reading literacy. Teachers can improve students' reading literacy by self-confidence in ICT high level tasks in a full ICT resource at home. Be-sides, it could fund that ICT for school related tasks and ICT availability in school were not helpful in students' reading literacy. Students like learning with their wish, they don't like learning with teachers' wish.

Most study only focus on student-level variables, such as Lee and Wu's study [9, 10] and Tse, Yuen, Loh, Lam, and Ng's study [11]. The study focus on both student-level variables and school-level variables. The urbanization of school area and the total number of students of school were positive with students' reading literacy. Most schools in large city were also many students, so these two results were similar. The gap of students' reading literacy was large, and government should pay attention to these schools in village or small town for their weak reading literacy. Computers for education in schools were negative with students' reading literacy. Stu-dents think computers in school were for learning, and learning was active as teachers wish and unfunny. So improve students' reading literacy was not just mean lots computers in school. Teachers should let students know the meaning of reading literacy and make learning with computers more fun.

References

1. Goldstein, H., Multilevel Statistical Models. In: Kendall's Libary of Statistics **3**. Arnold, London, (1995)
2. Searle, S.R., Casella, G., McCulloch, C.E., Variance Components, Wiley, New York (1992)
3. Castelli, A., Jacobs, R., Goddard, M., & Smith, P.C., Health, policy and geography: Insights from a multi-level modelling approach, Social Science & Medicine **92**, 61-73 (2013)
4. Abrahantes, J.C., Molenberghs, G., Burzykowski, T., Shkedy, Z., Abad, A.A., & Renard, D., Choice of units of analysis and modelling strategies in multilevel hierarchical models, Computational Statistics & Data Analysis **47**, 536-563 (2004)
5. OECD., PISA 2009 key findings, PISA. Re-trieved June 1, 2013 from http://www.oecd.org/pisa/pisa2009keyfindings.htm, (2013)
6. Sweet, R. and Meates, A., ICT and Low Achievers: What Does PISA Tell Us? In A. Karpati (Ed.), Promoting Equity through ICT in Education: Project, Problems, Prospects. Budapest, Hungary: OECD and Hungarian Ministry of Education (2004)
7. OECD, PISA 2009 at a glance, Paris: OECD, 2011.
8. Aşkar, P. and Olkun, S., The use of ICT in schools based on PISA 2003 Data, Eurasian Journal of Educational Research **19**, 15-34 (2005)
9. Lee, Y.H. & Wu, J.Y., The effect of individual differences in the inner and outer states of ICT on engagement in online reading activities and PISA 2009 reading literacy: Exploring the relationship be-tween the old and new reading literacy, Learning and Individual Differences **22(3)**, 336-342 (2012).
10. Lee, Y.H. & Wu, J.Y., The indirect effects of online social entertainment and information seeking activities on reading literacy. Computers & Educa-tion **67**, 168-177 (2013)
11. Tse, S.K., Yuen, A.H.K., Loh, E.K.Y., Lam, J.W.I., & Ng, R.H.W., The impact of blogging on Hong Kong primary school students' bilingual reading literacy, Australasian Journal of Educational Technology **26(2)**, 164-179 (2010)
12. OECD, Learning to Change: ICT in Schools, Paris:OECD (2001)

MATEC Web of Conferences **71**, 03008 (2016) DOI: 10.1051/matecconf/20167103008
ICCPE 2015

A robust cloud access scheme with mutual authentication

Chin-Ling Chen[1,2,a], Yong-Yuan Deng[3], Kun-hao Wang[2] and Chun-Long Fan[4]

[1]*Department of Computer Science and Information Engineering, Chaoyang University of Technology, Taiwan*
[2]*School of Information Engineering, Changchun University of Science And Technology, China*
[3]*Department of Information Management, Chaoyang University of Technology, Taiwan*
[4]*School of Computer Science,Shenyang Aerospace University, China*

Abstract. Due to the progress of network technology, we can access some information through remote servers, and we also can save and access lots of personal data in remote servers. Therefore, to protect these data and resist un-authorized access is an important issue. Some researchers proposed authentication scheme, but there still exist some security weaknesses. This article is based on the concept of HDFS (Hadoop Distributed File System), and offers a robust authentication scheme. The proposed scheme achieves mutual authentication, prevents re-play attack, solves asynchronous issue, and prevents offline password guessing attack.

1 Introduction

1.1 Background

Due to the rapid development of hardware net-work environment, a variety of services and applications began to attract personal attention in today's social environment. In earlier, because of the smaller scale of services, fewer numbers of users, a single server is sufficient to provide all the necessary services.

With the rise of integrated services, single company can provide diverse services, and even operate as cross-industry alliance type. User can use single account to access various services. On the other hand, the huge amount of users is impossible to use single server to satisfy various services. Therefore, we need multi-server to handle these services, and that's a concept of cloud. [7, 8, 11]

1.2 Cloud architecture

There are three kinds of architectures about user and server in cloud architecture [9].

1.2.1 All users connect to a single server for authentication and communication

This service model is simple and easy to set up. But the number of users provided service con-tents increasingly, it will be no longer a single server can handle it, and such architecture is vulnerable to be braked by malicious people.

1.2.2 All users through the gateway and lead to a different server

Such service model can effectively solve the excessive burden of a single server problem, but this will significantly reduce the reliability of the system architecture, and it is also vulnerable to be braked by malicious people.

1.2.3 All users process identity verification with front server and backend server

Such service mode can also effectively solve the excessive burden of a single server problem, and the user's real identity is stored in the backend server. It's hard to be braked by malicious peo-ple, the architecture is similar to HDFS (Hadoop Distributed File System). [6, 10]

2 Literature review

2.1 Review of Nivethitha et al.'s scheme [5]

2.1.1 Registration stage of Nivethitha et al.'s scheme

Step 1: User selects his/her identity ID, password PW, and then sends (ID, PW) to register server through secure channel. Registration server forwards the same message to backend server.

Step 2: Register server generates one time password OTP, and use the symmetric key OTP to encrypt the

[a]Corresponding author: clc@cyut.edu.tw

PW, $C1 = E_{OTP}(PW)$, $C2 = C1 \bmod 26$, and stores $C2$ in its database securely, then sends OTP to backend server through secure channel.

Step 3: Backend server calculates $C3 = E_{PW}(OTP)$, and then stores $(ID, C3)$ in its database securely.

2.1.2 Authentication stage of Nivethitha et al.'s scheme

Step 1: User inputs his/her identity ID', then sends it to register server. Register server forwards the same message ID' to backend server.

Step 2: Backend server checks its database to get $(ID, C3)$, which stored during registration stage, and then sends $C3$ to register server. Register server forwards the same message $C3$ to user.

Step 3: User calculates $OTP' = D_{PW}(C3)$, and then sends OTP' to register server.

Step 4: Register server calculates $C1 = C2 \bmod 26$, and use the decryption key OTP to decrypt PW, $PW = D_{OTP'}(C1)$. Afterward, uses the PW to encrypt OTP', $C4 = E_{PW}(OTP')$, and then sends $C4$ to backend server.

Step 5: Backend server checks $C4 \overset{?}{=} C3$. If pass the verification, the backend server gets ID which stored in registration stage, and then sends ID to register server.

Step 6: Register server checks $ID \overset{?}{=} ID'$. If pass, it generates OTP^{new} and calculates $C1^{new} = E_{OTP^{new}}(PW)$, $C2^{new} = C1^{new} \bmod 26$, and then updates $C2$ into $C2^{new}$ in its database securely for next access.

Later, Mrudula et al. [4] pointed out that Nivethitha et al.'s scheme can't prevent offline password guessing attack. When an attacker intercepts the messages ID', $C3$ and OTP' which transferred between user and register server during authentication stage, he/she can perform dictionary attack to guess a password PW', and calculate $OTP'' = D_{PW'}(C3)$. If OTP'' is equal to OTP', it means he/she gets the correct password successfully, and can be a legal user for future access.

2.2 Review of Mrudula et al.'s scheme

2.2.1 Registration stage of Mrudula et al.'s scheme

Step 1: User selects his/her identity ID, password PW and a random number a, uses a hashing function $h(\ )$ and exclusive OR operation $\oplus$ to calculate $APW = h(a \oplus PW)$, and then sends (ID, APW) to register server through secure channel.

Step 2: Register server generates OTP, and calculates $CID = ID \oplus h(OTP)$, $C1 = APW \oplus h(OTP)$, $C2 = C1 \bmod 26$, and then sends (CID, APW, OTP) to backend server through secure channel.

Step 3: Backend server calculates $C3 = h(APW) \oplus OTP$, $ID = CID \oplus h(OTP)$, and then stores $(ID, C3)$ in its database securely.

Step 4: Register server generates a random number ru, calculates $HID = h(ID \| X)$ and R, $R = ru \oplus h(X \| ID)$, stores $(HID, C2, R)$ in its database securely, and then sends ru to user through secure channel.

2.2.2 Authentication stage of Mrudula et al.'s scheme

Step 1: User inputs his/her identity ID', then sends it to register server. Register server forwards the same message ID' to backend server.

Step 2: Backend server checks its database to get $(ID, C3)$ which stored during registration stage, and then sends $C3$ to register server. Register server forwards the same message $C3$ to user.

Step 3: User calculates $OTP' = C3 \oplus h(APW)$, $C4 = OTP' \oplus h(ru \| ID')$, and then sends $C4$ to register server.

Step 4: Register server calculates $ru = R \oplus h(X \| ID)$, $OTP' = C4 \oplus h(ru \| ID')$, $C1 = C2 \bmod 26$, $APW = C1 \oplus h(OTP')$, $C5 = h(HPW) \oplus OTP'$, and then sends $C5$ to backend server.

Step 5: Backend server checks $C5 \overset{?}{=} C3$. If pass, backend server gets ID from $(ID, C3)$, which stored during registration stage, and then sends ID to register server.

Step 6: Register server checks $ID \overset{?}{=} ID'$. If pass, it generates ru^{new}, updates ru into ru^{new} in its database securely for next access, and calculates $C6 = h(APW) \oplus ru^{new}$, then sends $C6$ to user.

Mrudula et al.'s scheme solves offline password guessing attack successfully due to they use a random number ru between user and register server during authentication stage. The random number ru changes after a success round, so an attacker can't perform offline password guessing attack simply. We find that their scheme exists another problem that is asynchronous issue.

When an attacker intercepts the final message between user and register server during authentication stage, they can't communicate anymore in the future. User and register server hold different random number ru, and which is decided from register server, there is no way to let user gets another random number again. Besides, both Nivethitha et al. and Mrudula et al.'s scheme doesn't achieve mutual authentication. [1, 2, 3]

3 The proposed scheme

3.1 Registration stage

Step 1: User selects his/her identity ID, password PW and a random number ru, calculates

$$HPW = h(PW) \quad (1)$$

and then sends (ID, HPW, ru) to register server through secure channel.

Step 2: After receiving the message, register server generates OTP and calculates

$$EID = ID \oplus h(OTP) \quad (2)$$

$$C1 = HPW \oplus h(OTP) \quad (3)$$

$$C2 = C1 \bmod 26 \quad (4)$$

and then sends (EID, HPW, OTP) to backend server through secure channel.

Step 3: Backend server calculates

$$C3 = h(HPW) \oplus h(OTP) \quad (5)$$

$$ID = EID \oplus h(OTP) \quad (6)$$

and then stores $(ID, C3)$ in its database securely.

Step 4: Register server generates a secret parameter x, calculates

$$HID = h(ID \oplus x) \quad (7)$$

$$C4 = ru \oplus h(ID) \oplus h(x) \quad (8)$$

and stores $(HID, C2, C4)$ in its database securely, then sends HID to user through secure channel.

Step 5: User stores (HID, ru) in his/her device securely.

3.2 Authentication stage

Step 1: User inputs his/her identity ID', and then sends it to register server. Register server forwards the same message ID' to backend server.

Step 2: Backend server checks its database to get $(ID, C3)$ which stored during registration stage, and then sends $C3$ to register server. Register server forwards the same message $C3$ to user.

Step 3: After receiving the message, user calculates

$$h(OTP') = C3 \oplus h(HPW) \quad (9)$$

$$C5 = h(OTP') \oplus h(ID' \oplus ru) \quad (10)$$

generates a new random number ru^{new}, calculates

$$C6 = ru \oplus ru^{new} \quad (11)$$

and then sends $(C5, C6, HID)$ to register server.

Step 4: Register server checks

$$HID \stackrel{?}{=} h(ID' \oplus x) \quad (12) \text{ Note}^{a}$$

If pass, the register server calculates

$$ru = C4 \oplus h(ID) \oplus h(x) \quad (13)$$

$$h(OTP') = C5 \oplus h(ID' \oplus ru) \quad (14)$$

$$C1 = C2 \bmod 26 \quad (15)$$

$$HPW = C1 \oplus h(OTP') \quad (16)$$

$$C7 = h(HPW) \oplus h(OTP') \quad (17)$$

and then sends $C7$ to backend server.

Step 5: Backend server checks

$$C7 \stackrel{?}{=} C3 \quad (18)$$

If pass, backend server gets ID which stored during registration stage. Backend server calculates

$$C8 = C3 \oplus h(ID) \quad (19)$$

and then sends $C8$ to register server.

Step 6: Register server checks

$$C8 \stackrel{?}{=} C7 \oplus h(ID') \quad (20)$$

If pass, register server calculates

$$ru^{new} = C6 \oplus ru \quad (21)$$

$$C9 = h(HPW) \oplus ru^{new} \quad (22)$$

$$C4^{new} = ru^{new} \oplus h(ID) \oplus h(x) \quad (23)$$

and then sends $C9$ to user.

Step 7: User checks

$$ru^{new} \stackrel{?}{=} C9 \oplus h(HPW) \quad (24)$$

If pass, user updates ru into ru^{new} for next access.

4 Security analysis

4.1 Achieve mutual authentication

During the login phase, the user, register server and backend server authenticate each other to make sure the legality of each roles. Register server uses $HID \stackrel{?}{=} h(ID' \oplus x)$ to authenticate user, backend server uses $C7 \stackrel{?}{=} C3$ to authenticate register server and user, register server uses $C8 \stackrel{?}{=} C7 \oplus h(ID')$ to authenticate backend server, and user uses $ru^{new} \stackrel{?}{=} C9 \oplus h(HPW)$ to authenticate register server and backend server. The proposed scheme achieves mutual authentication.

4.2 Prevent replay attack

During login phase, an attacker may intercept the message $C5 = h(OTP') \oplus h(ID' \oplus ru)$, $C6 = ru \oplus ru^{new}$, $HID = h(ID \oplus x)$ that sent to register server, and sends it again in the future. The attacker can't success because of $C5 = h(OTP') \oplus h(ID' \oplus ru)$ includes a random value ru is changed in every round.

When the attacker sends $C5 = h(OTP') \oplus h(ID' \oplus ru)$ to register server again, register server computes $h(OTP') = C5 \oplus h(ID' \oplus ru)$ but can't get the original $h(OTP')$ because of the two $h(ID' \oplus ru)$ are different, and can't be recovered by exclusive-or operation. Therefore, the proposed scheme can prevent replay attack.

---

[a] $A \stackrel{?}{=} B$: determine if A equal to B

4.3 Solve asynchronous issue

To prevent replay attack, user and register server update a new random value after a successful authentication for next round. When an attacker intercepts the final message between user and register server, it will cause asynchronous problem. Because of user and register server hold different random value, they can't communicate anymore.

In the previous scheme, the random value decided by register server, the asynchronous problem will happen, and there's no way to solve this situation. In our proposed scheme, even the final message $C9$ has been intercepted by an attacker, user can run the registration phase to send (ID, HPW, ru) again to solve the asynchronous issue.

4.4 Prevent offline password guessing attack

When an attacker intercepts the messages ID', $C3 = h(HPW) \oplus h(OTP)$, $C5 = h(OTP') \oplus h(ID' \oplus ru)$, $C6 = ru \oplus ru^{new}$, $HID = h(ID \oplus x)$ and $C9 = h(HPW) \oplus ru^{new}$ between user and register server, he/she still can't do anything to get the password. The proposed scheme uses hash function and exclusive-or calculation to protect all the messages. It is computationally impossible to get the password HPW, the proposed scheme can prevent offline password guessing attack.

5 Security and complexity comparison

The following table 1 and table 2 are security comparison and complexity comparison respectively. Although the proposed scheme has higher complexity of calculation, it's not a problem for today's hardware and network environment. The proposed scheme achieves more security requirements than the previous schemes, especially mutual authentication.

5.1 Security comparison table

Table 1. Security comparison

	Nivethitha et al.'s scheme	Mrudula et al.'s scheme	The proposed scheme
Achieve mutual authentication	No	No	Yes
Prevent replay attack	Yes	Yes	Yes
Solve asynchronous issue	Yes	No	Yes
Prevent offline password guessing attack	No	Yes	Yes

5.2 Complexity comparison table

Table 2. Complexity comparison

	Nivethitha et al.'s scheme	Mrudula et al.'s scheme	The proposed scheme
Registration stage (User)	N/A	$1T_{XOR} + 1T_{HASH}$	$1T_{HASH}$
Registration stage (Registration server)	$1T_{ENC} + 1T_{MOD}$	$3T_{XOR} + 4T_{HASH} + 1T_{MOD}$	$5T_{XOR} + 5T_{HASH} + 1T_{MOD}$
Registration stage (Backend server)	$1T_{ENC}$	$2T_{XOR} + 2T_{HASH}$	$2T_{XOR} + 3T_{HASH}$
Authentication stage (User)	$1T_{ENC}$	$2T_{XOR} + 2T_{HASH}$	$4T_{XOR} + 5T_{HASH}$
Authentication stage (Registration server)	$2T_{ENC} + 1T_{MOD}$	$5T_{XOR} + 5T_{HASH} + 1T_{MOD}$	$12T_{XOR} + 10T_{HASH} + 1T_{MOD}$
Authentication stage (Backend server)	N/A	N/A	$1T_{XOR} + 1T_{HASH}$

T_{XOR}: time complexity of exclusive-or operation
T_{HASH}: time complexity of hash operation
T_{MOD}: time complexity of mod operation
T_{ENC}: time complexity of symmetric encryption/decryption

6 Conclusions

In the cloud environment, lots of users' sensitive data are stored in the remote servers. Besides legal users, malicious people also want to get these data. There must have a robust authentication protocol to protect these data. Some previous researchers proposed the authentication protocol for HDFS cloud architecture, but they still get some security defects. The proposed protocol solves these defects successfully, and also satisfies more security requirements.

Acknowledgement

This research was supported by the Ministry of Science and Technology, Taiwan, R.O.C., under contract number

MOST 105-2221-E-324-007, MOST 105-2622-E-212-008 -CC2 and MOST103-2632-E-324-001-MY3.

References

1. K. Hamlen, M. Kantarcioglu, L. Khan, & ham B. Thuraising. Security issues for cloud computing. *International Journal of Information Security and Privacy*. Apr–Jun; **4(2)**:39-51 (2010)
2. K. Hwang, S. Kulkarni, & Y. Hu. Cloud security with virtualized defense and reputation - based trust management. *Eighth IEEE International Conference on Pervasive Intelligence and Computing, (PICom2009)*. Chengdu, China. 717-722. (2009)
3. G. Kousiouris, G. Vafiadis, & T. Varvarigou. A frontend, Hadoopbased data management service for efficient federated clouds. *IEEE Third International Conference on Cloud Computing Technology and Science (CloudCom)*. 511-516. (2011).
4. S. Mrudula, M.R.M. Chandra, S.V. Chandra. A Secure and Light Weight Authentication Service in Hadoop using One Time Pad, *2nd International Symposium on Big Data and Cloud Computing (ISBCC' 15)*, 81-86 (2015).
5. S. Nivethitha, A. Gangaa, & S. Shankar. Authentication Service in Hadoop Using one Time Pad, *Indian Journal of Science & Technology*, **7**, 56-62 (2014).
6. O.M. Owen. Integrating Kerberos into Apache Hadoop Kerberos. *Conference 2010*, 26–27 Oct, MIT, USA (2010).
7. P.K. Rahul, & K.T. Gireesh. A Novel Au thentication Framework for Hadoop, *International conference on Intelligence and Evolutionary Algorithms in Engineering Systems (ICAEES 2014)* , **324**, 333-340. (2015).
8. G.S. Sadasivam, K.A. Kumari, & S. Rubika. A Novel Authentication Service for Hadoop in Cloud Environment, *IEEE International Conference on Cloud Computing in Emerging Markets (CCEM)*, 1-6, India (2012)
9. S.G. Sudha, K.K. Anitha, & S. Rubika. A novel authentication service for hadoop in cloud environment. *IEEE International Conference on Cloud Computing in Emerging Markets (CCEM)*. 16 (2012).
10. G. Turkington. *Hadoop Beginner's Guide*. PACKT Publishing (2013).
11. Q. Zhou, X. Deqin, T. Hunming, & R. Chunming. TSHC: Trusted Scheme for Hadoop Cluster, *Fourth International Conference on Emerging Intelligent Data and Web Technologies (EIDWT)*, 9-11 (2013).

MATEC Web of Conferences **71**, 03009 (2016) DOI: 10.1051/matecconf/20167103009
ICCPE 2015

Generalized Correlation Coefficient Based on Log Likelihood Ratio Test Statistic

Hsiang-Chuan Liu[a]

Department of Informatics and Biomedical Engineering, Asia University, Taichung 41354, Taiwan, ROC
Graduate Institute of Educational Information and Measurement, National Taichung University of Education, Taichung 40306, Taiwan

Abstract. In this paper, I point out that both Joe's and Ding's strength statistics can only be used for testing the pair-wise independence, and I propose a novel G-square based strength statistic, called Liu's generalized correlation coefficient, it can be used to detect and compare the strength of not only the pair-wise independence but also the mutual independence of any multivariate variables. Furthermore, I proved that only Liu's generalized correlation coefficient is strictly increasing on its number of variables, it is more sensitive and useful than Cramer's V coefficient, in other words, Liu generalized correlation coefficient is not only the G-square based strength statistic, but also an improved statistic for detecting and comparing the strengths of deferent associations of any two or more sets of multivariate variables, moreover, this new strength statistic can also be tested by G^2.

1 Introduction

For any quantity study, two kinds of statistics must be considered, one is the significant statistic, the other is the strength statistic, in other words, any significant test statistic should always be coupled with an appropriate statistic of strength, because if the association strength is weak, but the sample size is large, then the result of the null hypothesis is always rejected by a significant test, that is meaningless [1]. For comparing the pair-wise independent among some sets of multivariate variable, if the four rules are satisfied; we always use Pearson Chi square statistic [2], χ_P^2, or G^2 to test the significance; (a) Cochran's rule [3] addresses that there should be no expected frequency value under 1 and no more than 0.2 of expected frequency value should be under than 5, (b) Wikens' rule [4] suggests that the total sample size should be at less five times the number of cells of the contingency table, (c) McHugh's rule [5] suggests that the number of cells of contingency table is not larger than 20, (d) Liu's rule [6] suggests that all values of the correlation coefficients of variables are positive. If these rules are not be satisfied, the interpretation of the significant test results will be difficult, we can exploit the improving methods, such as combining some cells with smaller expected frequency of the contingency table, Yates' correct method or Fisher's exact test. And then, we can use Cramer's V coefficient [1] to test its strength.

For comparing the mutual independent of some sets of multivariate variable, if the above-mentioned four rules are satisfied; Sokal & Rohlf [7] suggested to use the log likelihood ratio statistic, G^2, rather than χ_P^2, since for a large dimensional contingency table, G^2 can be neatly decomposed into smaller components, it can be not done exactly with χ_P^2, and G^2 is the usual statistic for log-linear analyses. The most traditional researches always use G^2 or χ_P^2 to test the pair-wise independence of multivariate variables, since it only needs smaller sample size than that of mutual-independence, but only mutual-independence is the complete independence of multivariate variables. In recent years, big data researches and developments are more emphasized, how to compare the relations among some deferent sets of multivariate variables are the more important issues, and the strength statistic becomes more important than its significant statistic, since their mutual independence samples size are large, the null hypotheses of significant testing are always be rejected. We can use G^2 or χ_P^2, to test the mutual independence of them whether is significant or not, and we know that Cramer's V is the χ_P^2 based strength statistic, but how to find the G^2 based strength statistic? That is a very important issue.

In this paper, I will point out that two well-known mutual information based strength statistics can only be used for testing pair-wise independence, one is Joe's generalized correlation coefficient [7], and the other is Ding's generalized correlation coefficient [8], but both of them can still be not used for mutual independence of the three or more multivariate variables, since the value of mutual information of three variables may be negative [10], and its generalized cases can be not defined, and I will propose a novel G^2 based strength statistic, called Liu's generalized correlation coefficient, this new strength statistic can be used for detecting and comparing not only pair-wise independence but also mutual independence of any multivariate variables.

Furthermore, I will prove that the proposed statistic not only has some good properties as which of Pearson's determinant coefficient, but also its strength is strictly increasing on variable number n, therefore, the new

[a] Corresponding author: lhc@asia.edu.tw

strength statistic is more sensitive and useful than Cramer's V_n.

2 Significance statistics for detecting the independence

2.1. For detecting the pair-wise independence of multivariate variables

For detecting the pair-wise independence we can use χ_P^2 and G^2, below;

Definition 1: Pearson Chi square statistic, χ_P^2, [2] is defined below;

$$\chi_P^2=\sum_{i=1}^{r}\sum_{j=1}^{c}\frac{\left(O_{ij}-E_{ij}\right)^2}{E_{ij}}=N\sum_{i=1}^{r}\sum_{j=1}^{c}p_{ij}\left(\frac{O_{ij}/N-p_{ij}}{p_{ij}}\right)^2\sim\chi^2_{(r-1)(c-1)} \quad (1)$$

Where O_{ij} is the observed value of row i column j, E_{ij} is the expected value of row column j, N is the sample size, p_{ij} is the probability of row i and column j, $i=1,2,\ldots r,\quad j=1,2,\ldots,c$,

Definition 2: Log likelihood ratio Chi square statistic, G^2 [7], is defined below;

$$G^2=-2\sum_{i=1}^{r}\sum_{j=1}^{c}p_{ij}\log\left(\frac{p_{ij}}{p_ip_j}\right)\sim\chi^2_{(r-1)(c-1)} \quad (2)$$

Where p_{ij} is the joint probability of row i column j, p_i & p_j are the marginal probability of row i and column j, respectively.

2.2. For detecting the mutual independence of multivariate variables

χ_P^2 & G^2 can be extended to the generalized case for detecting the mutual independence below,

Definition 3. Generalized χ_P^2 and, G^2 are defined below, respectively,

(i) $$\chi^2_{P(n)}=\sum_{i_1=1}^{r_1}\sum_{i_2=1}^{r_2}\cdots\sum_{i_n=q}^{r_n}\frac{\left(O_{i_1i_2\ldots i_n}-E_{i_1i_2\ldots i_n}\right)^2}{E_{i_1i_2\ldots i_n}}\sim\chi^2_{\prod_{k=1}^{n}(i_k-1)} \quad (3)$$

(ii) $$G_n^2=-2\sum_{i_1=1}^{r_1}\sum_{i_2=1}^{r_2}\cdots\sum_{i_n=q}^{r_n}p_{i_1i_2\ldots i_n}\log\left(\frac{p_{i_1i_2\ldots i_n}}{\prod_{k=1}^{n}p_{i_k}}\right)\sim\chi^2_{\prod_{k=1}^{n}(i_k-1)} \quad (4)$$

3 The Chi square based strength statistics

Definition 4: The Chi square based strength statistics, Cramer's V [1, 6], is defined below;

(i) For pair-wise independent:

$$V=\sqrt{\chi_P^2/N\min\left(r-1,c-1\right)}, \quad (5)$$

(ii) For mutual independent:

$$V_n=\sqrt{\chi^2_{P(n)}/N\min\left(r_1-1,r_2-1,\ldots,r_n-1\right)} \quad (6)$$

where $0\le V,V_n\le 1$

Note that: if $\chi^2_{P(n)}=\chi^2_{P(n+1)}$ and

$$X_{n+1}\ne X_i, i=1,2,\ldots,n, \min\left(r_1-1,r_2-1,\ldots,r_n-1\right)=\min\left(r_1-1,r_2-1,\ldots,r_{n+1}-1\right), \quad (7)$$

then $V_n=V_{n+1}$, in other words, Cramer's V is not strictly increasing on its number of variables.

4. Two mutual information based strength statistics

4.1. Joe's generalized correlation coefficient

Definition 5: Joe's generalized correlation coefficient of $\left(X_1,X_2\right)$ [8] is defined below;

$$GCC_J\left(X_1,X_2\right)=\sqrt{1-\exp\left[-2\times I\left(X_1:X_2\right)\right]} \quad (8)$$

where

$$0\le GCC_{J-2}\left(X_1,X_2\right)\le 1,\ I\left(X_1:X_2\right)=H\left(X_1\right)+H\left(X_2\right)-H\left(X_1,X_2\right) \quad (9)$$

Note that Cover and Thomas [10] pointed out that the value of $I\left(X_1:X_2:X_3\right)$ may be negative, therefore, the generalized cases can be not defined.

4.2. Ding's generalized correlation coefficient

Definition 6: Ding's generalized correlation coefficient of $\left(X_1,X_2,\ldots,X_n\right)$ [9] is defined below;

$$GCC_D\left(X_1,X_2,\ldots,X_n\right)=\frac{\sum_{i=1}^{n}H\left(X_i\right)-H\left(X_1,X_2,\ldots,X_n\right)}{\left[\prod_{i=1}^{n}H\left(X_i\right)\right]^{\frac{1}{n}}} \quad (10)$$

Note that if $n>2, H\left(X_i\right)=H\left(X\right), i=1,2,\ldots,n$, then

$$GCC_D\left(X_1,X_2,\ldots,X_n\right)=n-1>1,$$

therefore, $GCC_D\left(X_1,X_2,\ldots,X_n\right)$ is not well defined, and it still can be not used for detecting the mutual independence of multivariate variables.

5. The G^2 based strength test statistic for testing independence

In this paper, a novel G^2 based strength test statistic is proposed as follows,

Definition 7: The G^2 based strength test statistic; Liu's generalized correlation coefficient of $\left(X_1,X_2,\ldots,X_n\right)$ is defined below;

$$\alpha_L(n) = GCC_L(X_1, X_2, ..., X_n) = \frac{n}{n-1}\left[1 - \frac{H(X_1, X_2, ..., X_n)}{\sum_{i=1}^{n} H(X_i)}\right] \quad (11)$$

Where n is the number of variables, N is the sample size, $H(X_1, X_2, ..., X_n)$ is the joint entropy of random variables $X_1, X_2, ..., X_n$, $H(X_i)$ is the marginal entropy of random variables $X_i, i = 1, 2, ..., n$.

Theorem 1: The Important properties of Liu' $\alpha_L(n)$ coefficient are listed below;

(i)
$$G_n^2 = -2\sum_{i_1=1}^{r_1}\sum_{i_2=1}^{r_2}...\sum_{i_n=q}^{r_n} p_{i_1 i_2 ... i_n} \log\left(\frac{p_{i_1 i_2 ... i_n}}{\prod_{k=1}^{n} p_{i_k}}\right) = 2\left[\sum_{k=1}^{n} H(X_k) - H(X_1, X_2, ..., X_n)\right] \quad (12)$$

(ii)
$$0 \le \alpha_L(n) = \frac{n}{n-1}\left[1 - \frac{H(X_1, X_2, ..., X_n)}{\sum_{i=1}^{n} H(X_i)}\right] \le 1 \quad (13)$$

(a) If $X_1, X_2, ..., X_n$ are mutual independent, then $\alpha_L(n) = 0$.

(b) If they are all equal, then $\alpha_L(n) = 1$.

(iii)
$$H(X_{n+1}) > 0, X_{n+1} \ne X_i, i = 1, 2, ..., n, \alpha_L(n) \in (0,1) \Rightarrow \alpha_L(n) < \alpha_L(n+1), \forall n \in N, n \ge 2 \quad (14)$$

(V)
$$G_n^2 = 2\left(1 - \frac{1}{n}\right)\left[\sum_{k=1}^{n} H(X_k)\right]\alpha_L(n) \sim \chi^2_{\prod_{k=1}^{n}(i_k - 1)} \quad (15)$$

Proof:(i)

$$G_n^2 = -2\sum_{i_1=1}^{r_1}\sum_{i_2=1}^{r_2}...\sum_{i_n=q}^{r_n} p_{i_1 i_2 ... i_n} \log\left(\frac{p_{i_1 i_2 ... i_n}}{\prod_{k=1}^{n} p_{i_k}}\right)$$
$$= -2\left[\sum_{i_1=1}^{r_1}\sum_{i_2=1}^{r_2}...\sum_{i_n=q}^{r_n} p_{i_1 i_2 ... i_n}\left(\log p_{i_1 i_2 ... i_n}\right) - \sum_{i_1=1}^{r_1}\sum_{i_2=1}^{r_2}...\sum_{i_n=q}^{r_n} p_{i_1 i_2 ... i_n}\left(\log \prod_{k=1}^{n} p_{i_k}\right)\right]$$
$$= -2\left[H(X_1, X_2, ..., X_n) - \sum_{k=1}^{n} H(X_k)\right]$$
$$= 2\left[\sum_{k=1}^{n} H(X_k) - H(X_1, X_2, ..., X_n)\right] \quad (16)$$

(ii) From the Entropy theory [10], we know that;

$$0 \le H(X_1, X_2, ..., X_n) \le \sum_{i=1}^{n} H(X_i) \quad (17)$$

Thus
$$0 \le 1 - \frac{H(X_1, X_2, ..., X_n)}{\sum_{i=1}^{n} H(X_i)} \le \frac{n-1}{n} \quad (18)$$

We can obtain

$$0 \le \alpha_L(n) = \frac{n}{n-1}\left[1 - \frac{H(X_1, X_2, ..., X_n)}{\sum_{i=1}^{n} H(X_i)}\right] \le 1 \quad (19)$$

(a) If $X_1, X_2, ..., X_n$ are mutual independent, then $\alpha_L(n) = 0$.

(b) If they are all equal, then $\alpha_L(n) = 1$.

(iii): According to Entropy theory [10], we know that

$$\sum_{I=1}^{n} H(X_i) \ge H(X_i)_{i=1}^{n} \ge H(X_{n+1} | X_i)_{i=1}^{n} \ge 0 \quad (20)$$

$$H(X_{n+1} | X_i)_{i=1}^{n} + H(X_i)_{i=1}^{n} = H(X_i)_{i=1}^{n+1} \quad (21)$$

Since $H(X_{n+1}) > 0$, then

$$H(X_{n+1} | X_i)_{i=1}^{n} > 0, \sum_{i=1}^{n+1} H(X_i) > \sum_{i=1}^{n} H(X_i), \quad (22)$$

We can obtain

$$\frac{\sum_{I=1}^{n} H(X_i)}{\sum_{I=1}^{n+1} H(X_i)} < \frac{\sum_{I=1}^{n} H(X_i)}{\sum_{I=1}^{n} H(X_i) + H(X_{n+1})} \le \frac{\sum_{I=1}^{n} H(X_i)}{\sum_{I=1}^{n} H(X_i) + H(X_{n+1} | X_i)_{i=1}^{n}}$$
$$\le \frac{H(X_i)_{i=1}^{n}}{H(X_i)_{i=1}^{n} + H(X_{n+1} | X_i)_{i=1}^{n}} = \frac{H(X_i)_{i=1}^{n}}{H(X_i)_{i=1}^{n+1}} \quad (23)$$

and

$$\frac{\sum_{I=1}^{n} H(X_i)}{\sum_{I=1}^{n+1} H(X_i)} < \frac{H(X_i)_{i=1}^{n}}{H(X_i)_{i=1}^{n+1}}, \quad \frac{H(X_i)_{i=1}^{n+1}}{\sum_{I=1}^{n+1} H(X_i)} < \frac{H(X_i)_{i=1}^{n}}{\sum_{I=1}^{n} H(X_i)} \quad (24)$$

Thus

$$\left(\frac{n}{n+1}\right)\frac{\sum_{I=1}^{n} H(X_i)}{\sum_{I=1}^{n+1} H(X_i)} < \left(\frac{n}{n+1}\right)\frac{H(X_i)_{i=1}^{n}}{H(X_i)_{i=1}^{n+1}}, \quad (25)$$

$$\frac{(n+1)H(X_i)_{i=1}^{n+1}}{(n+1)\sum_{I=1}^{n+1} H(X_i)} < \frac{nH(X_i)_{i=1}^{n}}{n\sum_{I=1}^{n} H(X_i)} \quad (26)$$

We have

$$\frac{nH(X_i)_{i=1}^{n} - \sum_{I=1}^{n} H(X_i)}{n\sum_{I=1}^{n} H(X_i) - \sum_{I=1}^{n} H(X_i)} > \frac{(n+1)H(X_i)_{i=1}^{n+1} - \sum_{I=1}^{n+1} H(X_i)}{(n+1)\sum_{I=1}^{n+1} H(X_i) - \sum_{I=1}^{n+1} H(X_i)} \ge 0 \quad (27)$$

Then
$$\frac{nH(X_i)_{i=1}^{n} - \sum_{I=1}^{n} H(X_i)}{(n-1)\sum_{I=1}^{n} H(X_i)} > \frac{(n+1)H(X_i)_{i=1}^{n+1} - \sum_{I=1}^{n-1} H(X_i)}{(n)\sum_{I=1}^{n+1} H(X_i)} \quad (28)$$

and

$$1 - \frac{nH(X_i)_{i=1}^{n} - \sum_{I=1}^{n} H(X_i)}{(n-1)\sum_{I=1}^{n} H(X_i)} < 1 - \frac{(n+1)H(X_i)_{i=1}^{n+1} - \sum_{I=1}^{n-1} H(X_i)}{(n)\sum_{I=1}^{n+1} H(X_i)} \quad (29)$$

DOI: 10.1051/matecconf/20167103009

Therefore

$$0<\alpha_L(n)=\frac{n}{n-1}\left[1-\frac{H(X_i)_{i=1}^{n}}{\sum_{i=1}^{n}H(X_i)}\right].$$

$$<\alpha_L(n+1)=\frac{n+1}{n}\left[1-\frac{H(X_i)_{i=1}^{n+1}}{\sum_{i=1}^{n+1}H(X_i)}\right],\forall n\in N,n\geq 2 \quad (30)$$

(v) It is trivial.

Note that the proposed strength statistic is strictly increasing on its number of variables, but Cramer's V is not. Therefore, Liu's generalized correlation coefficient is more sensitive and useful than Cramer's V, and it can be tested by G^2.

6. Conclusion

In this paper, I pointed out that both Joe's and Ding's generalized correlation coefficients can be not used for detecting the mutual independence of three or more multivariate variables, further, I proposed a novel strength statistic based on the significant statistic G^2, called Liu's generalized correlation coefficient, it can be used to detect and compare the strength of not only the pair-wise but also the mutual independence of any multivariate variables, moreover, I proved that only the new strength statistic is a strictly increasing function on its number of variables, it is more sensitive and useful than Cramer's V, moreover, this new strength statistic can also be tested by G^2.

This work is partially supported by the National Science Council grant (NSC 100-2511-S-468-001).

References

1. H. Cramer, *Mathematical Methods of Statistics* (Princeton: Princeton University Press, 1946)
2. K. Pearson, Philos. Mag **50** (1900)
3. W.G. Cochran, Biometrics **10** (1954)
4. T.D. Wickens, *Multiway contingency tables analysis for the social sciences* (Erlbaum, NJ, 1989)
5. M.L. McHugh, Biochem. Medica **23**, 2 (2013)
6. H.-C. Liu, Proceeding of 2015 Global Chinese Conference on Educational Information and Assessment (2015)
7. R.R. Sokal, F.J. Rohlf, *Biometry: The principles and practices of statistics in biological research* (WH Freeman and company, San Francisco, 1994)
8. H. Joe, J. Am. Stat. Assoc **84** (1989)
9. J .Ding, W. S. Wang, Y. L. Zhao, J. Sichuan. Univ: Eng. Sci. Ed **34,** 3 (2002)
10. T. Cover, and J. Thomas, *Elements of Information Theory* (John Wiley and Sons, 1991)

Effects of exercise intervention on the functional fitness of community elderly

Tsui-Er Lee[a]

Asia University, Office of Physical Education, 41354 Taichung City, Taiwan

Abstract. This study aims to discuss the effects of the 12-week exercise intervention on the functional fitness of community elderly. Total 25 community elderly aged above 65 are proceeded functional fitness tests. The data are analyzed with Descriptive Statistics. The results show that the male elderly significantly enhance the upper limb strength (musculus biceps brachii left arm curling, musculus biceps brachii right arm curling), static balance (open-eye single leg standing), and agility and dynamic balance (chair stand detour) and the female elderly make remarkable progress on the upper limb strength (musculus biceps brachii left arm curling, musculus biceps brachii right arm curling), cardiorespiratory endurance (high knees in place), static balance (open-eye single leg standing), and agility and dynamic balance (chair stand detour) after the 12-week regular exercise intervention. With functional fitness tests, the research findings reveal that regular exercise presents the effectiveness on distinct physical fitness of community elderly, while the aerobic endurance and upper/lower body flexibility still require improvement. Apparently, it is suggested to practice courses with stages and match fitness test results to adjust types of exercise and exercise intensity in order to achieve the effect. Keywords: regular exercise, the elderly, type of exercise.

1 Introduction

Population is a primary element of a nation, and the number of population and the change of age structure are the key factors in the national development. Population of young adults aged 15-64 (also called working-age population) achieved the peak in 2015 and started to decrease in 2016. Furthermore, aged population would exceed young population in 2016 because of decreasing young population and increasing aged population. In comparison with 2014, the Population Projection result revealed that young population and population of young adults would reduce about a half by 2061 (52.2% and 47.9%, respectively), while aged population would increase 1.6 times [1]. It proves that the constant increase of aged population structure has aging become an inevitable trend in Taiwan.

The issues related to the elderly and health conditions therefore should be emphasized. The physical fitness and physiological function, including nerves, muscles, skeleton, cardiopulmonary function, and hormone regulation function, of humans after 25 years old would degenerate with increasing ages [2]. Such a process would continue to the death and is irreversible. Degeneration might result in decreasing activity of the elderly to affect the daily activity and cause inconvenient life. A lot of research indicated that maintaining the habit of moderate, non-exciting, and low-intensity regular exercise could enhance strengths and promote physical function, like cardiopulmonary function [3]. Aiming at the elderly aged 55-75, Pei-chen Chung & Fen-yu Tseng preceded the 8-week exercise intervention and found out the obvious improvement of lower body strength, upper body strength, cardiorespiratory endurance index, lower body flexibility, and agility and action balance as well as obvious decrease of cholesterol and low-density lipoprotein [4]. Regular exercise not only could enhance the physical fitness of the elderly to have better quality of life, but could also prevent incidence from occurrence.

It is more important for the elderly being able to independently live and safely complete daily activity [5]. For this reason, the elderly exercise prescription design should cover strengths, aerobic endurance, flexibility, agility/dynamic balance, and body mass index as the total performance of human physical and mental characteristic function. Regular physical activity could also strengthen physical fitness. After the 12-week elderly walk training, the walk speed was obviously enhanced and the problem of move with difficulty caused by physical degeneration was also released [6]. Having 50 senior university members be the participants in the 14-week integrated exercise training, including group exercise training, home-based exercise training and health promotion, and exercise management courses, Wei-yu Chang also pointed out the positive improvement of BMI, waistline, heart rate of rest, and functional fitness of the elderly after the integrated exercise training, which could benefit the health promotion of the elderly [7]. As a result, cultivating regular exercise habits, promoting fitness

[a] Corresponding author: vivian@asia.edu.tw

DOI: 10.1051/matecconf/20167103010

standards, and establishing healthy lifestyles are the primary issues currently.

In addition to regular exercise, it is also important to promote health related physical fitness education and knowledge and instill in favorable healthy lifestyles, expecting to overall promote the health of a community by the active participation of the citizens and the intervention of professional knowledge. Community citizens could also enhance the friendship by the weekly group training and continue the exercise motivation to achieve the national health index. This study aims to discuss the effect of the 12-week regular exercise intervention on the functional fitness of community elderly.

2 Methodology

2.1 Research subject

Total 25 healthy community elderly (15 females and 10 males) aged above 65 are selected as the research population. In regard to time and manpower limitation, purposive sampling is applied to this study.

2.2 Experiment procedure

The participants are proceeded the 24-week fitness exercise training and cognition courses, containing weekly group course and twice-a-week home-based exercise. Functional fitness tests are preceded before and after the intervention. The participants follow the principles of maintaining normal lifestyles, except for the participation in the courses and tests in this study, and do not engage in other exercise. The weekly group course should be fully participated and exercise records need to be filled in to record the physical activity. All examiners in this study have the national fitness examiner certificate approved by Sports Administration, Ministry of Education, Taiwan, in order to ensure the test accuracy.

2.3 Experiment design

This study focuses on the integrated exercise training courses, including warm-up, main exercise cool-down, and stretching, aiming to promote distinct fitness, such as stretching exercise, aerobics, strength exercise, and walking, for the intervention training. The 60min exercise and 30min sports and health cognition courses tend to establish correct sport concept and exercise methods in order to achieve daily work-out habits. ACSM suggested that the elderly should accumulate at lease more than 150min moderate physical activity every week [8]. The participants in this study maintain daily lifestyles and diet habits during the 12-week exercise intervention, which is practiced three days a week on every other day. The 90min exercise training is arranged by the professional exercise instructors on one day, containing 60min gradual and overload strength training 50% 1RM (10~15 times) and 30min low-impact aerobics led by coaches to train physical big muscles. The exercise intensity is up to the rating of perceived exertion (RPE) level 12-15, i.e. at least reaching the moderate exercise intensity of "a bit difficult" [9]. The other two days focus on the independent training of participants, mainly walking, to achieve the moderate exercise intensity of "a bit difficult" requested by the perceived exertion.

2.4 Test item

The participants in this study receive functional fitness tests before and after the training. The test items are based on the fitness test proposed by Sports Administration, Ministry of Education, in 2014 to evaluate the daily physical activity ability of the elderly [5]. The evaluation refers to the following 8 test items of body composition, 30sec chair stand (lower limb strength), musculus biceps brachii arm curling (upper limb strength), open-eye single leg standing (static balance), back scratch (upper limb flexibility), chair sit-and-reach (lower limb flexibility), chair stand detour (agility/dynamic balance), and 2min high knees in place (cardiorespiratory endurance).

2.5 Data processing

SPSS20.0 is used for the data analysis, and the basic data and functional fitness of the community elderly are established with Descriptive Statistics. Two-sample t test is applied to analyze the difference before and after the intervention measure, and the research significance is set $\alpha = .05$.

3 Result

3.1 Basic data of participants

Total 25 healthy community elderly aged above 65 (15 females and 10 males) are selected as the research subjects. The mean ages of female and elderly are 76±5.33 and 77±9.92, respectively. After the 12-week exercise intervention, the full participation rate reaches above 80%. Table 1 and Table 2 show the pretest/posttest basic data of the male and female elderly, including blood pressure (diastolic pressure, systolic pressure), height (cm), weight (kg), body mass index (BMI), waistline, hipline, and waist-hip ratio. The pretest/posttest analyses show the mean systolic pressure of the male elderly 127.20mmHg but 126.10mmHg after the intervention and the mean diastolic pressure 69.10mmHg but 72.70mmHg after the intervention, revealing the blood pressure in the normal range. The weight appears 61.49kg but 60.90kg after the intervention, decreasing 0.59kg, without significant differences. The body mass index (BMI), waistline, hipline, and waist-hip ratio do not present remarkable changes. The mean systolic pressure of the female elderly shows 122.40mmHg but 123.00mmHg after the intervention and the mean diastolic pressure 69.33mmHg but 70.87mmHg after the intervention, revealing the blood pressure in the normal range. The weight appears 58.17kg but 58.29kg after the intervention, increasing 0.12kg, without notable differences. The body mass index (BMI), waistline, and waist-hip ratio do not show remarkable changes, while the hipline achieves significant differences.

Table 1. Pretest/posttest basic data of the male elderly (n=10).

Item	Pretest	Posttest	*t*	*p*
Systolic pressure	127.20 ±16.06	126.10 ±13.92	0.58	0.58
Diastolic pressure	69.10 ± 7.28	72.70 ±10.96	-1.46	0.18
Height (cm)	164.80 ± 7.22	164.50 ± 7.23	1.00	0.34
Weight (kg)	61.49 ±11.38	60.90 ±10.82	1.15	0.28
BMI	22.73 ± 4.42	22.59 ± 4.26	0.86	0.41
Waistline	89.20 ±10.72	86.50 ±11.18	1.65	0.13
Hipline	96.55 ± 7.17	96.50 ± 6.70	0.05	0.97
Waist-hip ratio	0.92 ± 0.06	0.89 ± 0.08	1.34	0.21

$^*p < .05$

Table 2. Pretest/posttest basic data of the female elderly (n=15).

Item	Pretest	Posttest	t	p
Systolic pressure	122.40 ±22.14	123.00 ±10.33	-0.133	0.896
Diastolic pressure	69.33 ± 9.96	70.87 ± 8.09	-0.700	0.500
Height (cm)	153.20 ± 4.72	152.53 ± 4.03	1.380	0.190
Weight(kg)	58.17 ± 8.69	58.29 ± 7.53	-0.120	0.910
BMI	24.80 ± 3.45	28.39 ± 8.50	-1.650	0.120
Waistline	89.50 ± 9.27	87.93 ± 9.30	0.990	0.340
Hipline	103.53 ± 9.38	99.70 ± 8.50	2.220	0.04*
Waist-hip ratio	0.87 ±0.052	0.88 ± 0.04	-1.060	0.310

$^*p < .05$

3.2 Current functional fitness of the elderly

To understand the current situation based on the fitness test before the intervention, the functional fitness 5-scale norm of seniors in Chiayi City, Taiwan, is referred for the analysis.

(1) Chair stand (evaluation of lower body strength): Females present the mean 16.53 times, with ordinary-favorable level, and males show the mean 15.2 times, with ordinary level.
(2) Musculus biceps brachii left arm curling (evaluation of upper body strength): Females reveal the mean 20.93 times, with good level, and males appear the mean 18.8 times, with favorable level.
(3) Musculus biceps brachii right arm curling (evaluation of upper body strength): Females show the mean 21.4 times, with good level, and males reveal the mean 19.3 times, with favorable-good level.
(4) High knees in place (evaluation of aerobic endurance): Females appear the mean 86.93 times, with ordinary-slightly bad level, and males present the mean 91.9 times, with ordinary-slightly bad level.
(5) Chair sit-and-reach (evaluation of lower body flexibility): Females reveal the mean 0.27, with slightly bad-bad level, and males appear the mean -3.6, with slightly bad level.
(6) Open-eye single leg standing (evaluation of balance): Female present the mean 12.78sec, with ordinary-favorable level, and males show the mean 21.09sec, with favorable-good level.
(7) Left back scratch (evaluation of upper body flexibility): Females appear the mean -13.93, with slightly bad-bad level, and males present the mean -16.6, with slightly bad-bad level.
(8) Right back scratch (evaluation of upper body flexibility): Females show the mean -4.93, with ordinary-slightly bad level, and males reveal the mean -13 with ordinary-slightly bad level.
(9) Chair stand detour (evaluation agility and dynamic balance): Females appear the mean 6.88 sec, with ordinary level, and males present the mean 6.26sec, with ordinary-favorable level.

3.3 Effects of intervention on functional fitness

Table 3 shows that the male elderly reveal notable differences on the upper limb strength (musculus biceps brachii left arm curling, musculus biceps brachii right arm curling), the lower limb strength (chair stand) does not reach the significance, and the static balance (open-eye single leg standing) and agility and dynamic balance (chair stand detour) achieve remarkable differences ($p<$ 0.5). In regard to flexibility, the upper limb flexibility (back scratch), lower limb flexibility (chair sit-and-reach), and cardiorespiratory endurance (high knees in place) do not reach significant differences.

Table 3. Male functional fitness pretest/posttest difference (n=10)

Item	Pretest	Posttest	*t*	*p*
Chair stand	15.20 ± 3.74	17.42 ± 4.91	-1.46	0.1790
Musculus biceps brachii arm curling (left)	18.80 ± 2.78	21.20 ± 3.19	4.27	0.002*
Musculus biceps brachii arm curling (right)	19.30 ± 3.65	22.10 ± 3.07	-5.25	0.001*
High knees in place	91.90 ±12.82	94.90 ±10.86	-2.13	0.0620
Chair sit-and-reach	15.20 ± 3.74	17.42 ± 4.91	-0.72	0.4870
Open-eye single leg standing	21.09 ±20.50	21.76 ±20.20	-2.97	0.016*
Back scratch (left)	-16.60 ±14.42	-17.90 ± 8.19	0.33	0.7500
Back scratch (right)	-13.00 ±10.32	-11.60 ±10.76	-2.26	0.0500
Chair stand detour	15.20 ± 3.74	17.42 ± 4.91	2.99	0.015*

$^*p < .05$

Table 4 reveals that the female elderly present remarkable differences on the upper limb strength (musculus biceps brachii left arm curling, musculus biceps brachii right arm curling), the lower limb strength (chair stand) does not achieve the significance, and the cardiorespiratory endurance (high knees in place), static balance (open-eye single leg standing), and agility and dynamic balance(chair stand detour) reach notable differences ($p < 0.5$). In terms of flexibility, both upper limb flexibility (back scratch) and lower limb flexibility (chair sit-and-reach) do not achieve significant differences.

Table 4. Female functional fitness pretest/posttest difference (n=15)

Item	Pretest	Posttest	*t*	*p*
Chair stand	16.53 ± 1.36	16.87 ± 4.58	-0.310	0.7600
Musculus biceps brachii arm curling (left)	20.93 ± 3.06	22.39 ± 3.97	-2.940	0.011*
Musculus biceps brachii arm curling (right)	21.40 ± 4.47	23.93 ± 4.59	-4.670	.000*
High knees in place	86.93 ±12.12	93.20 ±15.92	-3.010	0.009*
Chair sit-and-reach	0.27 ± 5.82	1.53 ± 6.84	-0.670	0.5170
Open-eye single leg standing	12.78 ±16.75	14.27 ±16.60	-2.280	0.039*
Back scratch (left)	-13.93 ±10.07	-11.87 ± 8.19	-1.050	0.3110
Back scratch (right)	-4.93 ± 7.47	-3.80 ± 7.18	-1.800	0.0940
Chair stand detour	6.88 ± 0.90	6.34 ± 0.92	4.232	0.001*

*$p < .05$

4 Conclusion

Senior Fitness Test (SFT) is utilized for evaluating the functional fitness of the elderly in this study. The research findings show remarkable improvement of the functional fitness of the female elderly, including upper limb strength, cardiorespiratory endurance, and balance, after the intervention of fitness exercise training and cognition courses. Although flexibility does not reach the significance, it reveals progress, comparing to the performance before exercise intervention. The result conforms to the research of Tai-yu Huang, Che-jung Chang, Mang-hui Lin & Chin-lung Fang [10]. In their research, 35 female elderly aged above 60 were proceeded 10-week water aerobic training, in which flexibility was improved but did not achieve the significance, while the rest functional fitness of upper/lower limb strength, aerobic endurance, and static and dynamic balance appeared notable improvement. Besides, the daily physical activity ability was promoted and the degeneration of physiological function resulted from increasing ages slowed down. Hallage et al. pointed out the remarkable effect of the 12-week aerobics intervention on the body mass index (BMI), waistline, chair stand, musculus biceps brachii arm curling, high knees in place, chair sit-and-reach, open-eye single leg standing, back scratch, and chair stand detour of female elderly [11]. Accordingly, aerobics could obviously improve the functional fitness of female elderly and was regarded as an effective training to prevent fitness from loss.

Regarding the male elderly, the results reveal the remarkable improvement of upper limb strength and balance. A one-year health promotion program for 148 elderly showed that aerobic and strength training and health education program and control could improve aerobic endurance and dynamic balance and promote fitness and health conditions [12]. Although lower limb strength, flexibility, and cardiorespiratory endurance did not achieve the significance, they presented progress, comparing to those before the intervention. Jhen-Yi Chen, Yen-Ju Lin, Yu-Ching Weng & Su-Chuan Yuan discussed the effect of the 12-week regular walking intervention on the fitness of community elderly with chronic diseases [13]. The results showed that, in addition to the promotion of cardiopulmonary function, the fitness of the elderly with chronic diseases was actually improved after exercise intervention. They suggested adding walking in the exercise intervention, as it was life -oriented exercise with better motivation to continue, in order to present more notable benefits on cardiopulmonary function.

From above research results, exercise intervention not only promotes the functional fitness of the elderly but also improves the health conditions. By comparing the physical activity of the elderly [14], Kim, Dogra, Al-Sahab & Tamim preceded the 16-week Taijiquan intervention in the elderly with and without fitness experiences. The results revealed that the elderly without experiences appeared remarkable differences on functional fitness after the intervention measure. The above statement is therefore proven. Health promotion could have the elderly present better confidence on exercise or daily activity. Aiming at intervening the daily life of weak elderly with exercise training program, Chou, Hwang & Wu [15] and Shimada, Uchiyama & Kakurai [16] revealed no significant effect of exercise intervention on the quality of life in groups, but increasing walk speed and improved balance to promote the daily activity performance. Consequently, the practice of exercise intervention programs in the elderly presents the benefits to enhance physical fitness, strengthen body, and promote independent living abilities for more active and safe aged life.

Acknowledgement

We are grateful for the project funding support of Ministry of Science and Technology (MOST 104-2629-E-468-002) and would like to show our appreciation.

References

1. National Development Council of the Executive Yuan of Taiwan, *The report of Republic of China from 2012 to 2060 census* (National Development Council of the Executive Yuan, Taiwan, 2014)
2. S. Roberts, R.A. Robergs, P.G. Hanson, *Clinical exercise testing and prescription: Theory and*

DOI: 10.1051/matecconf/20167103010

application (CRC Press, Boston, Massachusetts, 1997)
3. C.L. Liu, J.Y. Hou, R.H. Tu, J. of National Pingtung University Physical Education Research **17**, 139-144 (2014)
4. P.C. Chung, F.Y. Tseng, Formosan Journal of Medicine **13(2)**, 133-140 (2009)
5. C.J. Jones, R.E. Rikli, Journal of Aging and Physical Activity **7**, 129-161 (1999)
6. J.M. VanSwearingen, S. Perera, J.S. Brach, D. Wert, S.A. Studenski, Physical Therapy **91(12)**, 1740-1751 (2011)
7. W.Y. Chang, J. of National Taichung University of Science and Technology Physical Education Research **10**, 45-54 (2014)
8. American College of Sports Medicine, *ACSM's guidelines for exercise testing and prescription* (8th ed.), (American College of Sports Medicine, Philadelphia, PA, 2010)
9. G. Borg, Scandinavian Journal of Work, Environment & Health **16**, 55-58 (1990)
10. T.Y. Huang, C.J. Chang, M.H. Nin, C.L. Fang, Journal of Exercise Physiology and Fitness **3**, 121-129 (2005)
11. T. Hallage, M.P. Krause, L. Haile, C.P. Miculis, E.F. Nagle, R.S. Reis, S.G. Da Silva, Journal of Strength and Conditioning Research **24(8)**, 2261-2266 (2010)
12. Obesity, fitness & wellness week, *Studies from University of Porto Add New Findings in the Area of Physical Fitness* (NewsRx.com, Atlanta, GA, July 30 2011)
13. J.Y. Chen, Y.J. Lin, Y.C. Weng, S.C. Yuan, Cheng Ching Medical Journal **11(1)**, 39-48 (2015)
14. T.H. Kim, S. Eke Dogra, B. Al-Sahab, H. Tamim, Alternative Therapies in Health and Medicine **20(3)**, 20-25 (2014)
15. C.H. Chou, C.L. Hwang, Y.T. Wu, Arch. Phys. Med. Rehabil. **93(2)**, 237-244 (2012)
16. H. Shimada, Y. Uchiyama, S. Kakurai, Clinical Rehabilitation **17(5)**, 472-479 (2003)

Thin film Encapsulations of Flexible Organic Light Emitting Diodes

Fa-Ta Tsai [1], Ching-Kong Chao [1], Pei-Sin Jhu [2] and Rwei-Ching Chang [2,a]

[1] *Department of Mechanical Engineering, National Taiwan University of Science and Technology, Taiwan*
[2] *Department of Mechanical and Computer-Aided Engineering, St. John's University, Taiwan*

Abstract. Various encapsulated films for flexible organic light emitting diodes (OLEDs) were studied in this work, where gas barrier layers including inorganic Al2O3 thin films prepared by atomic layer deposition, organic Parylene C thin films prepared by chemical vapor deposition, and their combination were considered. The transmittance and water vapor transmission rate of the various organic and inorgabic encapsulated films were tested. The effects of the encapsulated films on the luminance and current density of the OLEDs were discussed, and the life time experiments of the OLEDs with these encapsulated films were also conducted. The results showed that the transmittance are acceptable even the PET substrate were coated two Al2O3 and Parylene C layers. The results also indicated the WVTR of the PET substrate improved by coating the barrier layers. In the encapsulation performance, it indicates the OLED with Al2O3 /PET, 1 pair/PET, and 2 pairs/PET presents similarly higher luminance than the other two cases. Although the 1 pair/PET encapsulation behaves a litter better luminance than the 2 pairs/PET encapsulation, the 2 pairs/PET encapsulation has much better life time. The OLED with 2 pairs/PET encapsulation behaves near double life time to the 1 pair encapsulation, and four times to none encapsulation.

1 Introduction

Due to the simplicity, organic light emitting diodes (OLEDs) can be easily built on not only glass substrates but also flexible substrates. However, the short lifetime and low efficiency of the flexible OLEDs are still the major problem to be solved. It is distributed to the penetration of oxygen and water vapor to the organic layers through the substrate and the encapsulated film [1-3].

The basic encapsulation method is used glass and epoxy adhesive in a N_2-filled glove box. However, the glass is not suitable for flexible encapsulation because of its rigid nature. The polymer substrates are the most promising candidate for the flexible encapsulation. The water vapor transmission rate (WVTR) of the polymer substrate is substantially below the requirement of OLED encapsulation. Therefore, polymer substrates coated barrier thin films to improve the WVTR are essential for flexible OLED encapsulation [4, 5].

Inorganic thin films have shown the potential to be the barrier layer for flexible OLED encapsulation [6]. Although these inorganic thin films have low WVTR, the defects of pinholes cannot be avoided because of the limitation of the vacuum deposition process. An organic-inorganic multi-barrier become an alternative to prevent the defects. [7-11]

In this work, to prevent the penetration of water vapor into the OLEDs, various encapsulated films for flexible OLEDs were studied, where gas barrier layers including inorganic Al_2O_3 thin films prepared by atomic layer deposition, organic Parylene C thin films prepared by chemical vapor deposition, and their combination were considered. The transmittance and water vapor transmission rate of the encapsulated films were tested. The effects of various encapsulated films on the luminance and current density of the OLEDs were compared. The life time experiments of the OLEDs with these encapsulated films were also conducted in the final.

2 Specimen Preparation

The structure of the OLED is shown in Fig. 1, where a flexible PET substrate, an ITO anode, a PEDOT:PSS (Clevios AI4083) hole injecting layer (HIL), a NPB hole transporting layer (HTL), an Alq_3 emitting layer (EL), an aluminum cathode, and a PET cover were stacked in sequence. The ITO-coated PET substrate with size 300×200×0.175 mm was treated by standard cleaning procedure including ultrasonic bath and oxygen plasma before deposition. To fabricate the OLED, a PEDOT:PSS film was deposited on the ITO/PET substrate by spin coating at the first. The specimen was followed by drying process at 70 °C for 10 min. And then, the NPB and Alq_3 were thermally evaporated with 1 Å/s deposition rate. And, an aluminum cathode was successively evaporated. Finally, the device was covered with the encapsulated film and packaged by UV glue, and cured by UV light for 1 hour.

To prepare the encapsulated films, Al_2O_3 thin films were fabricated on the PET substrates by ALD. The

[a] Corresponding author: author@e-mail.org

DOI: 10.1051/matecconf/20167104001

precursor and purging gas of the ALD reactor are controlled by switching valves and flow past the substrate surface and into a mechanical pump. The substrate temperature at 60 °C, and the precursor pulse time with 10 msec were considered, where the chamber pressure is controlled at 9×10^{-3} Torr. The precursor H_2O, purging N_2 gas, precursor TMA, and purging N_2 gas are controlled to flow into the chamber in sequence. The Al_2O_3 film sequentially grows with 10 msec precursor of H_2O, 6000 msec of N_2 purging, 10 msec precursor of TMA, and 6000 msec of N_2 purging. The film thickness is controlled as 50 nm, consequently near 500 cycles were needed.

Parylene C thin films with 100 nm thickness were deposited by a chemical vapor deposition system. Parylene C powders were evaporated at 150 °C and pyrolyzed at 650 °C. The deposition temperature was room temperature and the deposition time was 1 hour.

In this work, five encapsulated types were considered, including none, PET, Al_2O_3 /PET, Parylene C /Al_2O_3 /PET, and Parylene C /Al_2O_3 /Parylene C /Al_2O_3 /PET, respectively. In the following discussion, the fourth and fifth types are simply denoted as 1 pair/PET and 2 pairs/PET.

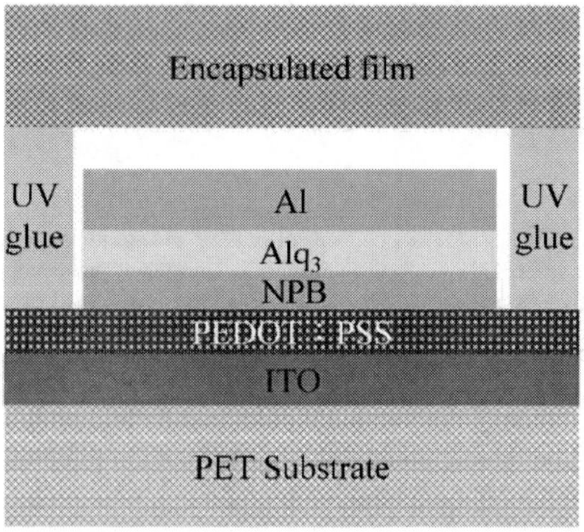

Figure 1. Sketch of the flexible OLED.

3 Result and Discussion

Optical transmittance of the encapsulated films were measured by using an UV/VIS/NIR spectrophotometer (BWTEK BTC112) with normal incidence in the visible wavelength ranged from 400 to 800 nm, taking the air as reference. The results are shown in Fig. 2, where the average transmittance of the PET substrate is 88.162%, PET coated Al_2O_3 is 87.05%, 1 pair/PET is 86.56%, and 2 pairs/PET is 84.65%. Obviously, the transmittance decreases as the barrier layer increases. However, there is only 1~5% transmittance loss by adding the barrier layers.

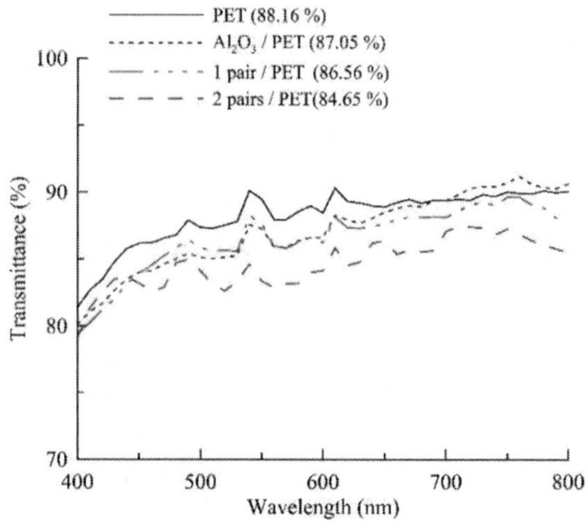

Figure 2. Transmittance of the encapsulated films.

The water vapor transmission rate (WVTR) of the encapsulated films is shown in Fig. 3. It shows the average WVTR of PET is 11.08 g/m^2day, where Al_2O_3 /PET is 3.44 g/m^2day, 1 pair/PET is 3.36 g/m^2day, and 2 pairs/PET is 2.77 g/m^2day. The results indicate 2 pairs/PET has the best water and vapor barrier performance, which is 4 times better than the bare PET. It also reveals that the PET substrate coated with a barrier layer will efficiently improve the water-vapor resistance.

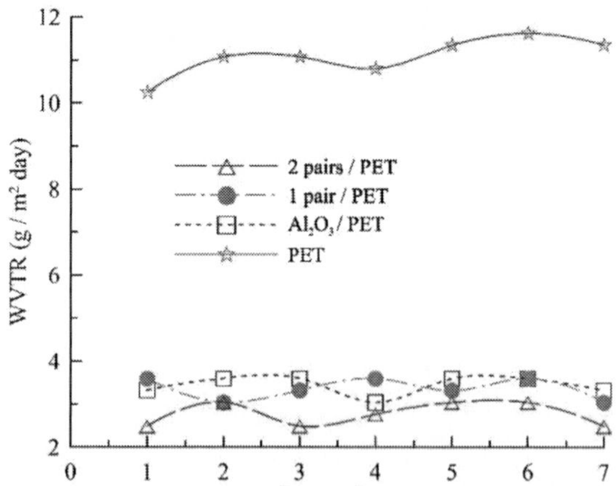

Figure 3. Water vapor transmission rate (WVTR) of the encapsulated films.

Luminance of the OLEDs with various encapsulated films is shown in Fig. 4. The results show that the OLEDs work as the applied voltage higher than 3V, and the luminance increases quickly as the voltage increases. It shows the luminance of the OLED without the encapsulated film is 111.82 cd/m^2 at 8V, and the luminance of OLED with PET, Al_2O_3 /PET, 1 pair/PET, and 2 pairs/PET are 162.18, 540.58, 557.23, and 544.93 cd/m^2 at 8V, respectively. It indicates the OLED with Al_2O_3 /PET, 1 pair/PET, and 2 pairs/PET presents similarly higher luminance than the other two cases. This phenomenon is consistent with the result of WVTR

DOI: 10.1051/matecconf/20167104001

shown in Fig. 3, revealing the barrier performance of the bare PET is as poor as no encapsulated film.

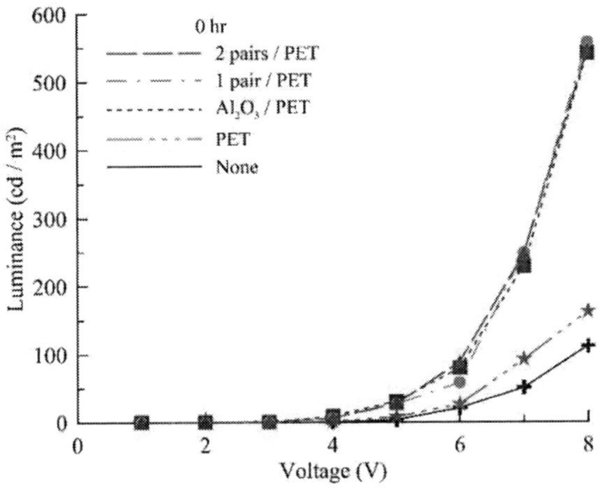

Figure 4. Luminance of the OLEDs with various encapsulated films.

The Current density vs. voltage diagram of the OLEDs with various encapsulated films is shown in Fig. 5, which is similar to the luminance diagram shown in Fig. 4. The Luminance and corresponding images of OLEDs with various encapsulated types at 8V are list in Table 1. Each luminant area of the OLED is 2×2 mm. The images indicate the OLED without encapsulation or with bare PET encapsulation presents lots of black points, and the case of 2 pairs/PET encapsulation has the best emitting patterns.

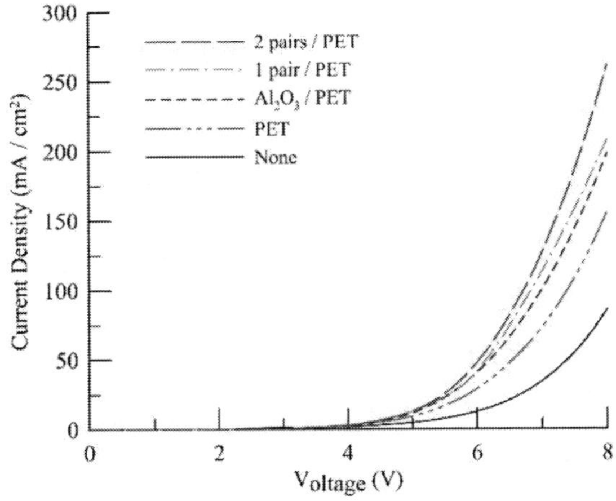

Figure 5. Current density vs. voltage diagram of the OLEDs with various encapsulated films.

Table 1. Luminance and corresponding images of OLEDs with various encapsulated films at 8 V.

Encapsulated type	None	PET	Al_2O_3 / PET	Parylene C / Al_2O_3 / PET	Parylene C / Al_2O_3 / Parylene C / Al_2O_3 / PET
Luminance (cd/m^2)	111.82±9.5	162.18±3.2	540.58±33.8	557.23±49.2	544.93±36.0
Image	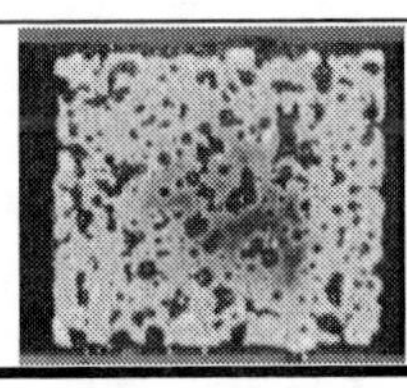	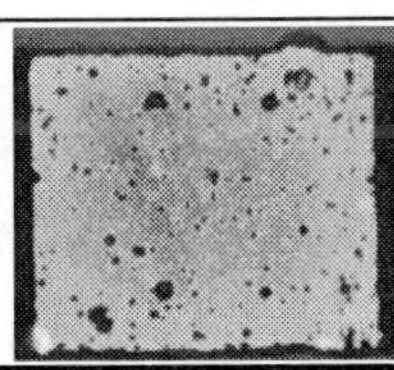	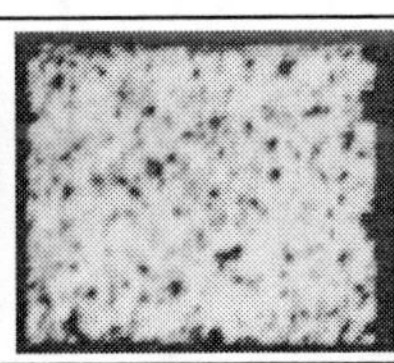	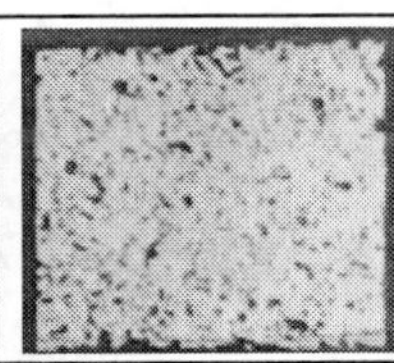	

The life time of the OLEDs with various encapsulated films is shown in Fig. 6. The life time of the OLED is indicated by the normalized luminance (L/L_0) that presents by the ratio of the luminance after usage hours (L) to the luminance at the finished moment (L_0). For the OLED without encapsulation, the normalized luminance is 0.26 for 12 hrs, 0.10 for 24 hrs, and 0.03 for 36 hrs. For the OLED with PET encapsulation, the normalized luminance are 0.37, 0.16, 0.08, and 0.02 for 12, 24, 36, and 48 hrs, respectively. The OLEDs with Al_2O_3 /PET and 1 pair/PET encapsulation behave the similar life time. The OLED with 2 pairs/PET presents the best life time, where the normalized luminance are 0.80, 0.66, 0.53, 0.43, 0.33, and 0.03 for 12, 24, 36, 48, 60, and 120 hrs, respectively. Note that, although the 1 pair/PET encapsulation behaves a litter better luminance than the 2 pairs/PET encapsulation, the 2 pairs/PET encapsulation has much better life time. The OLED with 2 pairs/PET encapsulation behave near double life time to the 1 pair encapsulation.

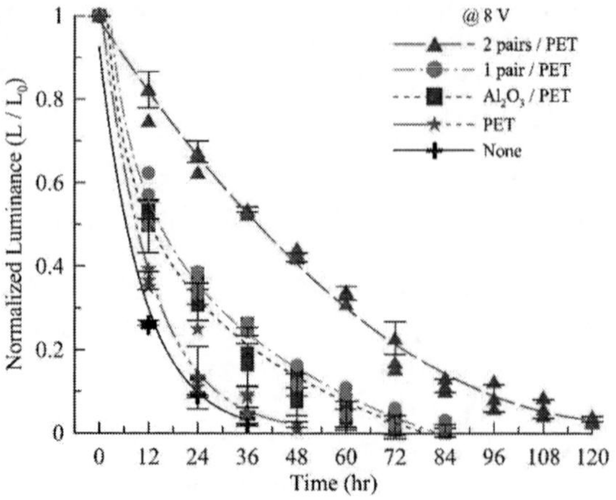

Figure 6. Life time of the OLEDs with various encapsulated films

4 Conclusions

Four types of encapsulated films to prevent the penetration of water vapor into the OLEDs were studied, including inorganic Al_2O_3 thin films prepared by atomic layer deposition, organic Parylene C thin films prepared by chemical vapor deposition, and their combinations. The results showed that the transmittance are acceptable even the PET substrate were coated two Al_2O_3 and Parylene C layers. The results also indicated the WVTR of the PET substrate improved by coating the barrier layers. In the encapsulation performance, it indicates the OLED with Al_2O_3 /PET, 1 pair/PET, and 2 pairs/PET presents similarly higher luminance than the other two cases. Although the 1 pair/PET encapsulation behaves a litter better luminance than the 2 pairs/PET encapsulation, the 2 pairs/PET encapsulation has much better life time. The OLED with 2 pairs/PET encapsulation behave near double life time to the 1 pair encapsulation, and four times to none encapsulation.

Acknowledgement

The support from the National Science Council through Grant NSC 102-2221-E-129-002-MY3 is gratefully acknowledged.

References

1. A. R. Cho, E. H. Kim, S. Y. Park and L. S. Park: Synth. Met. **193,** 77 (2014).
2. L. Ke, H. Liu, M. Yang, Z. Jiao and X. Sun: Chem. Phys. Lett. **623,** 68 (2015).
3. X. Wu, J. Liu and G. He: Org. Electron. **22,** 160 (2015).
4. P. F. Carcia, R. S. McLean, M. H. Reilly, M. D. Groner and S. M. George: Appl. Phys. Lett. **89,** 319151 (2006).
5. P. Mandlik, J. Gartside, L. Han, I. C. Cheng, S. Wagner, J. A. Silvernail, R. Q. Ma, M. Hack and J. J. Brown: Appl. Phys. Lett. **92,** 1033091 (2008).
6. J. H. Choi, Y. M. Kim, Y. W. Park, T. H. Park, J. W. Jeong, H. J. Choi, E. H. Song, J. W. Lee, C. H. Kim and B. K. Ju: Nanotechnol. **21,** 475203 (2010).
7. Y. C. Han, E. Kim, W. Kim, Y. G. Im, B. S. Bae and K. C. Choi: Org. Electron. **14,** 1435 (2013).
8. E. Kim, Y. C. Han, W. Kim, K. C. Choi, Y. G. Im and B. S. Bae: Org. Electron. **14,** 1737 (2013).
9. A. Singh, F. Nehm, L. Muller-Meskamp, C. Hosbach, M. Albert, U. Schroeder, K. Leo and T. Mikolajick: Org. Electron. **15,** 2587 (2014).
10. Y. Q. Yang, Y. Duan, Y. H. Duan, X. Wang, P. Chen, D. Yang, F. B. Sun and K. W. Xue: Org. Electron. **15,** 1120 (2014).
11. H. Klumbies, P. Schmidt, M. Hahnel, A. Singh, U. Schroeder, C. Richter, T. Mikolajick, C. Hosbach, M. Albert, J. W. Bartha, L. Leo and L. Muller-Meskamp: Org. Electron. **17,** 138 (2015).

MATEC Web of Conferences 71, 04002 (2016) DOI: 10.1051/matecconf/20167104002
ICCPE 2015

Development of an intelligent adapter for metered dose inhalers

Mingrong Zhang [1,a], Songhao Wang [1] and Yu-Ching Yang [1]

[1] *Mechanical Engineering Department, Kun Shan University, 71003 Tainan City, Taiwan*

Abstract: To better coordinate the interaction of inhalation and aerosol release, an intelligent adapter (IA) was developed for metered dose inhalers (MDIs). The adapter included three main units: a signal acquisition device, a micro-control-unit (MCU), and an actuation mechanism. To fully study the effectiveness of the intelligent adapter, an inhalation simulation experiment was done, and two bands of MDI were used for the experiment. The results indicated that, when inhalation, the intelligent adapter can press down the MDI automatically; moreover, this intelligent adapter could achieve an aerosol-release time Tr of 0.4 s for MDI A and 0.60 s for MDI B, which compares very well with the existing pure mechanical systems at 0.8 s and 1.0 s.

1 Introduction

The advantages of inhalation therapy includes direct impact on the respiratory airway, rapid onset of action, reduced dose and reduced adverse effects compared with injection and enteral therapies[1]. Furthermore, there are increasing potential applications for other diseases rather than only respiratory disease. For example, on June 29, 2014, the USA Food and Drug Administration (FDA) approved inhaled insulin for diabetes mellitus [2].

However, the inhalation technique is one of the most important factors that influences deposition of the inhaled drugs [3-5]. In the research of Piyush et al. [6], 300 bronchial asthma or COPD patients using different types of inhalation devices participated in an observational study. The maximum operation errors were observed in subjects using MDI (94.3%) followed by DPI (82.3%). For dry powder inhaler (DPI), a higher peak inspiratory flow rate is required, which prevents the application in children, the elderly and seriously ill patients [1,7]. Metered dose inhaler (MDI) is inspiratory flow independent, and coordination between inhalation and activation MDI is required [8]. However, MDI delivers larger dose (approximately 50% of the released drug) to the alveolar airways than DPI (approximately 30%) does [9], and the coordination could be improved by using new technology.

This paper demonstrates an intelligent adapter, capable of auto-spray. Moreover, design considerations for coordination of the inhalation and the aerosol-release time is also discussed.

2 Design and construction of an intelligent adapter

The intelligent adapter consisted of three main units: a signal acquisition device, a micro-control-unit (MCU), and an actuation mechanism.

2. 1 Signal acquisition device

The signal acquisition device (signal switch) was used to detect the action of inhalation. For the intelligent adapter, the signal switch included an impeller and a hall sensor as shown in figure 1. The hall sensor was installed at the top of the device, and the magnet was installed at one of the leaves on the impeller. Because of gravity, the magnet was always at the bottom to prevent malfunctioning due to small vibrations. During inhalation, the airflow passes through the small hole into the impeller chamber and causes the impeller to rotate. When the magnet rotates to

[a] Corresponding author:153155269@qq.com

the top, the hall sensor acquires the inhalation signal. Then, the rotation speed of the impeller was calibrated to the inhalation flow rate by the micro control unit.

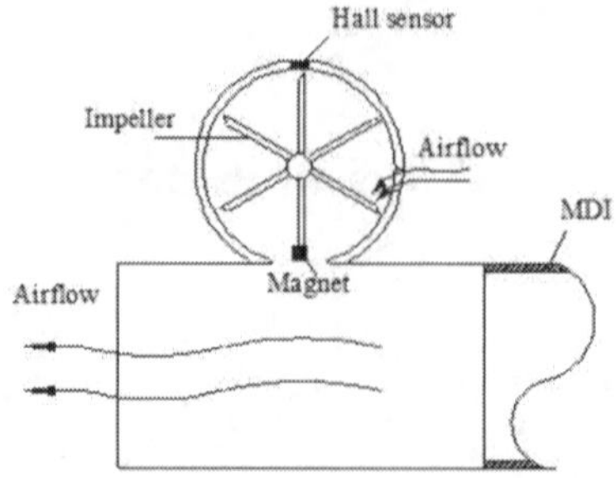

Figure 1. The structure of signal acquisition devices, an impeller and a hall sensor

2.2 Micro-control-unit

For controlling purposes, a micro-control-unit (MCU) (PIC 18F4550), with 32kbytes of program memory and a USB connection was used. The block diagram and printed-circuit-board of the control circuit are presented in figure 2. It included a sensing circuit to calculate the inhalation process, a power supply circuit for the MCU and a motor to drive the actuation device. The UART(Universal Asynchronous Receiver/Transmitter) module was connected to a PC or laptop computer through the USB port.

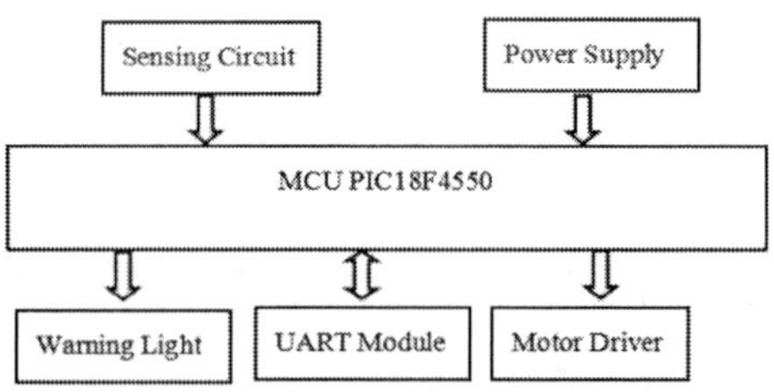

(a)

(b)

Figure 2. The block diagram (a) and PCB (b) of the control circuit

2.3 Actuation mechanism

A gear transmission mechanism was used to actuate the pressurized medicine canister because of its very high torque amplification efficiency. As shown in figure 3, three pairs of the gear were used to increase the torque from the small DC motor. In this study, the modulus of each gear was 0.6. The number of the teeth for each gear was: g1=g3=g5=10 and g2=46, g4=50, g6=75. Therefore, the torque could be amplified 172 times, and the mean push-down force could reach approximately 65N.

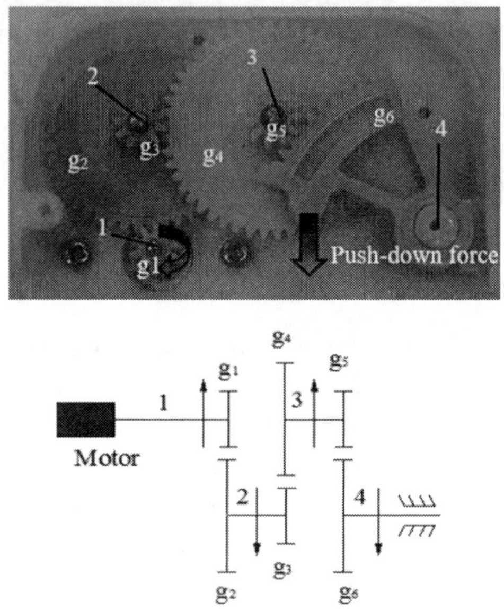

Figure 3. The gear transmission mechanism

2.4 The key parameters of the intelligent adapter

The feasibility for inhaled drugs depends mainly on aerosol release of MDI and the inhalation. The key technology of the intelligent adapter is to controll the aerosol-release time. For the intelligent adapter, the MDI was intalled on it. The aerosol-release time (T_r) is divided into three periods: switch-on time (T_s), mechatronics-hysteresis (T_m) and intentional-delay (T_i). T_s is the period between the starting of inhalation and the connecting of the signal switch. For impeller and hall sensor signal switch, the inhalation flow rate that can just drive the impeller to rotate a whole circle is defined as switch-on flow-rate (Q_S), the corresponding time is the switch-on time.T_m is the period between signal switch on and the time when the aerosol starts to release when T_i is equal to zero. T_i is the time that can be adjusted by the computer programmng, and to meet a suitable T_r.

3. Experiment

The experiments were conducted to measure T_r. For the experiments, an inhalation simulator was designed to generate the inhalation waveforms by vacuum pump. A simulation upper respiratory system (as shown in figure 4) including mouth-throat, main bronchus and adapter intalled with MDI, was used to simulate the aerosol delivery. Additional camera system was used to record the aerosol deliver process and the movement of the adapter, then using the recording time to caculate T_r.

Two brands of MDI were selected for this study. MDI A delivers fenoterol at a single dose of 100 μg. CFC-free MDI B delivers salmeterol and fluticasone propionate single doses of 25 μg and 250 μg, respectively.

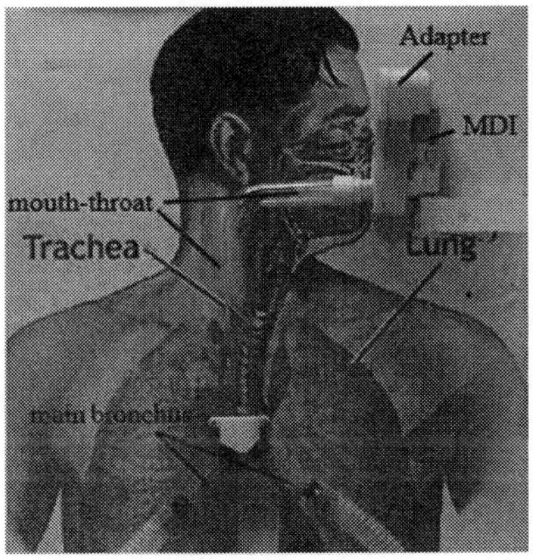

Figure 4. The upper respiratory system

4. Results and Discussion

Different signal acquisition devices have different switch-on time and switch-on flow-rate. Some autohalers with pure mechanical methods actuate the canister, however, the activating flow rate is above 30 L/min [10]. With the help of modern mechatronics, the starting flow rate can be reduced to only 10 L/min. Ts for the autohaler is approximately 0.6 s compared with 0.2 s for the adapter.

Tm is related with the bands of the MDI. In this study, Tm are 0.2 and 0.4 s for MDI A and MDI B respectively. Assuming the autohaler had the same mechatronics -hysteresis as this study, Tr is shown in figure 5. The intentional-delay is zero. The total aerosol-release time are 0.4 s or 0.6 s in this study and 0.8 s or 1.0 s for the autohaler for different bands of MDI. In the literature, the suggested aerosol-release times are 0-0.2 s [11] or 0.68 s for slow-deep inhalation [9]. Therefore, based on the results of this paper, it is better to start to actuate as soon as the system obtains the signal for some MDI. If Tr is small, it can be adjusted to a suitable value by the computer. Moreover, almost all the MDIs can use this intelligent adapter and it is reusable. The elddly, the child and the serious can operate it easily. However, for autohaler, the actor and the MDI are integrated and one-off; and the switch-on flow rate is higher than that of the intelligent adapter, it's not well for the weakness people.

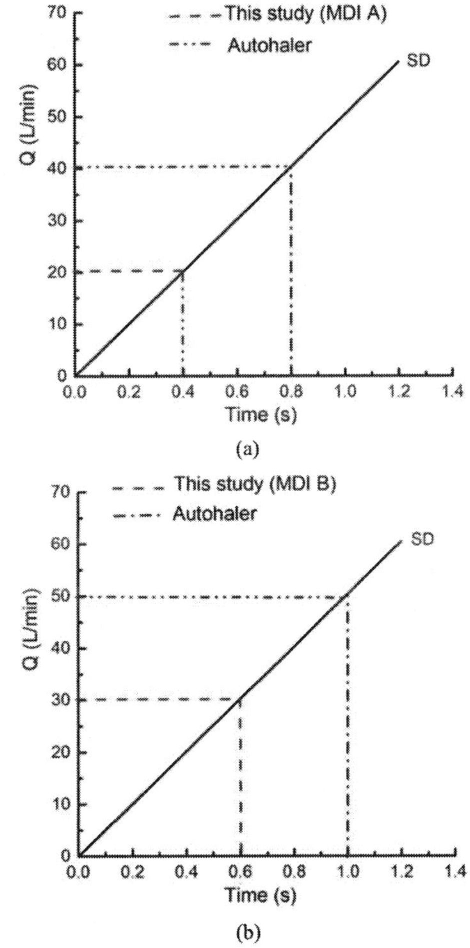

Figure 5. Comparison of aerosol-release time T_r for MDI A (a) and MDI B (b)

5. Conclusion

To help patients coordinate the interaction of inhalation and aerosol release, an intelligent metered dose inhaler has been developed. It is composed of an intelligent adapter and a metered dose inhaler. The added on mechatronic unit can detect the flow of inhalation and actuate the medicine canister at the specified release time.

The major advantages of the system are the low switch-on flow-rate (10 L/min), short switch-on time, adjustable aerosol-release time, and low cost. Moreover, the feasibility of detecting and recording inhalation flow rates can help doctors monitor the process of treatment and train asthma or COPD patients to properly use MDI.

References

1. R. Dalby, J. Suman, ADV. DRUG DELIVER REV., **55**, 779-791(2003)
2. J.J. McGrath, International Business Times, (2014)
3. G. M. Cochrane, CHEST,**117**, 542-550 (2000)
4. D. Hochrainer, H. Hölz, C. Kreher, L. Scaffidi, M. Spallek, H. Wachtel, J Aerosol Med., **18**, 273-282 (2005)
5. K. Nahar, N. Guta, R. Gauvin, S. Absar, EUR J PHARM, **49**, 805-818 (2013)
6. P. Arora, L. Kumar, V. Vohra, R. Sarin, RESP MED ,**108**, 992-998 (2014)
7. H.W. Johannes, D.D. Nigel, M.W. Joyce, G.D. Sunalene, N.L. Peter, J PEDIATR, **135**, 28-33 (1999)
8. J. Smith, J Aerosol Med Pulm Drug Deliv, **23**, 25-37 (2010)
9. P. W. Longest, G. Tian, L.W. Ross, M. Hindle, Pharm Res, **29**, 1670-1688 (2012)
10. D. Donnell, EUR REV MED PHARMACO, **5**, 7-16 (2001)
11. M.E. Broeders, J. Molema, W.C. Hop, H.T. Folgering, J Aerosol Med, **16**, 131–141(2003)

MATEC Web of Conferences 71, 04003 (2016)
ICCPE 2015
DOI: 10.1051/matecconf/20167104003

Design of Cam Mechanism with Negative Radius Roller-Follower

Jung-Fa Hsieh[1,a] and Wen-Chang Hsieh[2]

[1] *Department of Mechanical Engineering, Far East University, Tainan, Taiwan*
[2] *Department of Business Administration, Yunlin University of Science & Technology, Yunlin,Taiwan*

Abstract. Cam mechanisms with a negative radius roller-follower allow the design of simple machines capable of exerting remarkably high forces. Furthermore, the concave-convex nature of the contact surface between the cam and the follower results in a low contact stress and the ready formation of a lubricating layer. This paper presents a simple method for the design and analysis of a cam mechanism with a negative radius roller-follower. In the proposed approach, conjugate surface theory is employed to derive a kinematic model of the cam mechanism. Analytical expressions for the pressure angle and principal curvatures of the cam profile are then derived. The validity of the proposed design methodology is demonstrated by machining a cam mechanism having a negative radius roller-follower with a radius of 100 mm.

1 Introduction

Cam elements have many advantageous features for mechanical transmission systems, including simplicity, reliability, repeatability and a low running noise. Contact pairs comprising a convex cylinder and a concave cylinder have a large contact area, and hence a low contact stress [1]. Thus, cam profiles generally have a convex face, while follower profiles have either a convex face or a flat face. In order to distinguish conventional construct, followers with a concave face are referred to as negative radius followers. Cams with negative radius roller-followers use the inner rings of commercial roller bearings as roller-followers. the use of such mechanisms is particularly valuable in applications where space is at a premium and very high forces are involved. Carra *et al.* [2] presented a design nomograph and synthesis procedure for finding the suitable combination of parameters to satisfy the specified motion law without under-cutting. Wu *et al.* [3] compared the contact stress of two disc cam mechanisms with a concave-faced follower and a flat-faced follower, respectively. The results showed that the cam mechanism with a concave-faced follower produced a lower maximum contact stress than that with a common flat-faced follower. Moreover, it was shown that the negative radius roller-follower caused the sliding friction between the cam and the follower to be converted to rolling friction. However, the literature lacks a systematic investigation into the design of cam mechanisms with negative radius roller-followers. Accordingly, this study presents a simple yet robust methodology for the design and analysis of such mechanisms. The proposed methodology comprises four steps, namely (1) establishing a kinematic model of the cam mechanism, (2) designing the cam profile, and (3) analyzing the pressure angle and principal curvatures of the cam.

2 Surface geometry

In order to analyze the contact stress in a cam mechanism, it is first necessary to determine the profile of the cam. As discussed above, cams with negative radius roller-followers use the inner rings of commercial roller bearings as followers. In other words, the cam fits within an inner ring. In synthesizing the cam profile, the cam mechanism can be represented using the simplified model shown in Fig. 1. As shown, the cam rotates with an angular speed w_1 in the clockwise direction about its center point, 0_0. As the cam rotates, the follower rises to a total lift height h over a specified rise angle, remains in a high dwell position over the high dwell angle, falls through a distance h over the return angle, and then remains in a low dwell position over the low dwell angle.

To determine the cam profile using conjugate theory, it is necessary to label all of the coordinate frames in the system, starting with the cam (marked as "0" in Fig. 1) and ending with the translating follower (marked as "2" in Fig.1). Note that parameter b_2 is defined as $b_2=r_b+s_2(\ _1)$, where r_b is the base circle radius of the cam and s_2 is the translating follower displacement.

The configuration of frame $(xyz)_2$ with respect to frame $(xyz)_0$ is given by

$$ {}^0\mathbf{T}_2 = {}^0\mathbf{T}_1\,{}^1\mathbf{T}_2 = \begin{bmatrix} C\theta_1 & -S\theta_1 & -b_2 S\theta_1 \\ S\theta_1 & C\theta_1 & b_2 C\theta_1 \\ 0 & 0 & 1 \end{bmatrix} \tag{1} $$

[a] *Corresponding author:* seznof@cc.feu.edu.tw

DOI: 10.1051/matecconf/20167104003

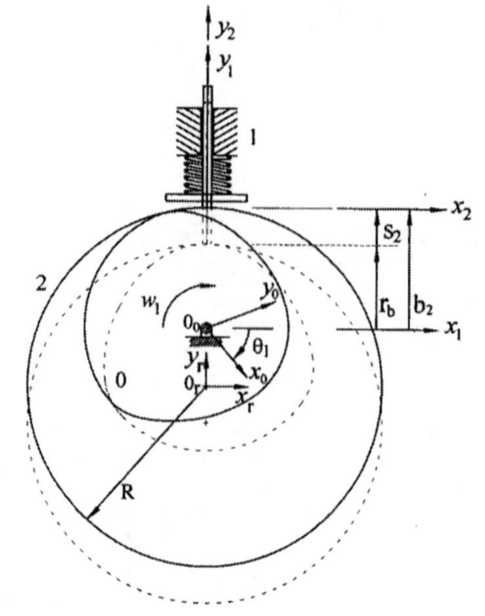

Fig.1. Simplified representation of cam and follower mechanism for synthesis of cam profile.

To express the surface of the roller in terms of frame $(xyz)_0$, it is first necessary to establish the frame of the roller-follower, namely $(xyz)_r$. The configuration of frame $(xyz)_r$ with respect to frame $(xyz)_2$ is given by the matrix

$$ {}^{2}\mathbf{T}_r = \begin{bmatrix} 1 & 0 & 0 \\ 0 & 1 & -R \\ 0 & 0 & 1 \end{bmatrix} \tag{2}$$

where R is the radius of the roller.

The surface equation, ${}^{r}\boldsymbol{S}$, and unit outward normal, ${}^{r}\mathbf{n}$, of the roller can be expressed with respect to frame $(xyz)_r$ as follows:

$$ {}^{r}S = \begin{bmatrix} RC\theta & RS\theta & 1 \end{bmatrix}^T \tag{3}$$

$$ {}^{r}n = \begin{bmatrix} RC\theta & RS\theta & 0 \end{bmatrix}^T \tag{4}$$

where θ is the polar angle.

In accordance with the principles of homogeneous coordinate transformation, the cam surface equation, ${}^{0}\boldsymbol{S}$, can be expressed as

$$ {}^{0}\mathbf{S} = {}^{0}\mathbf{T}_2\,{}^{2}\mathbf{T}_r\,{}^{r}\mathbf{S} = \begin{bmatrix} RC(\theta+\theta_1)+(R-b_2)S\theta_1 \\ RS(\theta+\theta_1)+(b_2-R)C\theta_1 \\ 1 \end{bmatrix} \tag{5}$$

In addition, the normal vector, ${}^{0}\mathbf{n}$, of the cam can be expressed as

$$ {}^{0}\mathbf{n} = {}^{0}\mathbf{T}_2\,{}^{2}\mathbf{T}_r\,{}^{r}\mathbf{n} = \begin{bmatrix} C(\theta+\theta_1) & S(\theta+\theta_1) & 0 \end{bmatrix}^T \tag{6}$$

Once the input-output relation of the cam mechanism has been defined, the conjugate points and cam profiles can be determined via the formulation

$$ {}^{0}\mathbf{n}^T \bullet \frac{d\,{}^{0}\mathbf{S}}{dt} = 0 \tag{7}$$

where ${}^{0}\mathbf{n}$ and ${}^{0}\mathbf{S}$ are the unit outward normal and surface equation of the cam profile with respect to frame $(xyz)_0$, respectively. The conjugate point (denoted as $\bar{\theta}$) is given by

$$ \bar{\theta} = -\tan^{-1}\left(\frac{\bar{F}}{\bar{E}}\right) \tag{8}$$

where $\bar{E}$ is defined as $\bar{E} = b_2 - R$ and $\bar{F}$ is defined as $\bar{F} = \frac{db_2}{d\theta_1}$.

Substituting Eq.(8) into Eq.(5), the profile of the cam, ${}^{0}\mathrm{S}$, can be obtained as

$$ {}^{0}\mathbf{S} = \begin{bmatrix} RC(\bar{\theta}+\theta_1)+(R-b_2)S\theta_1 \\ RS(\bar{\theta}+\theta_1)+(b_2-R)C\theta_1 \\ 1 \end{bmatrix} \tag{9}$$

3 Pressure angle

As shown in Fig. 2, the pressure angle is defined as the angle between the direction of motion (i.e., velocity) of the follower and the direction of the force transmission axis [5]. The pressure angle at the contact point p between the cam and the roller is given by

$$ \tan\psi = \frac{\left\| {}^{1}\mathbf{n}_p \times {}^{1}\mathbf{v}_p \right\|}{\left| {}^{1}\mathbf{n}_p \bullet {}^{1}\mathbf{v}_p \right|} \tag{10}$$

where the unit normal vector of the contact point, ${}^{1}\mathbf{n}_p$, and the direction of velocity of the contact point, ${}^{1}\mathbf{v}_p$, are both defined with respect to the fixed frame $(xyz)_1$.

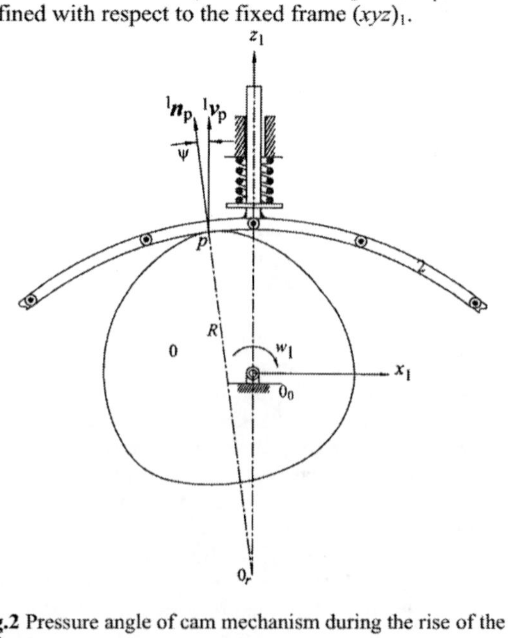

Fig.2 Pressure angle of cam mechanism during the rise of the follower

The unit normal vector of the contact point on the roller-follower surface with respect to frame $(xyz)_1$ is given by

$$ {}^{1}\mathbf{n}_p = {}^{1}\mathbf{T}_2\,{}^{2}\mathbf{T}_r\,{}^{r}\mathbf{n}_p = \begin{bmatrix} C\bar{\theta} & S\bar{\theta} & 0 \end{bmatrix}^T \tag{11}$$

Meanwhile, the location of the contact point, with respect to frame $(xyz)_1$. is given by

$$ {}^1\mathbf{S}_p = {}^1\mathbf{T}_2\,{}^2\mathbf{T}_r\,{}^r\mathbf{S}_p = \begin{bmatrix} RC\bar{\theta} & RS\bar{\theta} - R + b_2 & 1 \end{bmatrix}^T \quad (12) $$

Since the roller-follower translates along the y_1 axis direction, the direction of velocity of the contact point on the roller-follower surface with respect to frame $(xyz)_1$. can be expressed as

$$ {}^1\mathbf{v}_p = \frac{d\,{}^1\mathbf{S}}{dt} = \begin{bmatrix} 0 & \frac{db_2}{dt} & 0 \end{bmatrix}^T \quad (13) $$

Substituting Eqs.(11) and (13) into Eq.(10), the pressure angle, ψ, can be derived as

$$ \tan\psi = \frac{db_2}{d\theta_1}\Big/(R - b_2) = \frac{db_2}{d\theta_1}\Big/R - (r_b + s_2) \quad (14) $$

Equation (14) shows that an increase of the base circle radius r_b leads to a lower pressure angle ψ. Notably, a higher maximum velocity causes an undesirable increase of the contact pressure angle. However, for negative radius (concave-faced) follower cams, the adverse effects of a higher pressure angle are generally insignificant.

4 Analysis of curvature radius

The curvature radius of the cam must be analyzed in order to prevent the occurrence of singular points. Note that the radius of curvature of the cam cannot be greater than the radius of the follower since unintended multiple-point contacts will otherwise occur [6]. According to basic differential geometry principles, the principal curvatures of the cam profile can be evaluated as follows:

$$ K_1, K_2 = H \pm \sqrt{H^2 - K} \quad (15) $$

where K_1 and K_2 are the principal curvatures, and K and H are defined respectively as

$$ K = \frac{LN - M^2}{EG - F^2} \text{ and } H = \frac{2FM - EN - GL}{2(EG - F^2)} \quad (16) $$

where $L = {}^0n \bullet \frac{\partial^2\,{}^0S}{\partial^2 u}$, $M = {}^0n \bullet \frac{\partial^2\,{}^0S}{\partial u \partial\theta_1}$, $N = {}^0n \bullet \frac{\partial^2\,{}^0S}{\partial^2\theta_1}$, $E = \frac{\partial^0 S}{\partial u} \bullet \frac{\partial^0 S}{\partial u}$, $F = \frac{\partial^0 S}{\partial u} \bullet \frac{\partial^0 S}{\partial\theta_1}$, and $G = \frac{\partial^0 S}{\partial\theta_1} \bullet \frac{\partial^0 S}{\partial\theta_1}$

The present study considers the case of a planar cam. Hence, one of the principle curvatures is equal to zero. Moreover, the curvature radius, ρ, of the cam profile is equal to the inverse of the principal curvature

5 Implementation

To validate the design methodology presented in Sections 2~4, a cam mechanism was designed with a negative radius roller-follower having a radius of 100 mm. The motions of the cam and follower were defined by a modified sine motion curve. The follower motion was specified as follows: 20 mm rise for cam rotation from 0 to 150^0, dwell for cam rotation to 180^0, 20 mm fall for cam rotation to 330^0, and dwell for the remaining 30^0. The radius base circle was specified as 40 mm.

Figure 3 shows the simulation results obtained for the cam profile. Figure 4 shows the variation of the pressure angle at the contact point as the cam rotates. Figure 5 shows the curvature radius of the cam profile. Figure 6 presents a photograph of the machined cam. Note that the process of generating the NC equations required to produce the driving cam on a 3-axis machine tool is described in [7]. Note also that the three holes lying along the x_0 axis of the cam are the workpiece origin (center hole) and clamping holes, respectively. It is seen that a good qualitative agreement exists between the machined profile and the designed profile (see Fig. 3). Thus, the feasibility of the proposed methodology is confirmed.

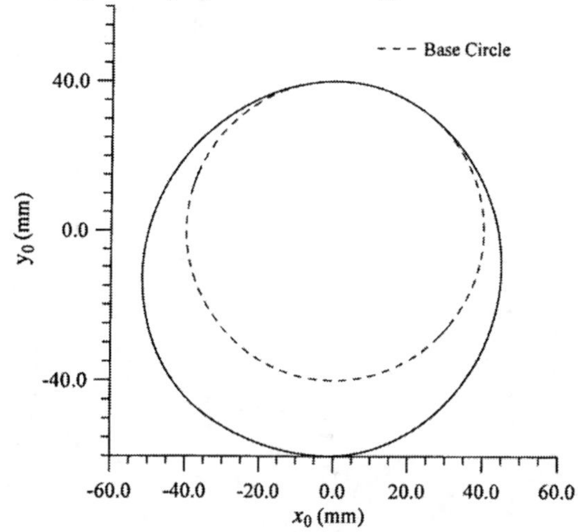

Fig.3 Simulated profile of cam with negative radius roller-follower

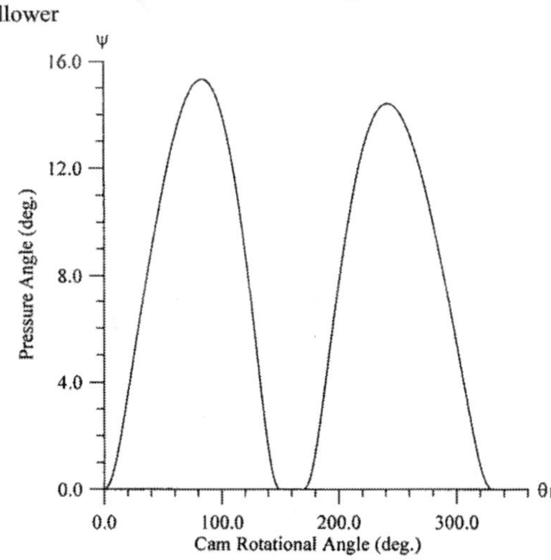

Fig.4 Variation of pressure angle with cam rotational angle

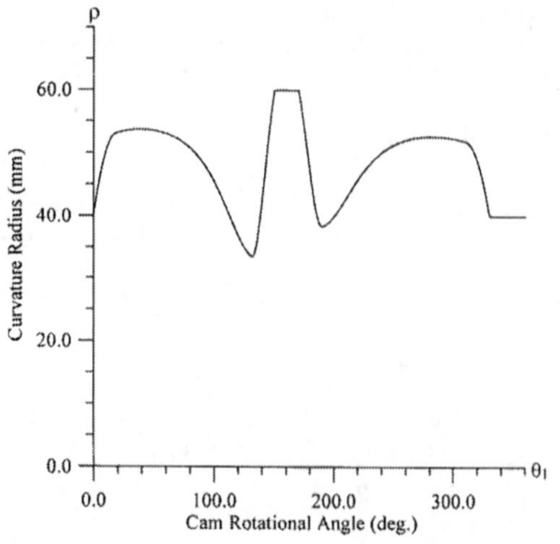

Fig.5 Variation of curvature radius with cam rotational angle

Fig.6 Photograph of machined cam

6 Conclusions

This paper has presented an integrated methodology for the design of a cam mechanism with a negative radius roller-follower. A kinematic model of the cam mechanism has been derived utilizing the homogeneous coordinate transformation method and conjugate surface theory. In addition, analytical expressions have been presented for the pressure angle and principal curvatures of the cam profile. The feasibility of the proposed approach has been demonstrated both numerically and experimentally.

References

1. Collins,J. A., *Mechanical design of machine elements and Machines*, New York: John Wiley & Sons Press (2003)
2. Carra, S., Garziera, R. and Pellegrini, M., Synthesis of cams with negative radius follower and evaluation of the pressure angle, *Mechanism and Machine Theory*, **39**, 1017-1032 (2004)
3. Wu, L.-I., Liu, C-H and Chen, T.-W., Disc cam mechanisms with concave-faced followers, *Proc. Instn.Mech,* Part C: *J. Mechanical Engineering Science*, **223**, 1443-1448 (2009)
4. McCarthy, J. Michael, Geometric Design of Linkage, New York: Springer-Verlag (2000)
5. Gonzalez-Palacios, M.A. and Angles, J., Cam Synthesis, Springer (1993)
6. Jones, J. Rees, Mechanisms-cam curvature and interference, *Engineering*, **218**, 460-463 (1978)
7. Hsieh, Jung-Fa, Application of homogenous transformation matrix to the design and machining of a Geneva mechanism with curved slots, *Proc. Instn. Mech. Engrs, Part C: J. Mechanical Engineering Science*, **221**, 1435-1443 (2007)

Analysis of the Real-time Compensation for Thermal Error at CNC Milling Machine

Tsung-Chia Chen[a] and Chia-Jung Chang

National Chin-Yi University of Technology, Dept. of Mechanical Engineering, 41170 Taichung, Taiwan (R.O.C.)

Abstract. This paper focuses on analyzing and discussing thermal errors at CNC milling machines. The thermal affection makes the deformation of machine tools and is the main problem of accuracy error over than 65%. Effectively improving or controlling thermal errors is helpful for the accuracy of machine. The key point of this paper is the position of tool center point. Firstly, 14 pieces of temperature sensors are used for checking the real field of temperature, and then four sensors with better linearity are chosen for real situations. The test bar and 5 pieces of non-contact sensors are utilized for clearly getting the displacement of the tool center point and head during the process. Based on the theory of MRA (Multiple Regression Analysis), the external zero-point is shifted to build the mathematical module. The database is input to the control board and the PLC is used for real-time compensation for machining. Finally, two work pieces (one compensated and one non-compensated) are tested and compensated one presents better precision.

1 Introduction

Taiwan is a famous country for manufacturing CNC machines in the world, especially is good at the quality and price. Regarding the rising sales in China, South Korean, and countries in south-east Asia during the recently years, the marketing of Taiwan is decreasing. By the way, like the advanced countries, Swiss, Japan, Germany has been developed machining of intelligent, compensation in-time that is bases on high quality and efficacy for many years. It means that there is more and more competitiveness in the world and Taiwan must follow such a trend. That's why the academia and manufacturers need to work with each other to reach the goal. Although the 3-axis center is the basic one, there is more space to improve and advance. In local famous companies, the technology could reach the geometry accuracy less than 0.015mm (based on ISO 230 rule). Regarding the mechanism of machine, it is built by complicated parts, transmission, and electronic elements. Utilizing tradition theories to model, control and analyze the error of kinematic isn't easily.

The heat is the biggest factor in making errors because of the heat deformation and displacement which will directly cause bad quality of work-piece. For this reason, high accuracy of CNC machines is only the basic standard. How to reach high reliability and stability for precision work-piece is the customer need.

Recently years for the static, dynamic, and the error of the heat on machine tools, that as following to reference. Mr. Ramesh, Mannan & Poo [1] indicated that the most important was to consider the accuracy of machine and focus on the error of geometry accuracy, heat, and shear force to explain lots of errors in detail. Finally, for utilizing sensors to gather great data for error compensation, the cheaper and easier way under the value of the precision and allowance is expected. The authors [2] discussed the error of heat and pointed out that when a machine worked continuously, the heat of moving parts and mechanism unit would cause errors. Using FEA and HTM models, there are three steps to model, measure and compensate for general compensation of heat in time, depending on PC system to control the compensation of machine directly. Although the technology needs improvement, especially in heat errors, it will become the standard equipment of advance machines in the future. To overcome limits to micron machines, Mr. Ha, Liu & Yang [3] used a capacitance sensor, which was developed to measure the geometry accuracy, to measure every axis and vertical straightness led to the coordination. Multiple degrees of freedom measurement have been realized to measure 5 components of geometry of work-piece errors at the same time. Totally 18 errors of geometry have been successfully measured by utilizing least squares method. Result in the error of horizontal decreases 1.2mm and the error of straightness to 0. Mr. Wu & Gong [4] used the thermocouple and non-contact displacement gauge to measure the temperature and thermal deformation in a gantry machine. Multiple Regression Analysis is utilized for building the mathematical model of temperature difference and thermal deformation. After the compensation simulation , the error of spindle that could be controlled in a certain range is acquired to highly prove the upgraded precision of work-piece. This method

[a] Corresponding author: ctchen@ncut.edu.tw

is predictive and accurate reliably. Mr. Lee and Yang [5] studied machine tools with the optimal statistics and evaluation, it be analysis by continuous and least squares regression. That could be applied to the actual variable selection effectively, and several related groups could decrease the variable problems to improve the multiple regression, so it is effectively shorter calculates and is more reliable of model. To develop a system of thermal compensation in time for the application of thermal model, Mr. Vanherck, Dehaes and Nuttin [6] used the artificial neural network model to access the relation of multiple regressions and effectively compensate the thermal deformation of multiple axes' machine tools. The back-propagation calculation is used for deducting the maximum deformation from 0.15mm to 0.015mm and finally machining two work-pieces. One is compensated, and the other is non-compensated. This way could shorten the error from 0.075mm to 0.016mm and the time for trying error by vanishing to decrease the thermal error. For the tool center point, Mr. Horejša, Mareša, and Novotnýa [7] used MLR and TTF to develop an advanced thermal compensation in horizontal 4-axis machine tools. Different from original 21 sensors, 4 are utilized by multiple regressions. This model is different from general TTF, which was built up based on the operation of spindle. It effectively decreases the error from 0.10mm to 0.01mm and shortens the displacement of TCP, better than the original 75%.

2 Theory and application of heat

2.1 Heat transmission

Generally, there are three ways to transmit heat.

1. Heat Conduction: Solid, heat conducts from higher temperature to lower.

$$q = -ka\frac{dX}{dT} \quad (1)$$

q is heat conduction, k is factor, a is the area of interface, dX is the distance, dT is temperature of each side.

2. Heat convection: The fluid or air, it delivers naturally to affect volume and density by itself.

$$q = hA\Delta T \quad (2)$$

q is heat convection, h is factor, A is sectional area, ΔT is temperature difference.

3. Heat irradiation: It delivers via electromagnetic wave and doesn't depend on anything.

$$q = \sigma A(T_1^4 - T_2^4) \quad (3)$$

σ: Stefan-Boltzmann constant ($5.669\text{x}10^{-8}\text{w}/(\text{m}^2\text{k}^4)$), A: sectional area, T_1 and T_2 is the temperature of difference for two irradiation.

2.2 Heat source of machine tools

There are three kinds of heat source for the heat deformation (Figure 1 & Figure 2). Firstly, the heat deformation of machine tools (roughly 65~70%), include spindle rotation, ball-screw expanded, heat displacement, and heat bend of machine tools. Secondly, the error of volume of machine tools (roughly 30%), include static bend, error of building, and error of mechanism. Finally, error of controlled system of machine tools (roughly 5%), it's from the error of interpolation of CNC controllers or the error of servo systems. Anyway, the above needs to be overcome for recent precision manufacturing industry.

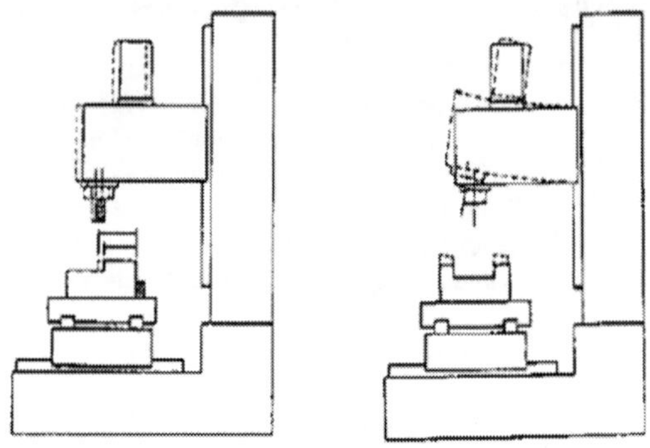

Figure 1. Heat deformation of machine tools.

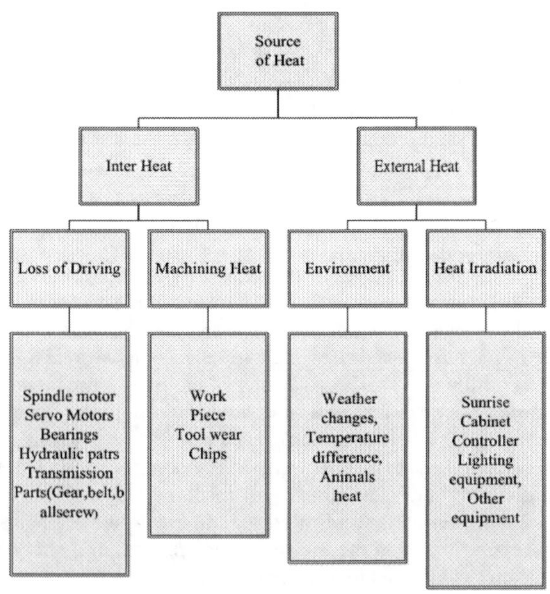

Figure 2. Factors in thermal deformation.

2.3 Model building of thermal error

There are two ways for building models.

2.3.1 Theory model

By analyzing the heat transfer, differential formula are utilized for getting the temperature of machine and then finding deformation based on the theory of elasticity, including data and numerical answers. It needs to transfer the numbers to show the mechanism and the heat source of machine. It is hard to explain the heat situation of machines accurately, because the mechanism is very complicated.

2.3.2 Actual model

It's an actual method to think that the machine system is a black box and to input the data of temperature about machine and other variables. Outputs are the deformation of machine tools or main factor of thermal errors in the system. It uses the parameters of system for the operation, like statistics model, exponential function, multiple regression, non-linear multiple regression, neural networks, expert system and fuzzy theory.

3 Experiment

3.1 Experiment plan

This paper uses 14 pieces of temperature sensors to measure the real field of temperature of the machine (Figure 3). Firstly, stick them on the machine to check the request. After the confirmation, fix 4 better relation sensors in the machine for gathering correct data as the following picture (Figure 4). Then, 4 pieces of non-contact displace sensors are used for checking the changeability of X, Y, Z axis.

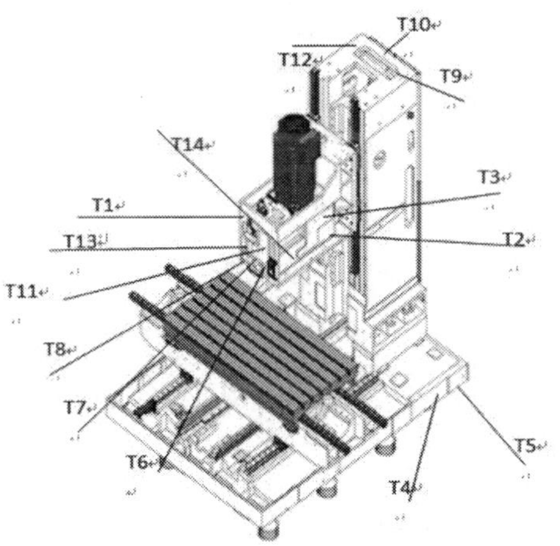

Figure 3. Position of sensors first.

After measuring the spindle running (7200RPM & 10800RPM) for 48hrs, it actually needs 4 pieces of sensors to fix in good relations (Table 1) and then start to gather data.

Table 1. Point sensors.

T_1	Nearby rear side bearing of spindle
T_2	Heat source of belt transmission in head
T_3	Nearby spindle motor (directly)
T_4	Base, temperature of environment

3.2 Equipment

The Qauser machine tools, MV204C type (Figure 5), with the specification of Heidenhain Itnc530 controller, 12000RPM spindle (belt type), THK linear way (X, Y baller, Z roller), and auto-lubrication system, is used

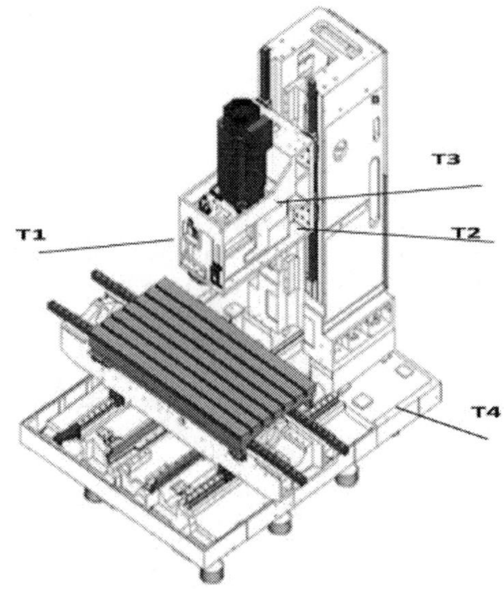

Figure 4. Position of needed sensors.

Figure 5. MV204C Quaser machine tools.

To compensate the microprocessor chip module (Figure 6), deliver through the PCB817 optical coupling electricity to PLC interface. (IO)

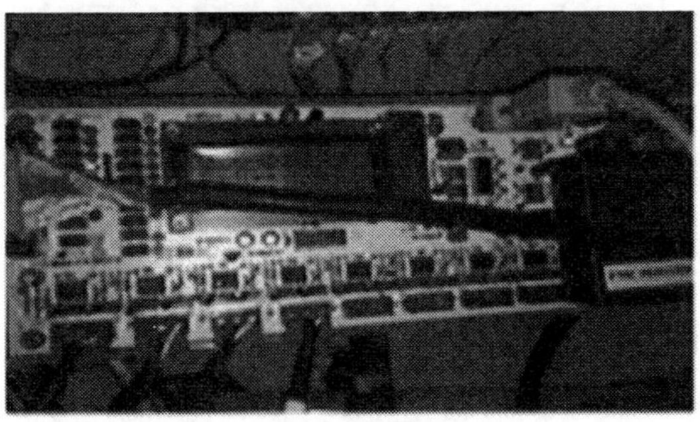

Figure 6. Compensation module.

4 Result of analysis and discuss

4.1 Displacement of temperature measuring

Firstly, use 14 sensors to fit on the machine. Spindle runs 7200 RPM (60%) and then the higher speed to 10800 RPM (90%). After 48hrs, the data show steady state. Then, choose 3 good related sensors, like above ways, to operate again and get the data as following (Figure 7).

4.2 Temperature changes in the amount of displacement

From Figure 7, the displacement of X and Y axis is firstly run less than 0.02mm, and the position to the tool center is extended (TCP drift). Factors should be the bearing of oil being solid and running unstable, friction of bearing, or not being built well. After running, the chart shows in steady state within 0.055mm, 90% RPM operation, the tool center point extends to 0.073mm and then becomes steady. The displacement of X axis is roughly 0.01mm, meaning that the head of X axis tilts right side 0.01mm (X1, X2) by thermal effect. The displacement of Y axis is roughly 0.016mm (Y1, Y2), meaning the head is the rear side of 0.016 by error of accuracy. T1 is near the rear side of bearing and it's the biggest heat source; T3 is the operation of spindle motor; T2 is the friction of belt between pulleys; and, T4 is the temperature of environment. (Air condition factory, temperature is controlled in 28°C)

4.3 Machining test

After several times of compensation, the displacement of X, Y axis is less than 0.02mm, and the point of grinding test bar is also within 0.02mm. Starting test, one is compensated and the other one is non-compensated.

Machining parameters contain 120mm tool length, R3 ball tool, material being aluminum, and size of work-piece 200mm X 200mm X 50mm. Radial machine, from center to outside, consumes time for more than two hours.

Off the compensation, the face of accuracy is 0.025mm (min.) and 0.021mm (max.). After the compensation, the accuracy is 0.014 (min.) and 0.001mm (max.). The effect rises 33% that the accuracy is absolutely more precise (as shown Figure. 8).

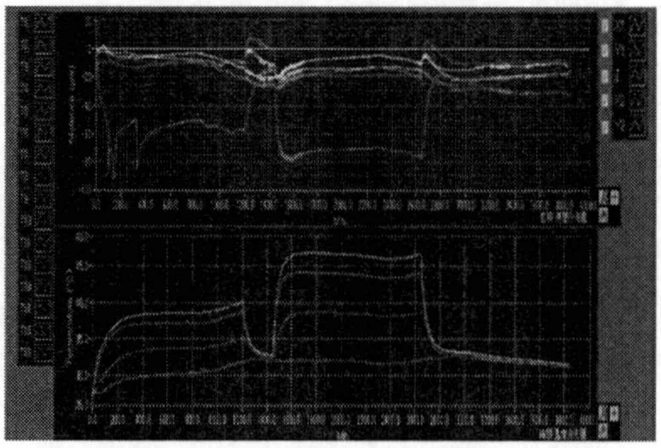

Figure 7. Temperature and displacement.

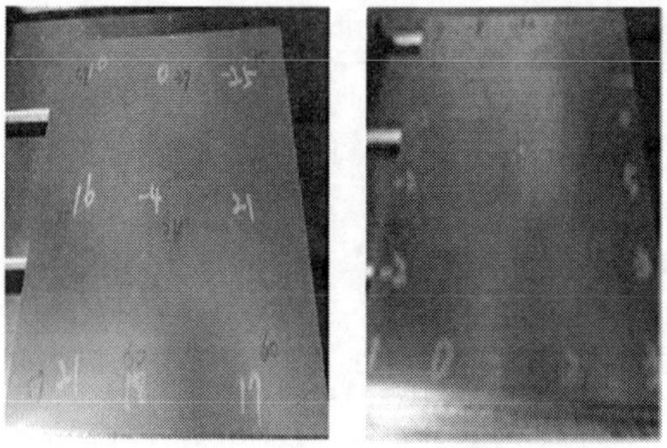

Figure 8. (a) Non-compensation. (b) Compensation.

5 Conclusions

After the experiment, the effects of heat on the error of accuracy for machine are very big, especially in deformation and displacement of tool point. The result of work-piece is apparent that it is concluded as below.

1. Mechanism of machine includes five parts of casting, where good heat balance and symmetrical design are very important. They could lower the error of heat, when they are well designed.
2. It's very important for the position of sensors. The experiment shows that too many sensors will lead to complex and unreliable result. Effect and lowest amount of sensors can get parameters easily and lower the cost.
3. When the data are acquired from the same type of machines, the geometry accuracy is focused to make correctly adjustment, for building the machine. When heat errors occur, the accuracy is in the tolerance.
4. The core technology of the FANUC, HEIDENHAIN controller cannot be acquired that it can merely be compensated by external way. It is expected to get the core interpolation technology for the direct compensation in order to develop good software.
5. If the machine is 4 or 5 axes machine, the result of compensation will become very hard, due to adding tilting or rotating axis. In the future, it is necessary to do more study and experiment at multiple axes machine tools.

Acknowledgements

Thanks to Quaser machine tools supporting the hardware, software and others and the PMC's assistance and technology transfer to finish this paper.

References

1. R. Ramesh, M.A. Mannan, A.N. Poo, International Journal of Machine Tools & Manufacture **40**, 1235-1256 (2000)

2. R. Ramesh, M.A. Mannan, A.N. Poo, International Journal of Machine Tools & Manufacture **40**, 1257-1284 (2000)

3. J.H. Lee, Y. Liu, S.H. Yang, International Journal of Machine Tools & Manufacture **46**, 1508-1516 (2006)

4. J.S. Wu, Y.T. GONG, Machinery Monthly **25(3)**, 48-58 (2003, In Chinese)

5. J.H. Lee, S.H. Yang, International Journal of Machine Tools & Manufacture **42**, 147-155 (2002)

6. P. Vanherck, J. Dehaes, M. Nuttin, Computers in industry **33**, 119-125 (1997)

7. O. Horejša, M. Mareša, L. Novotnýa, Procedia CIRP **4**, 67-72 (2012)

MATEC Web of Conferences 71, 04005 (2016) DOI: 10.1051/matecconf/20167104005
ICCPE 2015

Using AlN-Coated Heat Sink to Improve the Heat Dissipation of LED Packages

Ming-Der Jean [1,*], Ji-Bin Jiang [1], Ming-San Xu [1], and Jia-Yi Chien[2]

[1]School of Mechanical and Automotive Engineering/FuJian University of Technology, Fuzhou 350118, China.
[2]School of Nursing/National Taipei University of Nursing and Health Science, Taipei 11219, Taiwan.

Abstract. This study optimizes aluminum nitride (AlN) ceramics, in order to enhance the thermal performance of light-emitting diode (LED) packages. AlN coatings are grown on copper/ aluminum substrates as a heat interface material, using an electrostatic spraying process. The effect of the deposition parameters on the coatings is determined. The thermal performance of AlN coated Cu/Al substrates is evaluated in terms of the heat dissipated and compared by measuring the LED case temperature. The structure and properties of the coating are also examined a scanning electron microscopy (SEM). In sum, the thermal performance of the LED is increased and good heat resistance characteristics are obtained. The results show that using AlN ceramic coating on a copper/aluminum substrate increases the thermal performance.

1 Introduction

Light emitting diodes (LEDs) represent the new generation light sources and have attracted considerable attention because they are used in solid lighting devices. LEDs are more efficient than most traditional light sources because they have a long lifetime, high reliability, high efficiency and are environmentally friendly, so they are widely used in many applications. Current LEDs only have an efficiency of about 20% and around 80% of the energy is used to generate heat at the junction for the LED modules so most of the electrical power is converted to heat rather than light [1-3]. Insufficient heat dissipation in LEDs reduces light emission efficiency and shortens a LED lamp's lifetime. Therefore, thermal management, which ensures the rapid dissipation of the heat that is generated in the LED chip through a heat sink, is crucial for LED modules. Various solutions have been proposed [4-7].

Fire Retardant 4, which uses a conventional metal PCB and silicon submount was commonly used in the past as the material for electrical circuit boards for LEDs. It is currently used structurally and has an isolation layer with very low thermal conductivity. This structure gives the junction between the LED chip and metal PCB, which is formed by soldering, a large thermal resistance, so heat that is generated by the chip is rapidly transferred to the heat sink, but this heat transfer is limited by the die attachments and thermal conductivity grease, each of which have a high thermal resistance. Several recent studies have tried to improve the thermal conductivity of the base substrates by using alumina ceramic, AlN ceramic, silicon and Al metal PCB [8-11]. Several studies have reported noticeable improvements in the performance of LED’ s when ceramics with high thermal-conductivity are used to coat substrates [12-15]. However, cost reduction is a continuous requirement for LED packages and LED manufacturers must reduce the cost of producing LED modules. Aluminum nitride (AlN) coatings are commonly used for heat dissipation in LEDs packaging devices because of their high optical transparency and their thermal and chemical stability [16-22]. Very few studies have reported the use of electrostatically sprayed AlN coatings as a thermal interface material. To meet the demands of the electronics industry, highly conductive materials with reasonable production costs and the optimization of AlN coatings are necessary.

In this study, AlN films are optimally deposited on Al/Cu substrates by electrostatic spraying. The effect of the parameters for the AlN coating process is then determined. The morphology, the microstructure and the thermal performance of the LED packages are also determined. An optimized AlN coating is identified that compares very satisfactorily with the experimental results.

2 EXPERIMENTAL DETAILS

2.1 Aluminum nitride coated heat sink

AlN coatings were deposited on Cu/Al substrates (55mm in diameter and 2/3/5mm in thickness) using an

* Corresponding author: mdjeam@foxmail.com

DOI: 10.1051/matecconf/20167104005

electrostatic spraying system. The fabricated LED packaging system is composed of the LED module, an aluminum substrate, AlN coatings and a heat sink that uses both Al and Cu substrates, a 1.2W LED light source with a metal core printed parallel circuit board and a starter. AlN is a unique ceramic material that has outstanding properties, such as high thermal conductivity (280 $Wm^{-1}K^{-1}$), low electrical resistivity and chemical stability. In this study, AlN coatings are studied to assess their effectiveness for thermal management. During the assembly process, the aluminum substrate was bonded with the starter and the thermal interface material was attached to the heat sink. The interfaces were coated with an AlN and these were optimized to minimize the contact thermal resistance. In order to understand the thermal performance of the AlN coated Cu/Al substrates in terms of the heat dissipated, a 1.2-watt LED light source was used. The junction temperature (Tj) of the LED chips that were fixed on the AlN coated Cu/Al substrates was evaluated by measuring the case temperature because the junction temperature is difficult to measure directly. In this study, the junction temperature is the main quality characteristic that is used to evaluate the LED package. The case temperature at various locations was directly measured using a thermographic camera (IRI-4010) that offers true 3D thermal mapping software and allows precise visualization, measurement and comparison of the thermal maps. The illumination device comprised four 0.3W LEDs that used InGaN-based chips and a base substrate with AlN ceramics, assembled in a LED package. The deposited surface morphology of the AlN coatings was determined using a field emission scanning electron microscope (JEOL JSM-6700F).

2.2. Experimental design

In this study, there are several possible control factors for this design. Seven (B-H) of these control factors have three levels, but factor A (substrate) has two. All levels are selected for proprietary reasons according to experience and electrostatic spraying reference information. The substrate materials (A) that were used in the experiments are Cu and Al, which have high thermal conductance and are commonly used in heat dissipation components. The assignment of the eight factors to an L18 orthogonal array and the subsequent test results are shown. Each trial of the OA had a total of 18 tests and each of the eight factor combinations had two repetitions. The complete list of factors and levels for this experiment is given. The levels for the control factors are the proposed specification limits for the control system design. In the orthogonal array, 18 tests were used to systematically optimize the various control factors with respect to thermal performance and quality..

3. RESULTS AND DISCUSSION

3.1. Surface morphologies

The surface morphologies of the AlN coatings are shown in Fig. 1. Several parts of the OM images of the surfaces of the coatings obtained in all 18 tests using the orthogonal array table are shown. Clearly, the microscopic hills, voids, valleys and poor surface flatness at the solid-air interface are the greatest barrier to thermal conduction. In terms of the surface roughness values (Sa) of the AlN coatings, as shown in Fig.1a-1b, sample 14 exhibits the highest value for surface roughness (4.26 um), which corresponds to a T_j value of 45.25℃. This is the worst. Sample 4 (1.73um) has the lowest T_j value for all of the L18 tests at 39.7℃. Figs 1c-1d show that the average volume of the voids on the surface of AlN coatings is measured to be 0.55% in sample 14 and 0.68% in sample 4. The respective volumes of AlN material in both are 6.49mm^2/mm^2 and 3.80mm^2/mm^2, which indicates that sample 4 has a large amount of AlN on the top surface of the films. In other words, the OM patterns confirm that the AlN content for sample 4 is greater than that for sample 14, so there is a decrease in the contact resistance, which results in a lower thermal resistance. The overall result is confirmed by a lower thermal temperature (T_j), with low surface roughness and a high content of AlN.

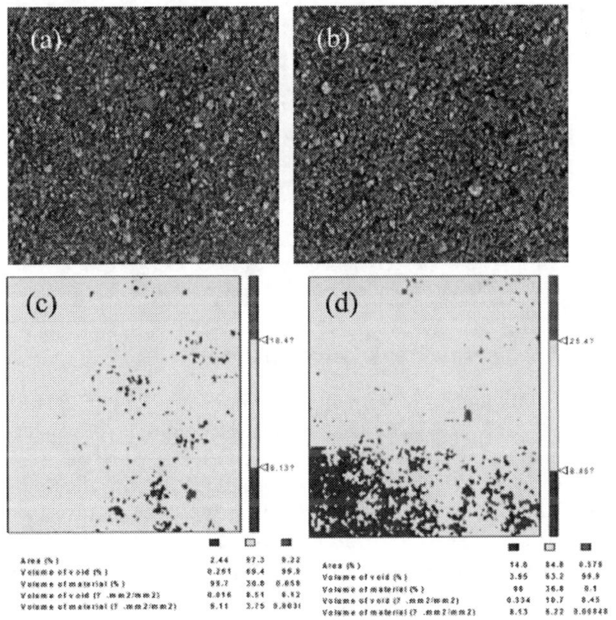

Fig.1. A comparison of the scanning surface morphology and the volume of the voids in the AlN coatings, using a TalyScan 150 instrument for the 4th and 14th tests.

3.2. Thermal Performance of the AlN Coatings

In order to test their performance, the AlN coatings were used as a heat sink for a 1.2W LED package that was attached to a parallel circuit board. During the thermal test, the LED was driven using a 700mA current when it reached a steady state. The observed cooling curves are shown in Fig.2. It is seen that when the driving currents are shut down after 60min, the temperature decreases over time in Figs. 2a -2b are from 54℃ to 43℃ and from 53 ℃ to 42 ℃, respectively. This shows that the temperature of the LED packages is reduced significantly when an AlN coated Al substrate is driven at 700mA. Figs.2c-2d show the surface curves for temperature (T_j) with a contour. These are plotted to determine the distribution of heat dissipation on an AlN coated Al substrate. It is seen that the central portion of the Cu/Al substrate exhibits the highest Tj value for AlN coatings. This may be due to solid-air interface defects in the crystal structure, which increase the thermal barrier and affect the heat conductance.

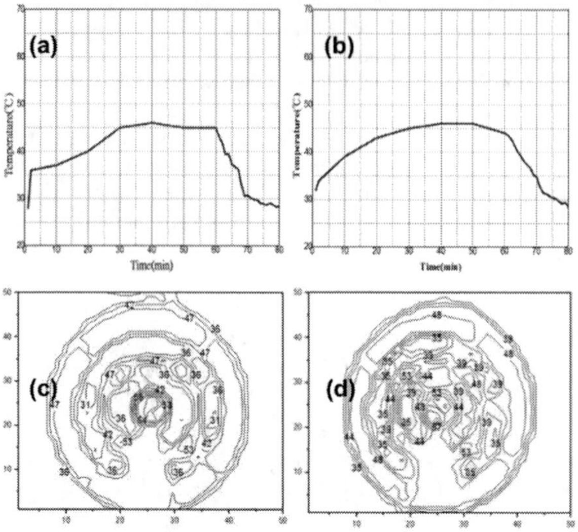

Fig.2. The observed cooling curves for a LED that is driven with a 700mA current when it reaches steady state.

3.3. Thermal resistance analysis

In this study, the thermal performance of a 1.2W LED on an AlN-coated Al substrate that is used as a heat sink and the influence of the parameters on the T_j and thermal resistance (R_{ja}) values are reported. Fig.3 shows that the R_{ja} value for the Cu substrate is greater than that for the Al substrate. The respective Ave. R_{ja} values are 16.2℃/w and 15.1℃/w. The R_{ja} value for the Cu substrate decreases slightly as the ratio for resin/AlN increases, but the R_{ja} value for the Al substrate decreases significantly, when the ratio of resin/AlN increases. However, the R_{ja} value for the Cu substrate has the highest value at a resin/AlN ratio of 1:3 (17.08±1.78℃/w), but the R_{ja} value for the Al substrate has the lowest value for a resin/AlN ratio of 1:4 (14.24 ±1.54℃/w), probably because there is a gradual increase in the concentration of resin particles, which is detrimental to a reduction in the total thermal resistance of the LED packages. The result is inefficient heat dissipation. These results show that the Cu/Al samples have better thermal performance with an optimal design for the electro-spraying of AlN coatings.

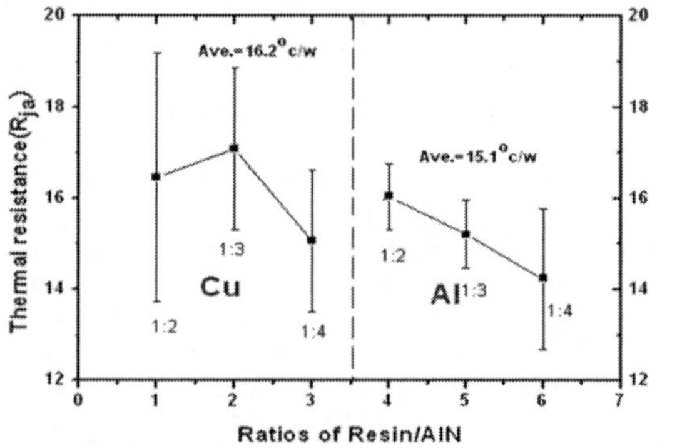

Fig.3. A comparison of the thermal resistance (R_{ja}) for a 1.2W LED on AlN-coated Cu and Al substrates.

4. Conclusion

In this study, the thermal performance of a 1.2W LED that uses an electro-sprayed AlN-coated Al substrate as a heat sink and the effect of parameters on the Tj and Rja values are reported. An L18 orthogonal array is used for the LED aluminium/copper plate and the effect of the deposition parameters on the coatings' thermal properties is determined. It is observed that the R_{ja} value for the Cu substrate is greater than that for the Al substrate. The respective Ave. R_{ja} values are 16.2℃/w and 15.1 ℃/w. These results demonstrate that the temperature of the LED packages is significantly reduced if there is an AlN-coated Al substrate and a current of 700mA. The Cu/Al substrate samples with AlN coatings also have better thermal performance when there is an optimal design for the electrostatic spraying process.

This work was financially supported by the National Natural Science Foundation of China (51575110), Natural Science Foundation of Fujian Province (2015J01628), Fujian Provincial Department of Education Research Projects (JA14211) and Fujian Province Industrial Technology Joint Innovation: Research and development of key technology of intelligent rotary die cutting machine (2014).

References

1. Shanmugan,S.; Mutharasu,D. A.; Haslan, H. International Journal of Electronics and Computer Science Engineering. 2, 1 (2014).
2. Kuo,P. K. G.; Auner,W.; Wu, Z. L. Thin Solid Films. 253 (1994).
3. Lee, J. W.; Cuomo, J. J.; Cho,Y. S.; Keusseyan, R. L. J. Amer. Ceram. Soci. 88, 7 (2005).
4. Huaiyu, Y.; Koh, S. H.; Zeijl, V. A.; Gielen,; W. J.; Guoqi, Z. A review of pas-sive thermal management of LED module. J. Semicon. **32** (2011).
5. Pan, T. S.; Zhang,Y.; Huang, J.; Zeng, B.; Hong, D. H.; Wang, S. L.; Zeng, H. Z.; Gao, M.; Huang, W.; Lin,Y. J. Appl. Phys. 112 (2012).

6. Heo, Y. J.; Kim ,H. T.; Kim, K. J.; Nahm, S.; Yoon, Y. J.; Kim, J.H. Applied Thermal Engineering. 50 (2013)
7. Lu, X.Y. Hua, T. C.; Wang, Y.P. Microelectronics Journal, 42 (2011)
8. Christensen, A.; Graham, S. Applied Thermal Engineering. 29 (2009).
9. Yung, K.C.; Liem, H.; Choy, H.S.; Lun, W.K. International Communications in Heat and Mass Transfer, 37 (2010).
10. Cheng, T.; Luo, X.B.; Huang, S.; Liu, S. International Journal of Thermal Sciences. 49(2010).
11. Kim,Y.P.; Kim, Y.S.; Ko, S.C.; Microelectronics Reliability. 56 (2016).
12. King, S. Y.; Tseng, J.; Zhao, .J. Applied Thermal Engineering. 52 (2013).
13. Yang, K.S; Chung, C.H.; Tu, C.W.; Wong, C.C.; Yang, T.Y.; Lee. M.T.; Applied Thermal Engineering.70 (2014)
14. Zheng, X.; Ren, Z.; Li, X.; Wang, Y. Applied Surface Science. 259 (2012)
15. Figueroa, U.; Salas, O.; Oseguera, J. Surface and Coatings Technology, 200 (2005)
16. Li, X.; Ren, Z.A.; Sun, D. Q. Materials Science and Engineering A,443 (2007)
17. Rada, N. M.; Triplett, G. E. Solid-State Electronics. 54,4 (2010)
18. Rada, N. Triplett, G. Graham, S. Kovaleski. S. Solid-State Electronics. 52,10 (2008)
19. Zheng, X.; Ren, Z.; Li, X.; Wang. Y. Applied Surface Science. 259,15(2012)
20. Zuo, Z.J.; North, M.T.; Wert, K.L. IEEE Transactions on Components and Packaging Technologies.24,2 (2001)
21. Shinde, S. High Thermal Conductivity Materials, Springer, New York (2006)
22. Shen, W.Q.; Zhu Y.W.; Wang G.L.
23. Appl. Sci. 5 (2015)

Application of damage locating vectors approach on monitoring shock isolation system

Quang Tuyen Le[1], Chiung Shiann Huang[1], and Wei Chih Su[2,*]

[1]Department of Civil Engineering, National Chiao Tung University, Hsinchu 30076, Taiwan
[2]National Center for High-performance Computing, National Applied Research Laboratories, Hsinchu 30076, Taiwan

Abstract. Structural damage detection based on the changes of dynamic properties is a major topic for structural health monitoring. In this paper, efforts are made to extend the flexibility-based damage localization methods, especially the damage locating vectors (DLVs) method, to the case of earthquake vibration, where the finite element model and mass matrices are not available. First, a new method using continuous Cauchy wavelet transform (CCWT) and ARX (autoregressive with exogenous input) model is applied to identify the modal parameters of a five-storey steel frame with seismic base isolation system LRB from its simulated acceleration responses under 10% and 100% of Chi-Chi Earthquake excitation (Taiwan, 1999). The DLVs, which determined from the change of flexibility matrix between two cases, are then used to monitor the shock isolation device in the structure through a weighted relative displacement index (WRDI). The proposed scheme is also proved to be superior to mode shape based methods (MAC, COMAC) in monitoring shock isolation system.

1 Introduction

In recent years, beside the popular applications in signal processing, wavelet transform also has demonstrated its excellent analysis capabilities in time-frequency domain that made many applications in system identification research. Narrow into the field of continuous wavelet transform, there are some studies really helpful for identifying modal parameters. Schoenwald[1] applied the continuous wavelet transform to the equation of motion of a single degree freedom system and identified the parameters in the equation of motion. Gouttebroze and Lardies[2] processed vibration responses of structures using Morlet mother wavelet and estimated natural frequencies and damping ratios of structures. Lardies & Gouttebroze[3] practiced their wavelet-based identification technique[2] to randomdec signatures of a TV tower. Huang & Su[4] proposed a procedure combining continuous wavelet transform with time series model ARX to estimate modal parameters of a structure from its free vibration responses and seismic responses using different mother wavelets (Shannon wavelet, Meyer wavelet, Morlet wavelet and Haar wavelet).

Approaches based on the change of flexibility formed an important group of methods for structural damage detection[5,6]. Bernal[7] recently proposed a damage locating vectors (DLV) method for damage localization, and this technique was experimentally studied by Gao & Spencer[8]. The generalized flexibility matrix change to detect the location and extent of structural damage has been introduced by Li et al.[9]. A method based on best achievable flexibility change with capability in detecting the location and extent of structural damage has also been presented by Yang & Sun[10]. The flexibility matrix and strain energy concepts of a structure have been used by Nobahari & Seyedpoor[11] in order to introduce a damage indicator for locating structural damage.

In this paper, a new method using continuous Cauchy wavelet transform (CCWT) and ARX (autoregressive with exogenous input) model is applied to identify the modal parameters of a five-storey steel frame with seismic base isolation system LRB from its simulated acceleration responses under 10% and 100% of Chi-Chi Earthquake excitation (Taiwan, 1999). The DLV's vectors, which determined from the change of flexibility matrix between two cases, are then used to monitor the shock isolation device in the structure through a weighted relative displacement index. The proposed scheme is also proved to be superior to mode shape based methods (MAC, COMAC) in monitoring shock isolation system.

2 Methodology

2.1 Identification of modal parameters using CCWT and ARX model

The dynamical responses of a linear structure are described by the equation of motion

$$[M]\{\ddot{x}\}+[C]\{\dot{x}\}+[K]\{x\}=\{f\} \qquad (1)$$

where $[M]$, $[C]$, and $[K]$ are the mass, damping and stiffness matrices of the structure system, respectively; $\{\ddot{x}\}$, $\{\dot{x}\}$, and $\{x\}$ are the acceleration, velocity and displacement response vectors of the system, and $\{f\}$ is

* Corresponding author: wichcv86@gmail.com

the input force vector. Equation 1 can be accurately discretized by the impulse invariant transformation as [12]:

$$\{z(t)\}=\sum_{i=1}^{I}[\Phi_i]\{z(t-i\Delta t)\}+\sum_{j=0}^{J}[\Theta_j]\{\overline{\mathbf{f}}(t-j\Delta t)\} \quad (2)$$

where Δt that represents the time increment.

$$\{z(t)\}=\begin{Bmatrix}\ddot{x}(t)\\ \dot{x}(t)\end{Bmatrix};\quad \{z(t-i\Delta t)\}=\begin{Bmatrix}\ddot{x}(t-i\Delta t)\\ \dot{x}(t-i\Delta t)\end{Bmatrix} \quad (3)$$

I, J denote the lags of output and input, respectively; $[\Phi_i]$, $[\Theta_j]$ are coefficient matrices related to [M], [C], [K] that need to be determined. Equation 2 is very similar to the time series ARX model with multiple variables.

The continuous Cauchy wavelet transform of a real signal of finite energy is

$$W_\psi f(b,a)=\frac{1}{\sqrt{a}}\int_{-\infty}^{+\infty}f(t)\psi_n^*\left(\frac{t-b}{a}\right)dt \quad (4)$$

where the standard Cauchy mother wavelet with order n is defined as:

$$\psi_n(t)=\left(\frac{i}{t+i}\right)^{n+1} \quad (5)$$

Treating $z(t-i\Delta t)$ and $\overline{\mathbf{f}}(t-j\Delta t)$ as vector functions, and applying the continuous Cauchy wavelet transform to Equation 2 yields

$$W_\psi z(a,\bar{b})=\sum_{i=1}^{I}[\Phi_i]W_\psi z(a,\bar{b}-i)+\sum_{j=0}^{J}[\Theta_j]W_\psi \bar{f}(a,\bar{b}-i) \quad (6)$$

where the translation parameter b is set to be $\bar{b}\Delta t$. Constructing Equation 6 for different $\bar{b}$ and rearranging the resulting equations gives

$$\left[Z^{(0)}\right]=\left[\hat{C}\right]\begin{bmatrix}[Z]\\ [F]\end{bmatrix} \quad (7)$$

where

$$[\hat{C}]=[[\Phi]_1\ \ [\Phi]_2 \ldots [\Phi]_I\ \ [\Theta]_0\ \ [\Theta]_1 \ldots [\Theta]_J] \quad (8a)$$

$$[Z]=\left[\left[Z^{(1)}\right]^T\ \left[Z^{(2)}\right]^T \ldots \left[Z^{(I)}\right]^T\right]^T \quad (8b)$$

$$[F]=\left[\left[F^{(1)}\right]^T\ \left[F^{(2)}\right]^T \ldots \left[F^{(J)}\right]^T\right]^T \quad (8c)$$

The coefficient matrix $[\hat{C}]$ is calculated by the least squares or the generalized inverse approaches. Hence:

$$\left[\hat{C}\right]=\left[Z^{(0)}\right]\begin{bmatrix}[Z]\\ [F]\end{bmatrix}^+ \quad (9)$$

where the superscript "+" denotes the generalized inverse operation. By adopting the concept behind the Ibrahim time domain system identification technique, Huang[12] proved that the modal parameters (natural frequencies, damping ratios and mode shapes) of the structure can be estimated from the eigenvalues and eigenvectors of [G], which is constructed from coefficient matrices as:

$$[G]=\begin{bmatrix}0 & I & 0 & 0 & \ldots & 0\\ 0 & 0 & I & 0 & \ldots & 0\\ \vdots & & & & & \\ [\Phi_I] & [\Phi_{I-1}] & & & & [\Phi_1]\end{bmatrix} \quad (10)$$

where I is an l*l unit matrix, and l is the dimension of measured $\{z(t)\}$.

2.2 DLVs approach

A simple damage localization method was proposed by Pandey and Biswas (1994), which consists in calculating the flexibility change matrix:

$$\Delta F=F_d-F_i \quad (11)$$

where F_i and F_d are the flexibility matrices for the intact and the damaged cases, respectively. The DLVs can be obtained from singular value decomposition (SVD) of ΔF. Each of DLVs is then applied to the intact model of structure. The stress σ_i in each structural element is calculated and a normalized cumulative stress (nsi) is obtained. If an element has zero normalized cumulative stress, then it is a possible candidate of damage.

In practice, the normalized cumulative stress induce by DLVs in the damaged elements may not be exactly zero due to noise and uncertainties. According to Hooke's law, under static condition:

$$\{f\}=K\{\Delta x\} \quad (12)$$

where K and $\{\Delta x\}$, respectively, are the stiffness matrix and relative displacement of the system subjected by force $\{f\}$. Equation 12 can be written as:

$$\{\Delta y\} = \hat{F}\{g\} \quad (13)$$

where

$$\{\Delta y\} = M^{1/2}\{\Delta x\} \quad (14a)$$

$$\hat{F} = M^{1/2}K^{-1}\left(M^{1/2}\right)^{T};\ \{g\} = \left(M^{-1/2}\right)^{T}\{f\} \quad (14b)$$

In the flexibility based-damage detection method, the DLVs are applied to structure as static force at the sensor locations to check whether stress area get zero value will be considered as damage position. So Equation 13 can be rewritten by:

$$\{\Delta y\} = \hat{F}_d\{L\} \quad (15)$$

where $\hat{F}_d$ is the flexibility matrix of damage structure, and $\{L\}$ is the damage locating vectors (DLVs).

Define the normalized relative displacement index of jth element as

$$nrdi_j = \left|\Delta y_{j+1} - \Delta y_j\right| / \left|\Delta y_{j+1} - \Delta y_j\right|_{\max} \quad (16)$$

one can compute a weighted relative displacement index (WRDI) as

$$WRDI = \sum_{j=1}^{q} \gamma_j nrdi_j \quad (17)$$

where γ_j =weights; q is the number of DLVs. Taking {PD} as the set of potentially damaged elements one has

$$PD = \left\{elements \middle| WRDI \le tol\right\} \quad (18)$$

In this paper, we select $\gamma = 1$, tol=0.1(WRDI)$_{\max}$ for a simple approach.

3 Application to shock isolation system

To confirm the accuracy and effectiveness of the proposed approach, SAP2000 was used to build up a finite element model of five-storey steel and simulate this frame (see Fig. 1) subjected to base excitations of the 10% 1999 Chi-Chi earthquake (Case 1) and 100% 1999 Chi-Chi earthquake (Case 2).

The five-storey steel frame under consideration was 3m long, 2m wide and 15m high. A seismic isolation system LRB is set up near base at the bottom of first storey to reduce the effect of earthquake excitation onto the structure. The simulation was conducted for 5% of the modal damping ratio. Assume plates were fixed on each floor, such that the total mass of steel frame was approximately 3.3516 tons. All types of data were saved at a sampling rate of 250Hz.

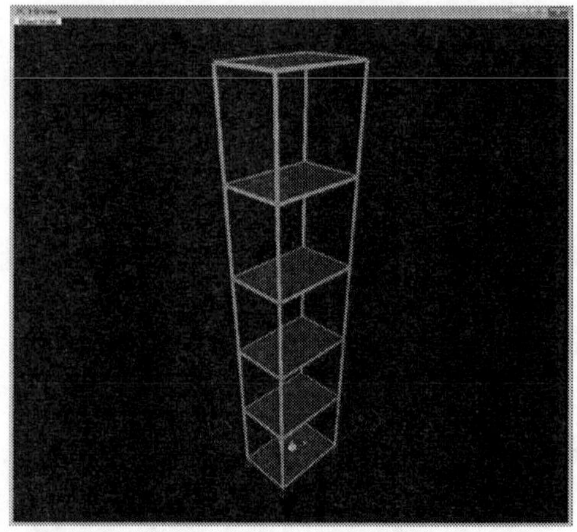

Fig. 1. Schematic diagram of a five-story shear frame.

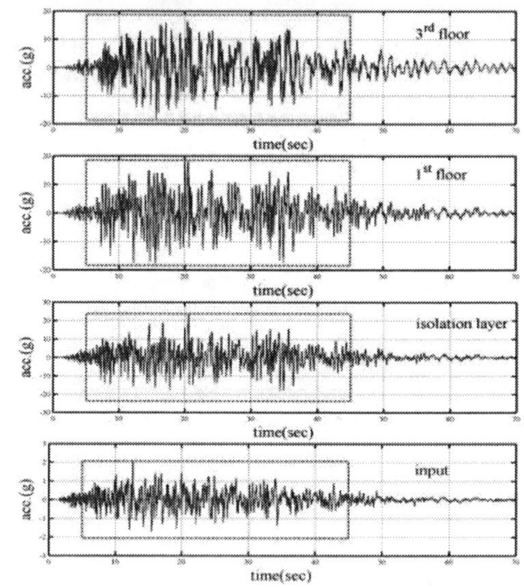

Fig. 2. Time history of base excitation and simulated acceleration responses of some floors of the frame in Case 1

By using full measurement responses of the frame in CCWT method, Table 1 and Table 2 respectively show the identified modal parameters of frame at two case obtained using Cauchy wavelets with different values of "n+1" and scale "a" corresponding to the judged natural frequencies.

Table 1. The identified modal parameters of frame under 10% 1999 Chi-Chi earthquake (Case 1)

Mode	1st	2nd	3rd	4th	5th	6th
f_n(Hz)	0.77	2.30	3.84	5.52	7.35	8.72
ξ(%)	0.05	0.05	0.05	0.05	0.05	0.05
Modal shapes	1.00	1.00	0.85	-0.66	-0.47	-0.21
	0.92	0.37	-0.33	0.83	1.00	0.56
	0.77	-0.45	-0.98	0.33	-0.82	-0.86
	0.56	-0.99	-0.23	-0.94	-0.01	1.00
	0.31	-0.94	0.86	-0.03	0.83	-0.98
	0.14	-0.56	1.00	1.00	-0.90	0.65

Table 2. The identified modal parameters of frame under 100% 1999 Chi-Chi earthquake (Case 2)

Mode	1st	2nd	3rd	4th	5th	6th
f_n (Hz)	0.75	2.23	3.66	5.39	7.19	8.62
ξ(%)	0.21	0.11	0.06	0.04	0.04	0.04
Modal shapes	1.00	-0.99	0.82	-0.73	-0.49	-0.23
	0.93	-0.41	-0.24	0.87	1.00	0.61
	0.79	0.40	-0.92	0.44	-0.73	-0.90
	0.61	0.96	-0.38	-0.98	-0.17	1.00
	0.41	1.00	0.68	-0.22	0.93	-0.91
	0.29	0.72	1.00	1.00	-0.70	0.49

The index of modal assurance criterion (MAC)[13] and coordinate modal assurance criterion (COMAC)[14] was computed to indicate the correlation between any two mode shapes of interest:

$$MAC(j) = \frac{\left|\sum_{i=1}^{n} \varphi_{ij,1}\varphi_{ij,2}\right|^2}{\left[\left(\sum_{i=1}^{n} \varphi_{ij,1}^2\right)\left(\sum_{i=1}^{n} \varphi_{ij,2}^2\right)\right]} \tag{19}$$

$$COMAC(i) = \frac{\left|\sum_{j=1}^{m} \varphi_{ij,1}\varphi_{ij,2}\right|^2}{\left[\left(\sum_{j=1}^{m} \varphi_{ij,1}^2\right)\left(\sum_{j=1}^{m} \varphi_{ij,2}^2\right)\right]} \tag{20}$$

to compare the identified mode shapes of undamaged frame in Case 1 with damaged frame in Case 2, $\varphi_{ij,1}$ and $\varphi_{ij,2}$ represent the deformation of the ith degree of freedom in the jth identified mode shapes, respectively.

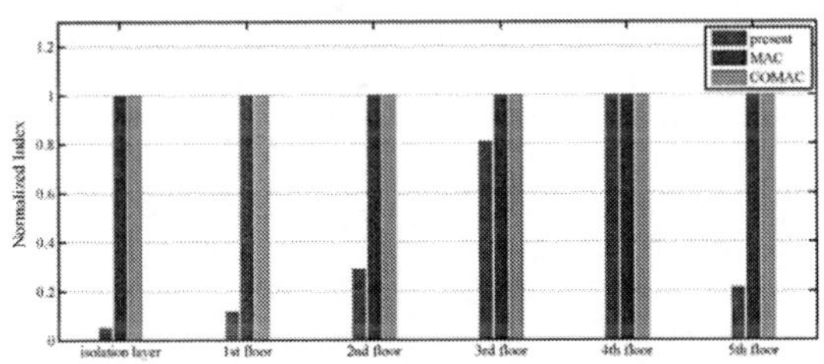

Figure 3.. Normalized index

When frame was subjected to 10% 1999 Chi-Chi earthquake excitation, no nonlinear behavior were observed, and it was treated as a reference structure. Based on result of WRDI index, we got the value of potential damage PD=1 which happened at the isolation layer of frame. This PD value correlated to the "present" diagram in Figure 3 and showed that there is damage in the seismic isolation layer only after frame subjected to 100% 1999 Chi-Chi earthquake excitation. The LRB system showed nonlinear behavior and the flexibility increase. This result also proved that the isolation system LRB had performed well its mission to absorb the power of the strong earthquake excitation onto the frame.

Figure 3 clearly demonstrates that the proposed approach is superior to MAC and COMAC methods in identifying the floor of the structure whose properties differ from those of the corresponding floor in the reference structure.

4 Concluding remark

The work develop a simple and efficient flexibility based approach, especially the damage locating vectors (DLVs) method, for monitoring the seismic isolation system on a structure. To demonstrate the feasibility of proposed method for actual application, the procedure was applied to a five-storey shear frame (with isolation system LRB set up at first floor) which subjected to 10% and 100% of 1999 Chi-Chi earthquake excitation. The proposed approach was validated by successfully identify the damage position through processing measured responses. The isolation system showed nonlinear behavior and absorbed the power of strong excitation of earthquake. Comparing the results obtained by the proposed approach with MAC and COMAC indexes revealed that the present approach is substantially superior to both in monitoring shock isolation system.

Acknowledgement

The authors would like to thank the Ministry of Science and Technology of Taiwan for financially supporting this research under Contract No. MOST 105-2622-M-492-001-CC2.

References

1. D. A. Schoenwald, In proceedings of the 32nd IEEE Conference on Decision and Control, (1993)
2. S. Gouttebroze & J. Lardies, Mechanics Research Communications **28**, 5 (2001)
3. J. Lardies & S. Gouttebroze, International Journal of Mechanical Sciences **44**, 11 (2002)
4. C.S. Huang & W. C. Su Mechanical System and Signal Processing **21**, 4 (2007)
5. S.W. Doebling & C.R. Farrar, Report LA-13070-MS. Los Alamos National Laboratory (1996)
6. A.K. Pandey & M. Biswas, Journal of Sound and Vibration, **169** (1994)
7. D. Bernal, Journal of Engineering Mechanics, **128** (2002)
8. Y. Gao & B.F. Spencer, Proceedings of the 1st International Workshop on Advanced Smart Materials and Smart Structures Technology (2004)
9. J. Li, B. Wu, O.C. Zeng & C.W. Lim, Journal of Sound and Vibration **329**, 22 (2010)
10. Q.W. Yang & B.X. Sun, Applied Mathematical Modelling, **35**, 10 (2011)
11. M. Nobabari & S.M. Seyedpoor, Structural Engineering and Mechanics, **46**, 2 (2013)
12. C.S. Huang, Journal of Sound and Vibration, **241**, 3 (2001)
13. R.L. Allemang & D.L. Brown, Proceeding of the 1st International Modal Analysis Conference, (1983)
14. N.A. Lieven & D.J. Ewins, In Proceedings of the 6th International Modal Analysis Conference, (1988)

MATEC Web of Conferences 71, 04007 (2016) DOI: 10.1051/matecconf/20167104007
ICCPE 2015

Fatigue Life Analysis of Cantilever Probe on Wafer Test

Te-Ching Hsiao[1,a], Hao-Yuan Chang[2] and Shyh-Chour Huang[1]

[1] *Department of Mechanical Engineering, National Kaohsiung University of Applied Sciences, Kaohsiung City 807, Taiwan (R.O.C.)*
[2] *Consultant, Sun Home Technology Co., Ltd*

Abstract. This research utilizes the finite element analysis software (ANSYS) to stimulate the different probe material quality (tungsten, SUS304 stainless steel, SUS316L stainless steel and SKD11 tool steel, respectively) during wafer tests. Under a room temperature of (25℃), the stress and fatigue life (cycles of probing test) of the cantilever probe were measured with an OverDriver (OD) of 20μm、40μm、50μm、60μm and 80μm, respectively. First, to obtain the magnitude of pinpoint shift of the probe under wafer test and the OverDriver is 50μm. And, calculate the fatigue life of the probe. Then, a probe model with the same characteristics as the experiment is created and the probe fatigue life analyzed with the ANSYS. After the reliability of the model is ascertained, the wafer tests of different probe materials are stimulated under different OverDriver circumstances to calculate its stress and fatigue life. The results indicate that the greatest stress measured during the wafer test of the tungsten, SUS304 stainless steel, SUS316L stainless steel and SKD11 tool steel cantilever probe are all smaller than the yield strength, and the fatigue life could reach over one hundred K cycles. When catalogued by the cantilever probe fatigue life during one hundred K cycles, the life span, in order, is tungsten < SUS316L stainless steel < SUS304 stainless steel < SKD11 tool steel.

1 Introduction

In recent years, semiconductor technology advances, from 1990s micron process evolve to today's nanometer manufacturing process, an integrated circuit also trends compact size and functional diversification. In order to improve IC yields, reduce costs, before wafer package, each die should be executed with wafer test, through touch between the probe card and grain specific pad, exposure to the amount of the sensing circuit electrical signals, and then filtered out in advance before packing poor electrical function of the wafer, in order to avoid defective products increased manufacturing costs.

When the wafer testing, pad dimension increased and contact with the probe then subjected to a fixed itinerary, and through a fixed voltage and current to measure its electrical signals. However, when the wafer and probe will cause pad surface scratches, without proper overdrive, may cause unable to measure electrical signals or scratches is too big, too deep, and the underlying cause pad micro-structural damage, arising from the subsequent packaging and reliability problems. In addition, use of the process, the probe was subjected to mush cyclic load, will cause the elastic fatigue, so that the probe tip offset, resulting in pads is overkill.

In 1998, Pandey and Higgins [1] propose a new P4 probe card technology (Photolithograhic Pattern Plated Process), this card is suitable for wafer testing high pin count and high speed, and when overdriving , stroke at 100μm, produced aluminum pads scratches size between 10μm ~ 25μm. In 2000 Maekawa [2] et. used tungsten probe as the test specimen, to compare the influence of different tip shapes (Radius Tip Probe and Flat Tip Probe) contact stability of resistance. The results show radius tip probe more stable by flat tip probe.

In 2001, Hotchkiss [3] and others to measure the number of different probe (1 ×, 2 ×, 3 ×, 4 ×), explore the influence of reliability the scratches on the drawn wire bonding structure fitted. The results show that when the proportion of scratches more than 20%, the combined strength and reliability of configuration of the wire will be significantly reduced. In 2003 Sauter [4], who experimented wafer testing with the thickness of 1.2 μm aluminum pad and found that the depth of the scratch of pad should not exceed 0.6μm, if for the 45μm diameter solder balls, solder ball indentation size can't exceed 25% of the solder ball bottom area 25~ 30%, otherwise it will affect the combined strength of wire bonding and reduce the reliability of the packaging process.

In 2005, Liu and Desbiens [5] etc. , who investigate the structure for aluminum pad, copper pad and tungsten probe, Tungsten-Rhenium probe and overdriving stroke analysis. Besides, use nonlinear finite element to simulate the probe tip contact with the pad surface, the article mentioned that the biggest scratch length calculation, the safe distance should remain must be 5 ~ 8μm. Further, the probe tip and the pad contact surface is smaller, the deformation of pads can be reduced.

In 1954, Coffin [6] and Manson [7] set up plastic strain-life theory, which was investigated the fatigue damage problem caused by thermal stress and the high

[a] Corresponding author: author@e-mail.org

stress load, advanced the empirical relationships between load times and the plastic strain in fatigue damage, according to the strain, to describe the fatigue was widely used, which is called Coffin-Manson relationship.

In 1998, Yasushi Miyano [8] et al proposed the following four basic assumptions used to predict the fatigue life, (1) at stationary strain rate, creep rate and fatigue damage, its destroyed structure is the same. (2) in the same time - the temperature, all breaking strength have the principle of superposition. (3) for a single load, it may use the laws of linear cumulative damage. (4) s tress ratio of linear fatigue have dependencies. With the above four assumptions, at a stationary strain rate, creep rate, frequency and stress ratio, with translational and rotational shift factor in the SN curve, to find an optimal master curve.

In this overdriving stroke is 50μm, the size of the scratch of aluminum pad, the fatigue life of the tungsten cantilever probe design will be calculated and calculate the maximum stress and fatigue life of the probe by finite element analysis software. After the simulation result and the experimental values are correspond, execute the analysis of maximum stress and fatigue life with different overdriving stroke (respectively 20μm, 40μm, 50μm, 60μm and 80μm), different probe material (respectively tungsten, low-carbon steel, SUS304 stainless steel, SUS316L stainless steel and SKD11 tool steel). The results can be used as a reference in probe design and through appropriate probe design to improve the life of the probe.

2 Theory of fatigue life

When structure repeated loads, will form a cyclic strain and stress, even if stress is less than the yield strength of the material, it will lead to the destruction of the structure, this is the fatigue damage. In the course of the fatigue study, in earlier studies focused on high cycle fatigue ($N > 10^4 \sim 10^5$) discussion, the corresponding fatigue curve that is the stress amplitude and cycle curve (S-N curve) to describe the fatigue properties of material. But as the study of low-cycle fatigue increased attention and needs to find in such a short life to explore the issue, S-N curve can't meet the requirements, and strain-controlled fatigue theory closer to the actual situation, and to avoid the stress under control unstable or out of control situations.

2.1 The relationship of Plastic stress - strain

Material by uniaxial cyclic loading, the plastic stress - strain relationship [14]:

$$\sigma = K'(\varepsilon_p)^{n'} \tag{1}$$

In this case, σ is the cyclically stress amplitude, ε_p is cyclically plastic strain amplitude, K' is cyclically strength coefficient, n' is the cyclically strain hardening exponent, generally n' is referred between 0.1 and 0.25[14]. The total strain is elastic strain and plastic strain sum and the equation of total strain can be obtained by the Hook's law.

$$\varepsilon = \varepsilon_e + \varepsilon_p = \frac{\sigma}{E} + (\frac{\sigma}{K'})^{\frac{1}{n'}} \tag{2}$$

Loading cycle repeated, $\Delta\sigma$ and $\Delta\varepsilon$ is the total amount of change of the stress and strain , so that (2) is rewritten as

$$\frac{\Delta\varepsilon}{2} = \frac{\Delta\sigma}{2E} + (\frac{\Delta\sigma}{2K'})^{\frac{1}{n'}} \tag{3}$$

Figure 1 is a material subjected to cyclic stress, the relationship between stress and strain of elastoplastic.

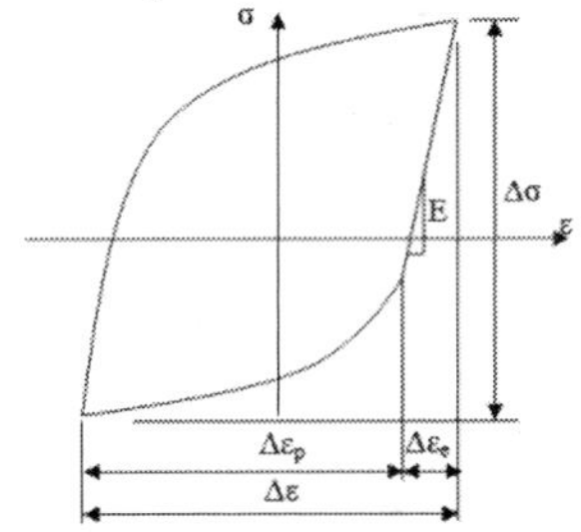

Figure 1. In cyclic stress, the relationship between stress and strain of elastoplastic.

2.2 Strain-Life Curve

In fatigue analysis, the strain-life approach is used of the material nature of fatigue and strain - life curve to evaluate material life; and in the elastic range, the fatigue life can be expressed as [14] :

$$\frac{\Delta\varepsilon_e}{2} = \frac{\Delta\sigma}{2E} = \frac{\sigma'_f}{E}(2N_f)^b \tag{4}$$

In this equation : $\Delta\varepsilon_e/2$ is true strain amplitude, $\Delta\sigma/2$ is true stress amplitude, σ'_f is fatigue strength coefficient, $2N_f$ is reversals to failure(1 rev = 0.5 cycle), b is fatigue strength exponent, in general, b is between -0.05 to -0.14 [9, 14].In the plastic strain range, the fatigue life can be expressed as [14] :

$$\frac{\Delta\varepsilon_p}{2} = \varepsilon'_f(2N_f)^c \tag{5}$$

In this equation: $\Delta\varepsilon_p/2$ is plastic strain amplitude, ε'_f is fatigue ductility coefficient, c is fatigue ductility exponent, in general, c is between -0.5 to -0.7[9, 14].

The total strain amplitude is the sum of elastic strain and plastic strain, it can be written as

$$\frac{\Delta\varepsilon}{2} = \frac{\sigma'_f}{E}(2N_f)^b + \varepsilon'_f(2N_f)^c \tag{6}$$

Coffin-Manson formula is estimating formula, equation (6) of the strain - fatigue life curve shown in Figure 2.

DOI: 10.1051/matecconf/20167104007

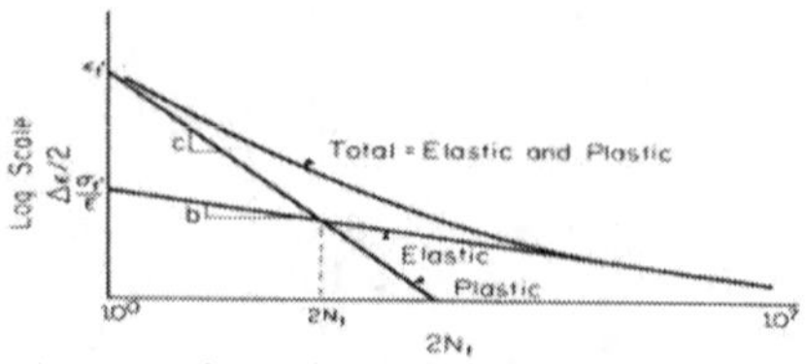

Figure 2. The strain - fatigue life curve

In recent years, experimental evidence accumulated by a number of researchers showed the correlation between cycle life and total strain amplitude ratio is better than the one between the plastic strain amplitude, but the fatigue life relationships in actual project need to go through another amendment. In this paper, use Morrow correction theory, the simulate fatigue life can approach real fatigue life, the corrected strain fatigue life is showed following equation：

$$\frac{\Delta\varepsilon}{2}=\frac{\sigma'_f-\sigma_m}{E}(2N_f)^b+\varepsilon'_f(2N_f)^c \quad (7)$$

In this equation: σ_m is average stress amplitude.

3 The parameters of test specimen

In wafer test, cantilevered probe direct contact with aluminum pads, and aluminum pads will be left permanent scratches, but also because of repeated use of the probe, and fatigue damage.

3.1 Probe geometry

In this paper, the geometry and dimensions of the cantilever probe are respectively shown in Figure 3 and Table 1. The more important factors which affect the structural strength of the probe are beam length, bending angle, shooting angle and probe diameter and so on.

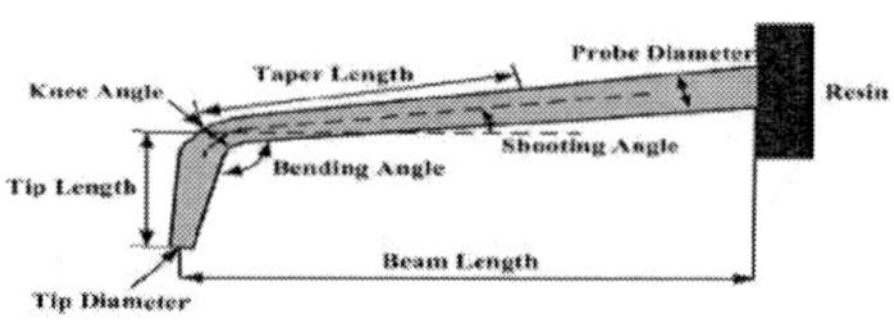

Figure 3. Cantilever probe schematic

Table 1. Cantilever probe geometry

Probe Diameter(μm)	150
Beam Length(μm)	2,500
Knee Diameter(μm)	45
Tip Diameter(μm)	23
Tip Length(μm)	279.4
Bending Angle(°)	103
Shooting Angle(°)	8
Taper Length(um)	1,750

3.2 pad geometry

The geometric shapes and sizes of aluminum pad and the silicon substrate were used as shown in Figure 4. When the wafer testing, fixed probe handle (at maximum diameter), then rises up the loading silicon substrate. After the pad in contact with the probe, and then rises to the set of overdrive, and silicon substrate returned to the original position. After an overdrive was completed, probing actions will leave permanent scratches in the welding pad.

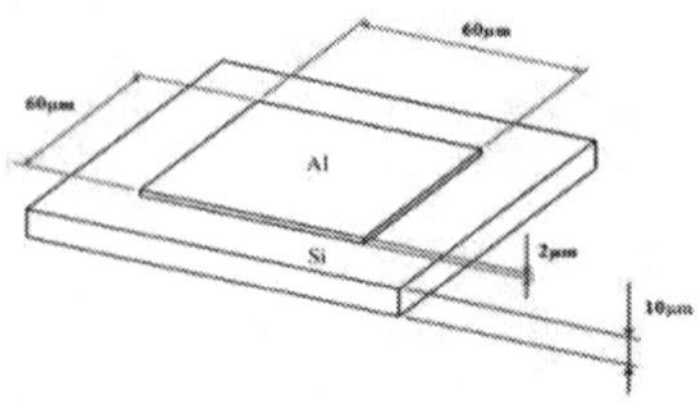

Figure 4. Geometry and dimensions of the pad

3.3 Related properties parameters of material

Nowdays, the most commonly probe material was used for wafer testing is tungsten, this article the starting point is angle of fatigue life, attempting to analyze other possible materials of metal of probe, where the choice of 304 stainless steel, 316L stainless steel and SKD11 tool steel, etc. The mechanical properties of each material shown in Table 2 [10], the analysis of fatigue about the material parameters required for as shown in Table 3 [10-13].

Table 2. The mechanical properties of material [10]

Materials	Tungsten	SUS304	SUS316L	SKD11
Density(kg/m^3)	19,300	8,000	7,800	8,400
ElasticModulus (GPa)	290	193	193	208
Poisson's Ratio	0.28	0.29	0.29	0.30
Yield Strength (MPa)	1770	405	365	585
Ultimate Strength (MPa)	2600	505	822	668

Table 3. The fatigue parameters of material [10-13]

Material	Tungsten	SUS304	SUS316L	SKD11
Fatigue Strength Coefficient(MPa)	1,020	1,100	997	890
Fatigue Strength Exponent	-0.09	-0.12	-0.11	-0.09
Fatigue Ductility Coefficient	0.5	0.34	0.47	0.51
Fatigue Ductility Exponent	-0.54	-0.49	-0.51	-0.59
Cyclic Strength Coefficient (MPa)	1,300	1,424	1,232	986
Cyclic Strain Harding Exponent	0.16	0.24	0.23	0.15

DOI: 10.1051/matecconf/20167104007

4 The probe tip offset analysis

When the wafer testing, the offset tip result from the probe material fatigue, Figure 5 is a schematic view of a probe offsets. Experiments that use tungsten cantilever probe in overdrive stroke under 50μm, the relationship between tip offset and cycles times as shown in Table 4; and Figure 6 is a relationship diagram about tip offset and cycles times, by the relationship between tip offset and cycles times, use the least squares method, we can calculate the equation related between the tip offset and Probing shown as.

$$Y=0.0167X+0.068 \tag{8}$$

In this equation, Y is the offset of tip(μm), X is probing times(10^3) 。

According to the relevant literature, in order to maintain the reliability when the chip package, when the wafer testing scratches length must be less than 20% of the length of the pad [3], and scratches must distance of 5 ~ 8μm border security. In response to the practice of tip offsets of probe, the tip of the probing the initial placement position control in one-third of the length, so calculate the tip offset arising due to fatigue and allow a maximum of 10μm. Then according to (5) calculation, the fatigue life of probe design is 5.95×10^5 times. Figure 7 is a tungsten probe and aluminum pad in a stroke of 50μm, the photo of the pad scratches.

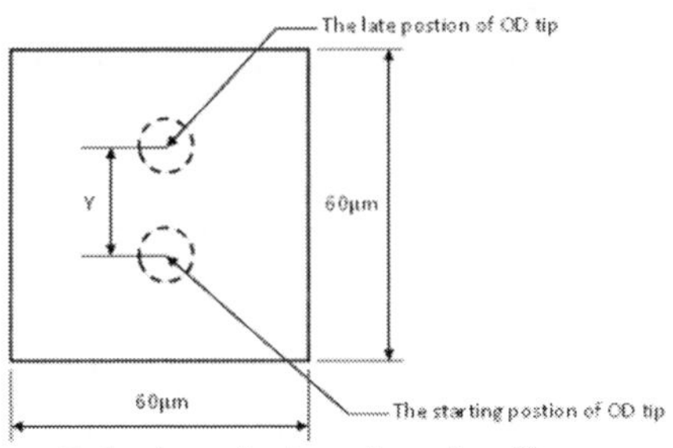

Figure 5. A schematic view of a probe offsets.

Table 4. The relationship between tip offset in Experimention and cycles times.

Cycles times	Tip offset(μm)
50×10^3	1.12
100×10^3	1.63
150×10^3	2.42
200×10^3	3.52

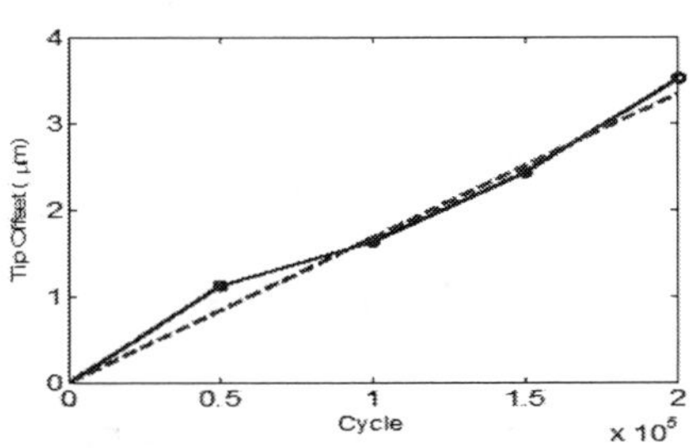

Figure 6. When OD=50μm, tip offset - cycles times curve

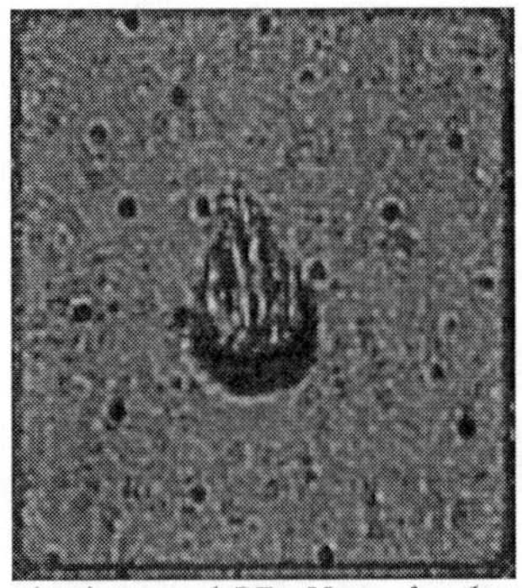

Figure 7. When aluminum pad OD=50μm, the photo of the pad scratches

5 The finite element model construction

5.1 CAD model construction

Wafer testing process is supported by the silicon substrate under the pad moves up, after the pads contact on the probe tip, to finish up the set of overdrive, then separated down, returned to the starting position, complete a cycle; pad of length, width and height respectively 60μm, 60μm and 2μm, shown in Figure 3.

5.2 Model grid and boundary conditions

Import the CAD model which was completed to the ANSYS analysis module, and go mesh, the model is cut to fit the size, the contact surface between the probe and the pad was more refined to 1μm or less, in order to achieve a more accurate analysis. Set the boundary conditions on the right side of fixed end in Fig. 8, the left side of the silicon substrate and aluminum pads move upward displacement overdrive required. The completion of meshing and boundary conditions set are shown in Figure 8.

Figure 8. The meshing of OD model and boundary conditions

5.3 Calculation process of fatigue analysis

After completing the model mesh, analyzed the using mechanical properties of the material in Table 2, input the finite element to the ANSYS analysis module. In calculating OD process, the mechanical behavior of the probe; after obtaining analysis information of probe on the mechanical behavior, in Table 3, the material parameters of fatigue and strain - fatigue (equation (1) to (7)) to calculate the fatigue life of the probe throughout the body, and retrieve the minimum value of the fatigue life as the probe design value of fatigue life. Figure 11 is the stress amplitude - life (S-N) curve of a tungsten probe that calculated with the information in Table 3.

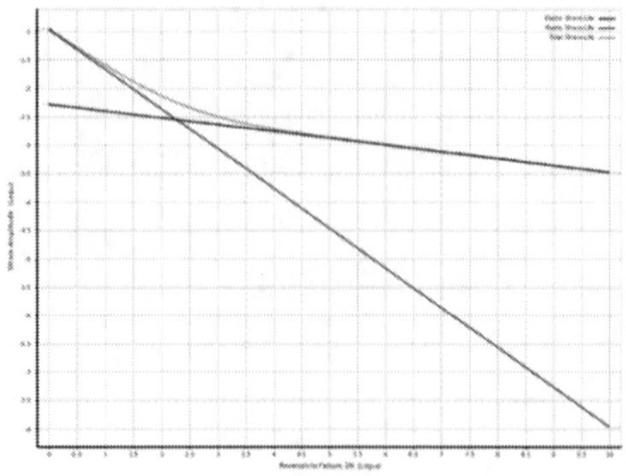

Figure 9. The stress amplitude - life (S-N) curve of a tungsten probe

6 The analysis result of finite element model

After model construction completed, run the simulation analysis of finite element in OD stroke is 50μm and calculate the value of fatigue life, and take the results to compare the experimental data, in determining the simulation results and experimental values are similar before further analysis.

6.1 Stress analysis of OD

After the ANSYS mesh model was establishd, we can performed OD analysis in OD stroke of 50μm, then we found that the maximum stress of the probe is 574.88MPa, the results shown in Figure 10, compare the material properties of tungsten in Table 2 and know the maximum stress of the probe is less than the yield stress of tungsten, it was found that the tip of the offset was resulted from the material fatigue, the obtained analysis from the scratch that on aluminum pad shown in Figure 11.

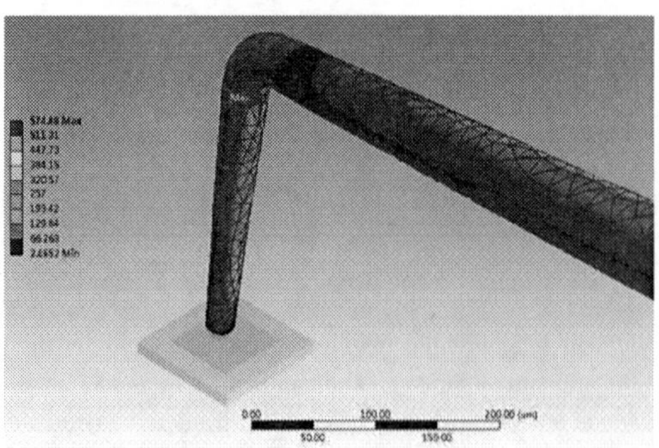

Figure 10. The diagram of the maximum stress of the tungsten probe in OD=50μm.

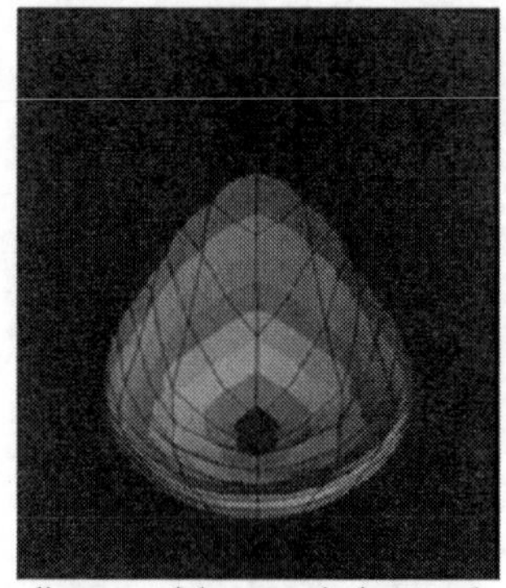

Figure 11. The diagram of the scratch that on aluminum pad in OD=50μm.

6.2 Fatigue analysis comparison

Processed the ANSYS analysis of fatigue with the material factor of tungsten metal in Table, and obtain the cycle life is 6.12×10^5 times, is similar to the experimental value is 5.95×10^5 times, which shows the reliability analysis model has good.

6.3 Simulation Results

After determining the analytical model, the idea that is based on expanse the type of materials to use, we try to analyze a material, that nobodys use on OD in past, for example SUS304 stainless steel, SUS316L stainless steel and SKD11 tool steel, in the same size for probe and pad and aluminum pad, processed the mechanical behavior of the probe and fatigue analysis.

Table 5 is the variety of materials of probe, in the maximum stress measured at each stroke of OD, the mechanical properties of the material in Table 5 and Table 2 is evident, in the OD stroke less than 60μm, in addition to the 316L stainless steel, other materials of probe that maximum stress is less than the yield stress. Table 6 and Figure 12 is the variety of materials of probe, according to the maximum stress in Table 5, calculated fatigue life of each stroke of OD.

Table5. The maximum stress of each stroke of OD with a variety of materials of probe (unit:MPa)

Material / OD	Tungsten	SUS304	SUS316L	SKD11
20μm	219.95	120.32	135.88	147.21
40μm	435.80	243.52	261.60	284.64
50μm	574.88	332.72	332.40	358.02
60μm	705.12	401.25	410.70	442.50
80μm	915.25	502.78	537.62	577.42

Table 6. The analysis result of fatigue of a variety of materials of probe (cycles)

OD(μm)	W	SUS304	SUS316L	SKD11
20	7.04×10^9	7.68×10^9	7.01×10^9	3.29×10^{10}
40	3.68×10^6	1.88×10^7	1.81×10^7	5.47×10^7
50	6.12×10^5	4.05×10^6	3.75×10^6	6.08×10^6
60	2.12×10^5	1.17×10^6	1.27×10^6	1.54×10^6
80	7.49×10^4	2.41×10^5	2.50×10^5	6.23×10^5

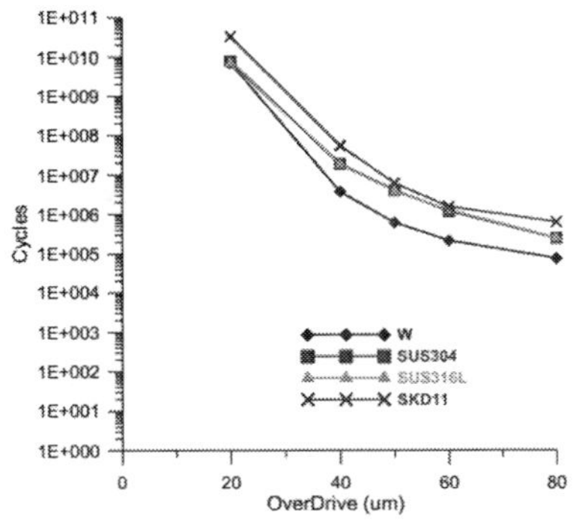

Figure 12. The analysis result of fatigue of a variety of materials of probe

7 Conclusion

In this study, we focus on the wafer testing process, analyze the stress is caused in different materials of probe and different stroke of OD and the fatigue life of the probe that under loading cycles. Results from the analysis are summarized several conclusions:

1. The overdriving experiments use tungsten probe that the allowed offset of tip is 10μm, the results show that when the OD stroke is 50μm, the number of OD will up to 595×10^3 times. In the simulation in the same set situation, the result will become 612×10^3 times, the results differ between expeniment and analysis is only 2.9%.

2. Analysis in different materials showed that the OD stroke is 80μm, the maximum stress of material the probe that are tungsten and SKD11 tool steel are less than yield stress, these two kinds of material of the probe did not produce plastic deformation, but when the material are SUS304 and SUS316L stainless steel, if the maximum stress will exceed yield stress, a part of the probe will produce plastic deformation.

3. As shown in Table 6 and Figure 11, the fatigue life of probe for each material from small to large are: Tungsten <SUS316L stainless steel <SUS304 stainless steel <SKD11 tool steel; wherein, SUS316L and SUS304 stainless steel fatigue life considerably. If you consider the fatigue life of the probe, you should use probes SKD11 tool steel for probes that can get the best effect.

References

1. R. Pandey, D. Higgins, IEEE International Test. Conference, 836-842, (1998)
2. S. Maekawa, M. Takemoto and Y. Kashiba, IEEE ECTC, 1152 - 1156, (2000)
3. G. Hotchkiss, G. Ryan, W. Subido, J. Borz, S. Mitchell, R. Rincon, R. Rolda and L. Guimbaolibot, IEEE ECTC, 1175-1180, (2001)
4. W. Sauter, T. Aoki, T. Hasida, H. Miyai, K.Petrarca, F. Beaulieu, S. Allard, J. Power, and M.Agbesi, IEEE ECTC,(2003)
5. Y. Liu, D. Desbiens, S. Irving, T. Luk, S. Edborg, D. Hahn and S. Park, IEEE ECTC Proceedings, Vol. **1**, 861-866, (2005)
6. L.F. Coffin, Transactions of ASME,**76**,931–950, (1954)
7. S.S. Manson, National Advisory Commission on Aeronautics, Report No. 1170, Cleveland: (1954)
8. Y. Miyano, M. Nakada, H. Kudoh, R. Muki, CFRP. J.Compos.Mater. **34**(7), 538-550,(1999)
9. W.W. Lee, L.T. Nguyen, G.S. Selvaduray, Microelectronics Reliability 40, 231-244, (2000)
10. Online Materials Information Resource Mat-Web in http://www.matweb.com/index.aspx
11. A. Bäumel. Jr. and T. Seeger, Elsever Science Publishers. (1990)
12. C. Boller, and T. Seeger, **1**(5), Elsevier Science Publishers, Amsterdam.(1987)
13. ANSYS WORKBENCH database.
14. Julie A Bannantine, Jess J Comer, James L Handrock, B. Englewood Cliffs, N.J. Prentice Hall (1990)

Development of Registration methodology to 3-D Point Clouds in Robot Scanning

Liang-Chia Chen[1,a], Dinh-Cuong Hoang[1], Toan-Thang Chu[2] , Hsien-I Lin[3]

[1]*Department of Mechanical Engineering, National Taiwan University, Taipei 10617, Taiwan*
[2]*School of Mechanical Engineering, Hanoi University of Science and Technology, Hanoi, Vietnam*
[3]*Graduate Institute of Automation Technology, National Taipei University of Technology, Taipei 10608, Taiwan*

Abstract. The problem of multi-view 3-D point clouds registration is investigated and effectively resolved by the developed methodology. A registration method is proposed to register two series of scans into an object model by using the proposed oriented-bounding-box (OBB) regional area-based descriptor. Robot 3-D scanning is often employed to generate set of point clouds of physical objects. The automated operation has to successively digitize view-dependent area-scanned point clouds from complex shaped objects by multi-view point clouds registration. To achieve this, the OBB regional area-based descriptor is employed to determine an initial transformation matrix and is then refined employing iterative closest point (ICP) algorithm. The developed method can be used to resolve the commonly encountered difficulty in accurately merging two neighbouring area-scanned images when no coordinate reference exists. The developed method has been verified through some experimental tests for its registration accuracy. Experimental results have preliminarily demonstrated the feasibility of the developed method.

Keywords: Robot; 3-D scanning; image registration; point clouds; reverse engineering; surface digitization

1 Introduction

A 3D model is a digital representation of a physical object, which can be used for many purposes. In manufacturing, producers apply 3D model to make new products, perform dimensional and comparative analysis of parts, or inspection robotics. For medical industry, 3D models play an important role as inputs in organ fabrication, creation of customized prosthetics, implants, and anatomical models. They are also used in architecture industry to demonstrate proposed buildings, in the movie industry as characters and objects for animated and real-life motion pictures.

Recently, automatic CAD model generating technology, known as many different names including 3-D scanning, 3D digitizers, laser scanning, industrial CT, LIDAR, and others, has an emerging interest. The common factor of all these devices is that they capture the geometry of physical objects and describe them through thousands or millions measured points called point clouds. This kind of technique is the most efficient way to generate a CAD model from a physical object that has any kind of complex or freeform shape. 3D scanning technology can be applied at any point in a typical manufacturing cycle, saving time, money and material. Because of the dominant advantages, 3D scanning systems have become more and more popular with numerous applications such as 3-D printing, reverse engineering, quality inspection, prosthetics, etc. Specially, reverse engineering (RE) is an efficient approach to significantly model the product and reduces the product development cycle [1-4].

According to the sensing principle being employed, the current solutions can be generally classified into two main categories, namely hand-guided and automated scanning techniques. Hand-guided scanning allows for acquiring arbitrary shapes. However, the effectiveness of this scanning method highly depends on the skills of the user and the scanning process is generally time consuming. To enhance the efficiency, the 6-axis robot arm integrated with 3-D imaging scanners has recently emerged as a technical developing trend for 3-D surface scanning for objects having arbitrary or complex geometry [5-7]. Both Callieri [5] and Larsson [6] presented a system for automated 3-D modeling consisting of a 3-D laser scanner, an industrial six axis robot and a turntable. The advantage of this system is that it automatically achieves the shape of the object from initial scanning and thus, if possible will scan the object using an orientation of the scanner which gives the most accurate result. Meanwhile, an autonomous 3-D modeling system, consisting of an industrial robot

[a] Corresponding author: lchen@ntu.edu.tw

and a laser striper, for efficient surface reconstruction of unknown objects was proposed by Simon Kriegel [7]. The system iteratively searches for possible scan paths in the local surface model and selects a next-best-scan (NBS) in its volumetric model.

The working principle of the above robotic 3-D scanning system is that the optical probe is attached on the robot end effector, in which the probe is controlled and scanned continuously by the robot in the working area and simultaneously posed with different gestures in the process of scanning. Shown in Figure 1, initially, the robot moves to a pre-defined position for the first probe scan. By doing so, the system uses the acquired image to determine the initial object orientation. According to the processed information, the position and gesture of robot for next view are further estimated and employed to generate the robot path for next movement. This process is repeated until the whole object is scanned. In the scanning process, the system controls the robot position and gesture to continuously scan and register point clouds of consecutive scenes of the scanned object in real time.

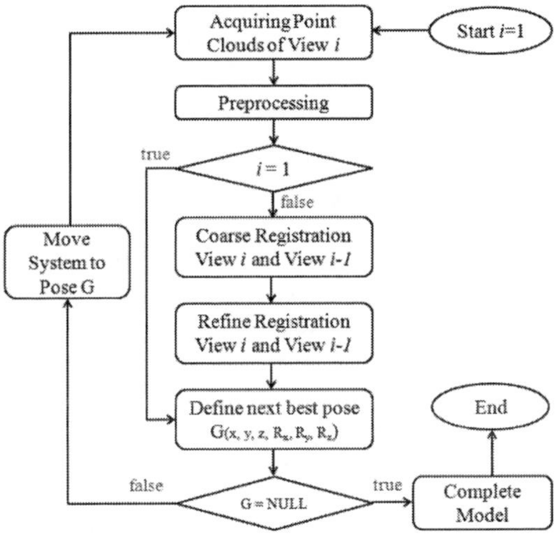

Figure 1. Overview of automated 3-D scanning process using a 6-axis robot.

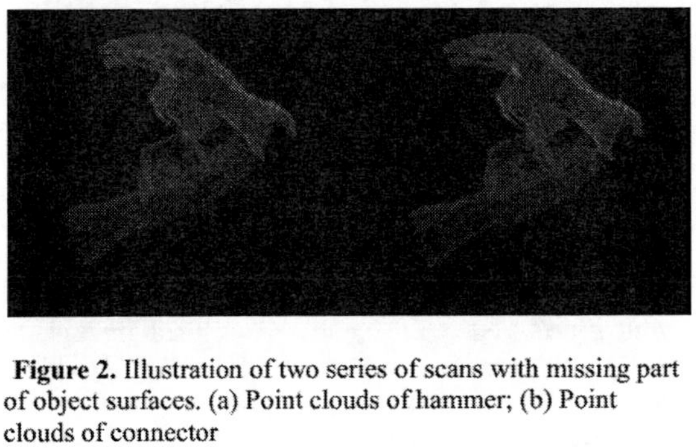

Figure 2. Illustration of two series of scans with missing part of object surfaces. (a) Point clouds of hammer; (b) Point clouds of connector

However, due to the lack of a global registration between successively scanned data, the existing scanning techniques still fail to robustly satisfy accurate object digitization with 100% surface coverage. Figure 2 show two cases with missing some part of object surfaces after performing automated object digitization using robot scanning. This issue is mainly caused by the fact that 3-D sensor cannot measure point data from the scanned object surface that is in contact with its supporting ground.

Thus, modeling scanning completion with 100% surface coverage still remains one of the most challenging problems in the current 3-D object digitization. Accurately registering point clouds of two series of scanned point clouds into the reconstructed object model without an established coordinate reference has been proven to be a difficult process [8-12]. To overcome the above difficulty, an automated robot scanning system integrated with a NTU-developed optical 3-D measuring probe is developed on a global registration method to accurately register two series of scans into an object model underlying 3-D imaging digitization by using the proposed oriented-bounding-box (OBB) regional area-based descriptor. The hardware and basic system configuration as well as data processing algorithms such as preprocessing, overlapping detection and robot path planning have been described in [12].

2 Global registration method based on the OBB regional area-based descriptor

2.1 Principle concept and flow chart diagram of the global registration method

To find the transformation between two series scans that are acquired from multi single scans, the OBB regional area-based descriptors of source point clouds are matched with the descriptor of the target point clouds. Each the proposed descriptor includes two components. The first component contains the information about the OBB of the point clouds. Each OBB is represented by a corner $C(x_C, y_C, z_C)$ and three vectors $CC_1(x_{max}, y_{max}, z_{max})$, $CC_2(x_{mid}, y_{mid}, z_{mid})$, $CC_3(x_{min}, y_{min}, z_{min})$ corresponding with the maximum, middle, and minimum dimensions of the OBB, respectively. The second component represents the distribution of the surface area of the object in the OBB. The similarity between the regional area based descriptor that represents the source point clouds and the regional area based descriptor that represents the target one is determined by using the normalized cross correlation.

To make the proposed method clear in its operation procedure, the following Algorithm 1 is used to describe the proposed method in its main 6 steps.

[a] Corresponding author: lchen@ntu.edu.tw

Algorithm 1

Input: Two series scans ***P*** and ***Q***
Output: Transformation matrix

1. Define and perform translation ***Q*** to ***P*** based on center of gravity of point clouds and estimation and enclosing the missing regions.
2. Determinate individual OBBs of two series scans.
3. Define and perform rough rotation Q based on applying initial three-axis alignment between ***P*** and ***Q***, which is an orientation alignment between two OBBs along the three principle axes.
4. Generate descriptors of **P** and **Q.**
   a. Calculate the OBB regional area-based descriptor of P.
   b. Rotate Q around the three axes of the coordinate system defined by the origin is center of gravity of Q and three unit vectors considered as the three directions of the OBB of point clouds P with every increment Δθ on each axis and generating the OBB regional area-based descriptors of Q.
5. Matching descriptors.
6. Transformation Estimation and Refinement.

2.2 Definition of translation matrix

Assuming that two series of scans to be registered are represented as two sets of unorganized points $\boldsymbol{P} = \{\boldsymbol{p}_1, \boldsymbol{p}_2, \ldots, \boldsymbol{p}_k\}$ and $\boldsymbol{Q} = \{\boldsymbol{q}_1, \boldsymbol{q}_2, \ldots, \boldsymbol{q}_k\}$, the initial transformation, which represents the coarse alignment between two series, can be estimated as follows.

If ***P*** and ***Q*** represent the same free-form shape, the least squares method can be used to calculate the initial rotation and translation by minimizing the sum of squares of alignment errors:

$$E = \sum_{i=1}^{k} \|\mathbf{e}_i\|^2 = \sum_{i=1}^{k} \|p_i - \mathbf{R}\cdot q_i - \mathbf{t}\|^2 \quad (1)$$

Where ***R*** is the rotation matrix, ***t*** is the translation vector, $\boldsymbol{p}_i$ and $\boldsymbol{q}_i$ are the corresponding points pair between the ***P*** and ***Q*** point clouds.

Let $\bar{p}$ be the centroid of the set of corresponding points in ***P*** $\{\boldsymbol{p}_1, \boldsymbol{p}_2, \ldots, \boldsymbol{p}_k\}$ and $\bar{q}$ be the centroid of the set of corresponding points in ***Q*** $\{\boldsymbol{q}_1, \boldsymbol{q}_2, \ldots, \boldsymbol{q}_k\}$. $\bar{p}$ and $\bar{q}$ can be defined as:

$$\bar{p} = \frac{1}{k}\sum_{i=1}^{k} p_i \quad (2)$$

$$\bar{q} = \frac{1}{k}\sum_{i=1}^{k} q_i \quad (3)$$

The new coordinates of points are denoted by:

$$p'_i = p_i - \bar{p} \quad (4)$$

$$q'_i = q_i - \bar{q} \quad (5)$$

Note that:

$$\sum_{i=1}^{k} p'_i = \sum_{i=1}^{k} (p_i - \bar{p}) = \sum_{i=1}^{k} p_i - k\bar{p} = 0 \quad (6)$$

$$\sum_{i=1}^{k} q'_i = \sum_{i=1}^{k} (q_i - \bar{q}) = \sum_{i=1}^{k} q_i - k\bar{q} = 0 \quad (7)$$

The registration error term is then expressed and rewritten as:

$$\mathbf{e}_i = p_i - \mathbf{R}\cdot q_i - \mathbf{t} = p'_i - \mathbf{R}\cdot q'_i - \mathbf{t}' \quad (8)$$

where

The sum $\mathbf{t}' = \mathbf{t} - \bar{p} + \mathbf{R}\cdot\bar{q}$ ors becomes:

$$\begin{aligned} E &= \sum_{i=1}^{k} \|p'_i - \mathbf{R}\cdot q'_i - \mathbf{t}'\|^2 \\ &= \sum_{i=1}^{k} \|p'_i - \mathbf{R}\cdot q'_i\|^2 - 2\mathbf{t}'\cdot\sum_{i=1}^{k}[p'_i - \mathbf{R}\cdot q'_i] + k\|\mathbf{t}'\|^2 \\ &= \sum_{i=1}^{k} \|p'_i - \mathbf{R}\cdot q'_i\|^2 + k\|\mathbf{t}'\|^2 \end{aligned} \quad (9)$$

The first term in Equation (9) does not depend on $\boldsymbol{t}'$. Therefore, the total error is minimized if $\boldsymbol{t}' = \boldsymbol{0}$, and then:

$$\mathbf{t} = \bar{p} - \mathbf{R}\cdot\bar{q} \quad (10)$$

Therefore, the initial translation matrix can be computed based on the difference between the centroid of ***P*** and the centroid of ***Q***.

In addition, the weight loss at missing parts of two scan series may result in a significant bias between the translation matrix being estimated (by the difference between the centroid of ***P*** and the centroid of ***Q***) and the real translation one. The shifting of two gravity centers (the red and green points shown in Fig. 3) being created using a virtual camera from a model having the centroid (white point) in Figure 3 (a) illustrates the influence of weight loss from the missing points to the centroid point position. Assume that the two point sets are separated as shown in Figure 3 (b). The accurate translation matrix to merge set ***Q*** (green point clouds) with set ***P*** (red point clouds) is defined through the difference between the current centroid of ***P*** and the original centroid of green point clouds (green points). Therefore, it is necessary to estimate the original centroid of ***P*** on the object for translation.

[a] Corresponding author: lchen@ntu.edu.tw

DOI: 10.1051/matecconf/20167104008

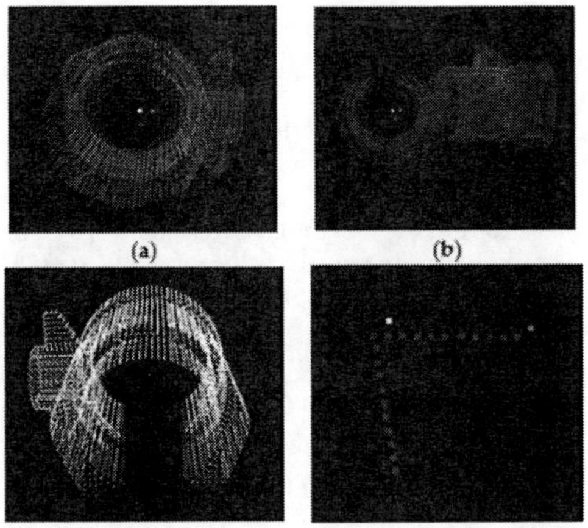

Figure 3. Improvement of translation: (**a**) The influence of weight loss from the missing points to the centroid point position; (**b**) Two point sets are separated; (**c**) Detection of the boundary points located nearby missing part; (d) The process of estimating the original centroid from two point sets being separated.

The boundary points located nearby missing part can be detected through the geometric relationship with the support plane and using a set of connected edges that satisfy two requirements, in which all edges have a similar orientation and each edge lacks an assigned triangle, either on the left or on the right side [7] (as shown in Figure 3 (c)). Then, the RANSAC algorithm [16] is used to fit a mathematical model to the boundary points in each subdivided box of the OBB (The total number of subdivided boxes in the OBB is m = k_1 x *1* x *1*). Figure 3(d) indicates the process of estimating the original centroid from two point sets being separated. The center of gravity of the scanned object O_W can be estimated as follows:

$$\frac{\overrightarrow{O_R O_W}}{\overrightarrow{O_W O_{RM}}} = \frac{\sum_{v=0}^{m} F_v}{\sum_{v=0}^{n} f_v} \tag{11}$$

where O_R is the centroid of red point clouds ***P***; $\boldsymbol{O_{RM}}$ is the centroid of missing points of ***P***; F_v is the surface area of object in each subdivided box from the missing points of ***P*** being computed based on the fitted mathematical model; and, f_v is the surface area of object in each subdivided box from red point clouds ***P***.

After that, applying the rough rotation matrix to the point set of ***Q***, the original centroid of green point clouds O_G can be estimated as follows:

$$\frac{\overrightarrow{O_G O_W}}{\overrightarrow{O_W O_{GM}}} = \frac{\sum_{v=0}^{m} F_v}{\sum_{v=0}^{n} f_v} \tag{12}$$

where O_W is the centroid of the scanned object computed in Equation (12); O_{GM} is the centroid of missing points of ***Q***, F_v is the surface area of object in each subdivided box from the missing points of ***Q*** being computed based on the fitted mathematical model; and, f_v is the surface area of the object in each subdivided box from green point clouds ***Q***.

As a result, the initial translation matrix can be computed based on the difference between the current centroid of ***Q*** and the original centroid estimated O_G.

2.3 Alignment of two series scans based ontheir OBBs' orientation

The oriented bounding box is a rectangular bounding box that covers entire object point clouds. Each OBB is represented by a corner $C(x_c, y_c, zc)$ and three vectors $\boldsymbol{CC_1}(x_{max}, y_{max}, z_{max})$, $\boldsymbol{CC_2}(x_{mid}, y_{mid}, z_{mid})$, $\boldsymbol{CC_3}(x_{min}, y_{min}, z_{min})$ corresponding with the maximum, middle, and minimum dimensions of the OBB. The orientations of the OBB can be determined by using the covariance-based method, which is proposed by Gottschalk, 1996 [14].

Each OBB is represented by three dimensions and three directions. The OBB matching process finds the view of the model that has the same dimensions with the object point clouds. The OBB of the point cloud of 1st series scans ***P*** is represented by three vectors $\mathbf{C_1C_{11}}(x_{1max}, y_{1max}, z_{1max})$, $\mathbf{C_1C_{12}}(x_{1mid}, y_{1mid}, z_{1mid})$, and $\mathbf{C_1C_{13}}(x_{1min}, y_{1min}, z_{1min})$. The OBB of the point cloud of 2nd series scans ***Q*** is represented by $\mathbf{C_2C_{21}}(x_{2max}, y_{2max}, z_{2max})$, $\mathbf{C_2C_{22}}(x_{2mid}, y_{2mid}, z_{2mid})$, and $\mathbf{C_2C_{23}}(x_{2min}, y_{2min}, z_{2min})$. Alignment of ***Q*** to ***P*** based on their OBBs' orientation leads to change of the coordinates of the point $\boldsymbol{q}_i(x_i, y_i, z_i)$ as follows:

$$\begin{bmatrix} x_i' \\ y_i' \\ z_i' \end{bmatrix} = \begin{bmatrix} \frac{C_2C_{21}}{|C_2C_{21}|}\frac{C_1C_{11}}{|C_1C_{11}|} & \frac{C_2C_{21}}{|C_2C_{21}|}\frac{C_1C_{12}}{|C_1C_{12}|} & \frac{C_2C_{21}}{|C_2C_{21}|}\frac{C_1C_{13}}{|C_1C_{13}|} \\ \frac{C_2C_{22}}{|C_2C_{22}|}\frac{C_1C_{11}}{|C_1C_{11}|} & \frac{C_2C_{22}}{|C_2C_{22}|}\frac{C_1C_{12}}{|C_1C_{12}|} & \frac{C_2C_{22}}{|C_2C_{22}|}\frac{C_1C_{13}}{|C_1C_{13}|} \\ \frac{C_2C_{23}}{|C_2C_{23}|}\frac{C_1C_{11}}{|C_1C_{11}|} & \frac{C_2C_{23}}{|C_2C_{23}|}\frac{C_1C_{12}}{|C_1C_{12}|} & \frac{C_2C_{23}}{|C_2C_{23}|}\frac{C_1C_{13}}{|C_1C_{13}|} \end{bmatrix} \begin{bmatrix} x_i - (x_{c_2} - x_{c_1}) \\ y_i - (y_{c_2} - y_{c_1}) \\ z_i - (z_{c_2} - z_{c_1}) \end{bmatrix} \tag{13}$$

Figure 4 shows a case with point clouds being derived from two series of scans before and after employing the rough rotation by alignment of two series scans.

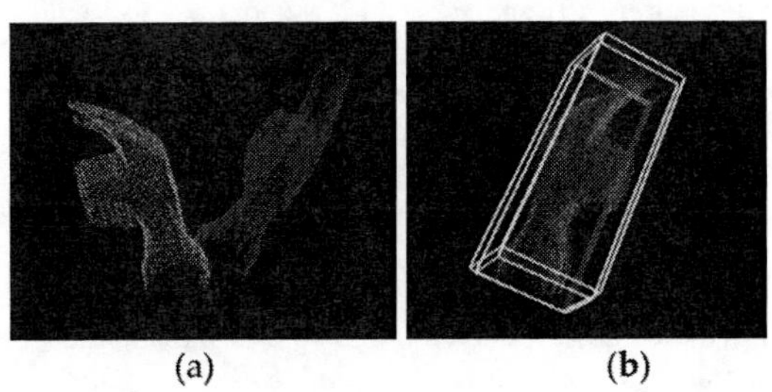

Figure 4. Illustration of the rough rotation estimate by alignment of two series scans. (a) Two series of scans before transforming by rough rotation matrix; (b) Two series of scans after transforming by rough rotation matrix.

[a] Corresponding author: lchen@ntu.edu.tw

2.4 Generating descriptors of point clouds

2.4.1 The OBB regional area-based descriptor

We assume that the three directions of the OBB are divided into k_1, k_2, and k_3 segments (as shown in Figure 5 (a)), the total number of subdivided boxes in the OBB is n = k_1 x k_2 x k_3. The surface area of the object in each subdivided box V_{ijk} ($i = 0, ..., k_1 - 1; j = 0, ..., k_2 - 1; k = 0, ..., k_3 - 1$) is denoted as f_v, where $v = k(k_1k_2) + j(k_1) + i$. f_v can be determined by subdividing the considered surface into the triangle mesh [15] (as shown in Figure 5 (c)). The distribution of S_v is defined as the regional area-based descriptor $FV = \{f_v, v = 0, \ldots, n\}$, shown in Figure 6 (d).

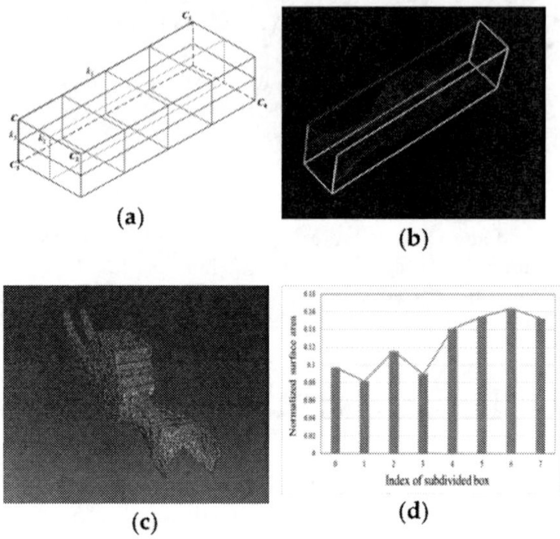

Figure 5. Illustration of the OBB regional area-based descriptor. (a) The OBB of the object is subdivided into smaller boxes; (b) The object point clouds and OBB; (c) Triangle mesh is generated over the point cloud data; (d) the regional area-based descriptor of the object.

2.4.2 Generating database and calculating descriptors

To define initial rotation matrix, we can consider the coordinate system $O_Gx_py_pz_p$ with the origin O_G defined in Equation (20) and three unit vectors v_1, v_2, and v_3, which are considered as the three directions of the OBB of point clouds ***P***. The regional area-based descriptor of point clouds ***P*** is $FV_P = \{f_{Pv}, v = 0, \ldots, n\}$ in the OBB of ***P***. Rotating ***Q*** around the three axes of the coordinate system $O_Gx_py_pz_p$ is performed with every increment $\Delta\theta$ on each axis. For each rotation, the regional area-based descriptor ***Q*** in the OBB of ***P*** is FV_Q can be then determined as described as Algorithm 2. To enhance the accuracy, the descriptor FV_P and FV_Q can ignore the subdivided boxes in which the missing part of point clouds ***P*** are located.

Algorithm 2. Generation of database

Input: Series scans $\boldsymbol{Q} = \{\boldsymbol{q}_1, \boldsymbol{q}_2, \ldots, \boldsymbol{q}_n\}$, the OBB of ***P***
Output: A set of descriptors of ***Q*** at different poses
Parameters: $(-\theta_x,\theta_x)$, $(-\theta_y,\theta_y)$, and $(-\theta_z,\theta_z)$ are ranges for rotation about the O_Gx_p axis, O_Gy_p axis, and O_Gz_p axis respectively.

```
for l = 0, 1, 2, ..., 2θx /Δθ
    for j = 0, 1, 2, ..., 2θy /Δθ
        for k = 0, 1, 2, ..., 2θz /Δθ
            θxl = - θx + l*Δθ, θyj = - θy + j*Δθ, θzk = - θz + k*Δθ
            R = Rx (θxl) Ry (θxj) Rz (θzk)
            Q --R--> Qi
            FVQi = {fQiv, v = 0, ..., n}
        end_for
    end_for
end_for
```

*2.4.3 Define rotation by searching the best matching pose between **P** and **Q** using the descriptors matching.*

The similarity between the regional area-based descriptors that represent the object point clouds ***P*** and ***Q*** is determined by using the normalized cross correlation. The normalized cross correlation between $FV_P = \{f_{Pv}, v = 0, \ldots, n\}$and $FV_Q = \{f_{Qv}, v = 0, \ldots, n\}$ is computed as follows:

$$C(FV_P, FV_Q) = \frac{\sum_{v=0}^{n}(f_{Pv} - \bar{f}_P)(f_{Qv} - \bar{f}_Q)}{\sqrt{\sum_{v=0}^{n}(f_{Pv} - \bar{f}_P)^2 \cdot \sum_{v=0}^{n}(f_{Qv} - \bar{f}_Q)^2}} \quad (14)$$

where $\bar{f}_P = \frac{1}{n+1}\sum_{v=0}^{n} f_{Pv}$ $\qquad \bar{f}_Q = \frac{1}{n+1}\sum_{v=0}^{n} f_{Qv}$

The rotation of Q is the initial rotation, in which the normalized cross correlation $C(FV_P, FV_Q)$ reaches its peak.

Although the parameters of transformation matrix can be obtained in the initial registration, the accuracy may not be satisfactory due to the error caused by the limited number of the rotating ***Q*** iterations. Fortunately, a refined registration like the iterative closest point (ICP) algorithm can be performed to achieve precise registration between multi-view point clouds.

3. Results and discussion

3.1 Case study on synthetic data

In this section we provide a discussion on influence of parameters to performance of the proposed method. To estimate the computation cost of the registration process, the experiments are processed on a computer with a Core i7 processor having 3.40 GHz and 8 GB RAM.

[a] Corresponding author: lchen@ntu.edu.tw

 DOI: 10.1051/matecconf/20167104008

In real application, it is very difficult to distinguish different error sources such as shape measurement error (noise, surface sampling, etc.), correspondence error (occlusion, outliers, etc.), and registration error. In order to evaluate the coarse registration errors, we have generated synthetic data, which is provided by the Industrial Technology Research Institute (ITRI) (as shown in figure 6). The dimensions of socket model and connector model are 45 x 25 x 25 mm^3 and 41 x 33 x 25 mm^3 respectively. Then, the point clouds corresponding with different views of the CAD model are extracted by a virtual camera. The motion between green and red point clouds is a displacement of 50 mm in each axis and a rotation of $\pi/4$ around a fixed axis. Therefore, we have precise knowledge of motion parameters rotation matrix **R** and translation vector **t** to serve as reference for error estimation and validation of the coarse registration methods. The measures used to determine the accuracy include rotation error and translation error. In order to evaluate the accuracy, the estimated transformation is compared to an expected one.

Results obtained shown that sampling does not affect considerably the accuracy of registration results. Thus, low resolution surfaces can be used in proposed approach to reduce the computation time. Meanwhile, as is shown in table 1, errors in proposed registration algorithms extremely depends on the rotation increment $\Delta\theta$ (rad). To enhance the accuracy, the rotation increment should become smaller. But the increment has a significant impact on the runtime of the proposed coarse registration process. As an example, the experiment with 1000 points in table 1, computation time in the case $\Delta\theta$ =0.01 (rad), at 126.439 (s), was much higher than that in the case $\Delta\theta$ =0.05 (rad), at only 1.063 (s).

Table 1. Experimental results using connector model obtained by proposed coarse registration method with different samplings and increments $\Delta\theta$.

Points	Increment $\Delta\theta$ (rad)	Translation error (mm)	Rotation error (rad)	Time (s)
	0.05	1.371	0.036	1.063
1000	0.02	1.371	0.017	15.328
	0.01	1.371	0.009	126.439
	0.05	1.352	0.039	7.244
5000	0.02	1.352	0.013	110.212
	0.01	1.352	0.012	904.237
	0.05	1.368	0.035	18.332
10000	0.02	1.368	0.015	203.795
	0.01	1.368	0.009	1762.824

OBB n = k_1 x k_2 x k_3. From table 2, it can be seen that if the OBB contains many subdivided boxes, it is more expensive to find the best rotation. For a fast registration procedure, less subdivided boxes are preferable. However, it is also required that the number of subdivided boxes is great enough to achieve expected rotation. For instance, in experiment using connector model of 1000 points, the rotation error rose to 0.057 (rad) for the OBB n = k_1 x k_2 x k_3 = 4 x 5 x 6, while it reached just 0.017 (rad) if the OBB n = k_1 x k_2 x k_3 = 7 x 8 x 9.

Besides, the ratio of the overlapped area is especially crucial. As shown in table 3, translation and rotation errors in proposed approach grow directly proportional to the percentage of non-overlapping region. This change is especially significant with over 40% of outliers. This is because it is difficult to estimate translation exactly without high level of overlapping. The excellent results are obtained for over 80% shape overlapping. In this case, the translation and rotation for tested models were at under 1.5 (mm) and 0.02 (rad), respectively. As described in the introduction section, the proposed registration method is developed to register two series of surface scans which exist certain degree of optical occlusion from each individual viewpoint of the probe scanning. To accurately register two series of scans under such the circumstance, it is important to have them overlapped as much as possible. However, realistically, the overlapped ratio really depends on the complexity of scanned object's geometry and how the object is placed on the rotation table. In general, the higher the overlapped ratio between two series of scans, the more accurate the registration can be achieved.

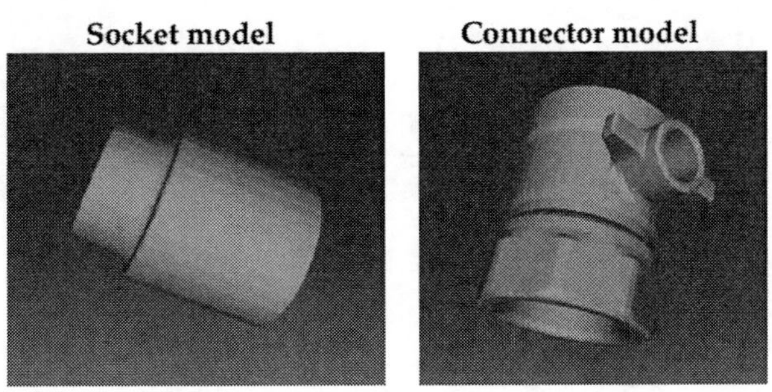

(a) (b)

[a] Corresponding author: lchen@ntu.edu.tw

MATEC Web of Conferences **71**, 04008 (2016) DOI: 10.1051/matecconf/20167104008
ICCPE 2015

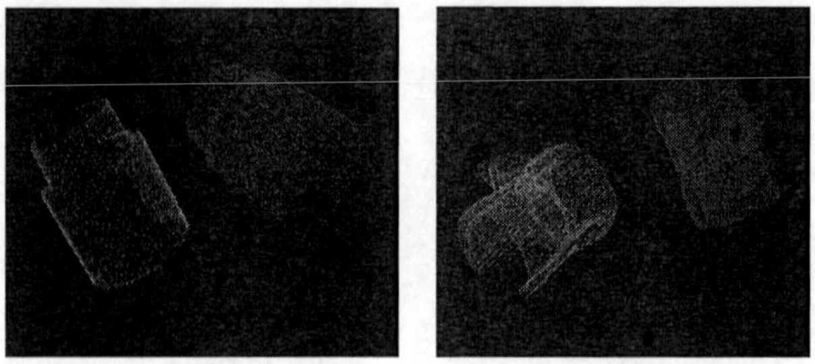

(c) (d)

Figure 6. The synthetic data used in experiments.

3.2 Case study on measured data

To verify the feasibility of the developed methodology, some experiments were performed and evaluated. In the experiments, the point clouds of a hammerhead were acquired from a 3-D optical scanner being developed by using random speckle pattern projection and triangulation measurement principle. The dimensions of the hammerhead are approximately 110 x 35.5 x 21 mm^3. Firstly, the hammerhead was placed on a fixed table, allowing for viewing the object from several positions around the object. However, not all the object surface could be detected due to optical occlusion. Therefore, after the 1st series of automated scans to obtain point clouds $\boldsymbol{P}$, the object was reoriented manually and the 2nd series of automated scans was continuously carried out to acquire point clouds $\boldsymbol{Q}$.

Rotating $\boldsymbol{Q}$ around the three axes of the coordinate system $O_q x_q y_q z_q$ with an increment of $\Delta\theta$ on each axis can be used to calculate the normalized cross correlation curves, shown in Figure 7. The width of the normalized cross correlation peak depends on the number of iterations and $\Delta\theta$. Meanwhile, the height of the normalized cross correlation depends on the number of OBB segments k_1 x k_2 x k_3 and $\Delta\theta$.

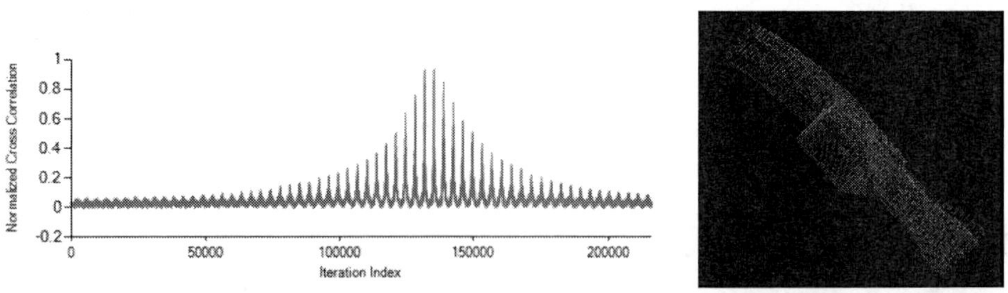

(a)

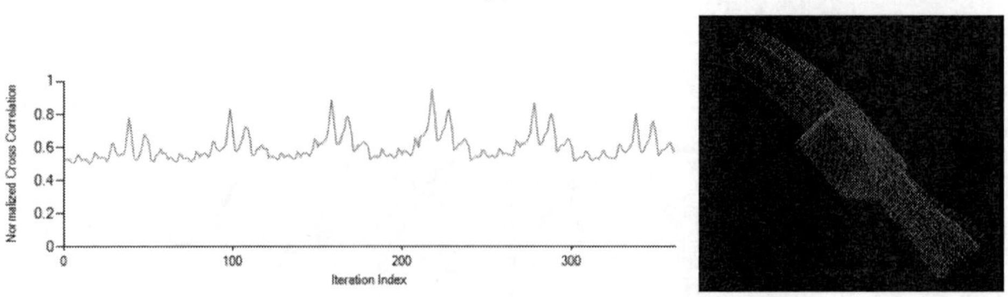

(b)

[a] Corresponding author: lchen@ntu.edu.tw

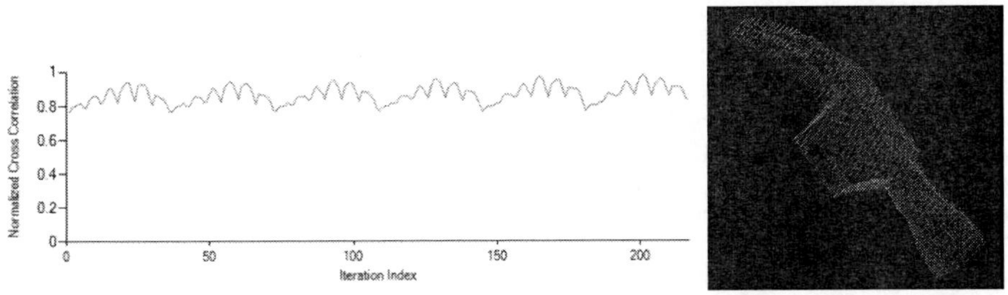

(c)

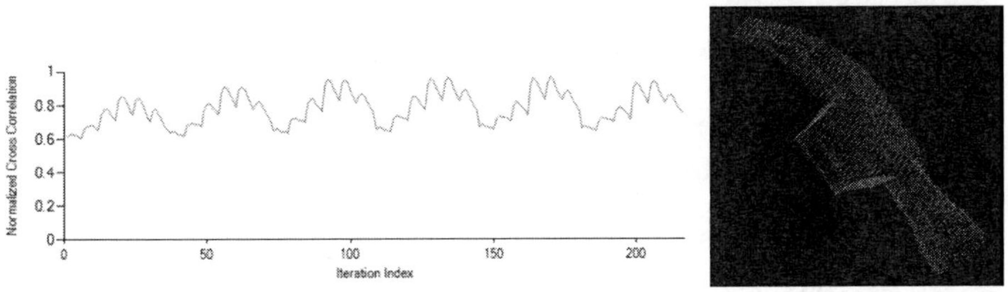

(d)

Figure 7. Normalized cross correlation curves given by Eq. (11): (a) the coarse registration with rotation $\Delta\theta = \pi/60$ without rough rotation estimate; (b) the coarse registration with rough rotation estimate before rotation $\Delta\theta = \pi/60$; (c) the coarse registration with rough rotation estimate before rotation $\Delta\theta = \pi/180$ (OBB segments: k_1 x k_2 x k_3=4 x 5 x6); (d) the coarse registration with the rough rotation estimate before a rotation increment $\Delta\theta = \pi/180$ (OBB segments: k_1 x k_2 x k_3=9 x 10 x11).

The performance of the registration can be estimated by using the distance from each overlapping point $\boldsymbol{q}_i$ in the point clouds $\boldsymbol{Q}$ to the fitting control point $\boldsymbol{p}_i$ that is the projected point of $\boldsymbol{q}_i$ onto the triangle mesh of $\boldsymbol{P}$. If d_i denotes the distance between a point $\boldsymbol{p}_i$ in $\boldsymbol{P}$ and its closest neighbor point $\boldsymbol{q}_i$ in $\boldsymbol{Q}$, the mean distance μ and standard deviation σ, which are used to evaluate the performance of the object registration, are computed as follows:

$$\mu = \frac{1}{n}\sum_{i=1}^{n} d_i \quad (15)$$

$$\sigma = \sqrt{\frac{\sum_{i=1}^{n}(d_i - \mu)^2}{n-1}} \quad (16)$$

In this experiment, the mean deviation value is 0.0347 mm and the standard deviation value is 0.0235 mm.

4 Conclusions

In this study, a global registration method is developed to overcome one of the most challenging problems remaining in current 3-D scanning systems for accurate image registration even for no reference coordinate existing between neighboring scanned surface patches. The key innovation in this work is the strategy in determining a robust registration transformation between two neighboring scans. The experimental results demonstrate the validity and applicability of the proposed approach. The registration error can be controlled within a few tens micrometers in one standard deviation for a size of object reaching 100 millimeters.

[a] Corresponding author: lchen@ntu.edu.tw

Acknowledgments

The authors would like to convey their appreciation for Grant support from the MOST of Republic of China under its grant with Reference number, MOST 103-2218-E-001-027-MY2. The appreciation is also conveyed to the Industrial Technology Research Institute (ITRI) for providing some testing samples.

References

1. T. Bodenmueller and G. Hirzinger, "Online Surface Reconstruction from Unorganized 3-D-Points for the DLR Hand-guided Scanner System," *Proceedings of 2nd International Symposium on 3-D Data Processing, Visualization and Transmission*, 285 - 292 (2004).
2. K. H. Strobl, W. Sepp, E.Wahl, T. Bodenmueller, M. Suppa, J. Seara, and G. Hirzinger, "The DLR Multisensory Hand-Guided Device: The Laser Stripe Profiler," In *Proceedings of ICRA*, **2** 1927-1932 (2004).
3. A. W. Fitzgibbon, G. Cross, and A. Zisserman, "Automatic 3-D model construction for turn-table sequences," In *3-D Structure from Multiple Images of Large-Scale Environments*, 155–170 (1998)
4. V. Fremont and R. Chellali, "Turntable-based 3-D object reconstruction," In *IEEE Conference on Cybernetics and Intelligent Systems*, 1277–1282 (2004)

5. M. Callieri, A. Fasano, G. Impoco, P. Cignoni, R. Scopigno, G. Parrini, and G. Biagini, "RoboScan: An Automatic System for Accurate and Unattended 3-D Scanning," In *IEEE 3-DPVT*, Thessaloniki, Greece, Sept. 805–812 (2004).
6. S. Larsson and J. A. P. Kjellander, "Path planning for laser scanning with an industrial robot," In *RAS*, 7, 615–624, (2008).
7. Simon Kriegel, Christian Rink, Tim Bodenmüller and Michael Suppa. "Efficient Next-Best-Scan Planning for Autonomous 3-D Surface Reconstruction of Unknown Objects", *Journal of Real Time Image Processing: Special Issue on Robot Vision*, 1-21 (2013).
8. Besl, P. J. and McKay, N. D., "A method for registration of 3-D shapes," *IEEE Transactions on Pattern Recognition and Machine Intelligence* **14**, 239–256 (1992).
9. C. J. R. Chua, "Point signatures: a new representation for 3-D object recognition," *Int. J. Comput. Vision* **25** 63–85 (1997)
10. A. E. Johnson and M. Hebert, "Using spin images for efficient object recognition in cluttered 3-D scenes," *IEEE Transactions on Pattern Analysis and Machine Intelligence*, vol. **21** 433-449 (1999).
11. F. Tombari, S. Salti, and L. Di Stefano, "Unique Signatures of Histograms for Local Surface Description," In *Proceedings of the 11th European Conference on Computer Vision Conference on Computer Vision*: Part III, 356–369 (2010)
12. Liang-Chia Chen , Dinh-Cuong Hoang, Hsien-I Lin, Thanh-Hung Nguyen, "An Automated Point Clouds Scanning and Registration Methodology for 3-D Object Reconstruction," In *The 39th National Conference on Theoretical and Applied Mechanics*, (2015).
13. L.-C. Chen, T.-H. Nguyen, and S.-T. Lin, "3-D Object Recognition and Localization for Robot Pick and Place Application Employing a Global Area Based Descriptor," In *International conference on Advanced Robotics and Intelligent Systems*, Taipei, Taiwan, May 29-31 (2015).
14. S. Gottschalk, M. C. Lin, and D. Manocha, "OBBTree: A Hierarchical Structure for Rapid Interference Detection," In *Proceedings of the 23rd Annual Conference on Computer Graphics and Interactive Techniques*, 171–180 (1996).
15. Z. C. Marton, R. B. Rusu, and M. Beetz, "On fast surface reconstruction methods for large and noisy point clouds," In *IEEE International Conference on Robotics and Automation*, 3218–3223 (2009).
16. M. A. Fischler and R. C. Bolles, "Random Sample Consensus: A Paradigm for Model Fitting with Applications to Image Analysis and Automated Cartography," *Commun. ACM*, **24** 381–395 (1981).
17. T. H. Nguyen, "3-D Object Recognition and Localization of Randomly Stacked Objects for Automation". Doctoral Thesis, National Taipei University of Technology, Taiwan, (2015).

[a] Corresponding author: lchen@ntu.edu.tw

MATEC Web of Conferences 71, 04009 (2016) DOI: 10.1051/matecconf/20167104009
ICCPE 2015

Fatigue and Model Analysis of the CNC Cylindrical Grinder

Jui-Chang Lin [1], Cheng-Jen Lin [2,a]

[1]*Department of Mechanical Design Engineering, National Formosa University, Huwei,Taiwan, R.O.C*
[2]*Institute of Mechanical and Electro-Mechanical Engineering, National Formosa University, Huwei,Taiwan, R.O.C*

Abstract. The purpose of this study is to lower deviation of workpiece by meeting high stability and rigidity to prevent the resonance in producing procedure of the CNC universal cylindrical grinding machine. Using finite element analysis software ABAQUS in grinder machine tools for numerical simulation of several analyses for the following: structural rigidity analysis, optimized design, vibration frequency analysis and fatigue damage analysis. This work aims on state of the transmission of outer diameter spindle to proceed in stress and fatigue life analysis by FE-SAFE Subroutine. The max values of equivalent stress and average amount of displacement in structural rigidity analysis are 0.67(Mpa) and 0.92(μ m). Optimization design effectively reducing extreme value of stress, the largest decline of about 5.43%. Modal analysis compared with the experimental, the average error percentage was less than 10% of parts. The whole structure error does not exceed 3%. The fatigue life of approximately 1,193,988 times, estimates into real life time can use more than sixty years, from the viewpoint of structural strength, spindle has a good high breaking strength is designed to be safe.

1 Introduction

In general, the influential factors of CNC machine accuracy are structural design, assembly errors and component machining problems. The structural vibration and deformation will affect CNC machine structure, the main reason is insufficient strength which is caused by improper reduction weight, some components of machine will shake during machine operation, in some situation the structural resonance may be happened, the machining quality must be influenced too. Therefore, experimental mode analysis and finite element analysis is necessary for structural design of CNC grinding machine. The improvement of the structural design errors, the avoidance of the vibration frequency and the structural rigid problems were an imperative task. FEM is one of the current analytic methods, which is obtained by numerical solution of engineering problems similar analysis.

FEM concept is to analyze the object which was meshed and cut into the appropriate size, number and shape, and then the basic theoretical principles (mainly elastic-plastic theory) combined with the material properties, the force conditions, and the boundary conditions, applying the limited cumulative iterations to solve the element and finally obtain the numerical solution. Finite element analysis software ABAQUS is widely used in automobiles, civil engineering, aerospace and mechanical industries. ABAQUS/CAE can analyze various mechanics problems from linear structure to non-linear material stress and strain including vibration and deformation. [1-4]

2 Theory

Mechanical systems are usually composed of several parts; the interaction is existed each other. A simpler physical model is represented in order to study the vibration characteristics of the actual mechanical system. Mechanics or mathematical theorems are applied to establish the physical model expression. A simple vibration system usually includes (1) spring or elastomer of stored potential energy; (2) the mass or inertia of stored kinetic energy; or (3) consumption of energy damping. When general structure of the vibration system can't be simplified to a point mass, the system will be treated as infinite degrees of freedom of the continuum. The FEM is used to solve equation of motion or equations of force for the complicated structure.

Vibration is the object which is treated as a reference point and moved gently back and forth with a small range of movement under the static equilibrium, so the vibration of the object is a regular cycle of movement. Four-bar linkage, slider-crank, swing and other periodic motion are not the so-called vibration. They only can be treated as dynamic system problems. In general, the vibration can be divided into two types: free vibration and forced vibration. Free vibration system is generated by the vibration exciting force in the initial conditions or after the abolition of the existing external vibration.

[a] Corresponding author: lcj@nfu.edu.tw

Forced vibration is caused by periodically or intermittently external forces. The two types of vibration can be a damping or undamped vibration, friction effect in the undamped vibration system is ignored; it can keep the vibration forever, but in reality the internal friction and external friction is exist, all objects are damping vibration.

Frequency is variable because of the structure shape, size, material properties and even the boundary conditions of the structure. Mode parameters is unique, constant, can be regarded as the characteristics of the system.

3 Static structural analysis for grinding machine

In this paper, FEM software was used to analyze the static structure of CNC cylindrical grinding ma-chine, including machine head, machine bed, and spindle. In the static analysis, the main structure analysis was made after the main structure assembly was finished, inspecting the stress changes and the deformation under the grinding load to find out the maximum strain of the mechanical structure, and enhance its structural rigidity by optimization.

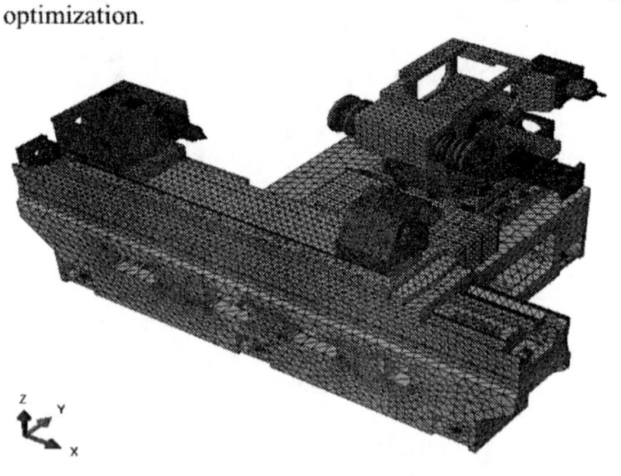

Figure 1. CNC grinding mechanical structure.

Table 1. The mechanical properties of machine material.

Material	Cast iron (FC300)
Density (ton/mm^3)	7.2×10^{-9}
Young's modulus (Gpa)	110
Poisson's ratio	0.28
yielding strength (MPa)	200

The Fig.1 showed the main structure of complex CNC cylindrical grinding machine including body, bed, head and other parts. The material of CNC cylindrical grinding machine is cast iron (FC300), the mechanical properties include density, Young's modulus, Poisson's ratio and yielding stress as listed in Table 1. The loading and support portion is added in the structure according to the actual situation and establishment of the basic conditions. Solving the results, the maximum displacement of the head of grinding machine is 0.016mm, and the maximum stress is 2.5MPa. However, the maximum stress at the grinding machine base is lower than the yielding strength of cast iron FC300 200MPa. The base structure of grinding machine does not produce permanent deformation owing to its grinding load, and reduce its precision and stability.

4 Grinding machine dynamical analysis and experiment

This study used finite element method to simulate the frequency response of complex CNC internal and external grinding machine, the frequency response was also actually measured to verify the accuracy. Therefore, the weakness of grinding machine structure will be known, in the working condition with spindle revolution, the resonance frequency should be avoided, and improved the machine structure.

4.1 Comparison between experiment and FEA

After modes experiment and modes analysis by FEA, the results showed that experiment and analysis was coincidence, the error was within 12 %, showed the comparison of modes analysis of whole machine in Fig.2. The rotational speed of complex CNC grinding machine is 30000 RPM, the corresponding frequency is 500 Hz. The figures of analysis model indicate that smaller parts have higher frequency, the first mode of the small most parts are larger than 500 Hz. It is evident that the most parts do not have vibration problems.

FEA Experiment

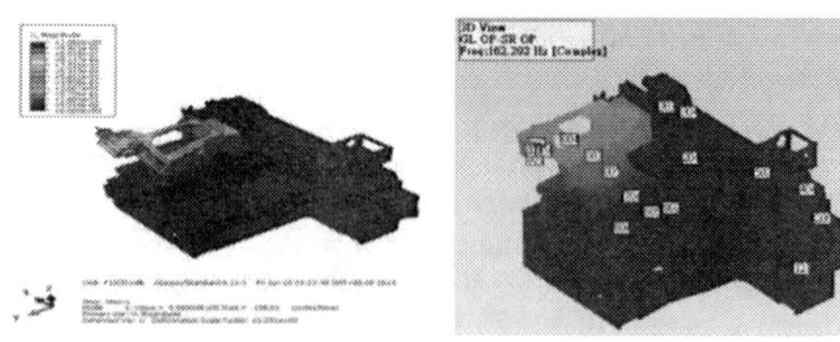

158.91 (Hz) 162.292 (Hz)

Figure 2. CNC grinding mechanical structure.

4.2 Fatigue estimate by finite element analysis

The finite element analysis (FEA) software ABAQUS was applied to analyze the fatigue of complex CNC grinding machine, including life of whole machine assembly structure. After modes of spindle analysis by FEA, the results showed in Fig.3. The fatigue life of approximately 1,193,988 times, estimates into real life can use more than sixty years, from the viewpoint of structural strength, spindle has a good high breaking strength is designed to be safe.

DOI: 10.1051/matecconf/20167104009

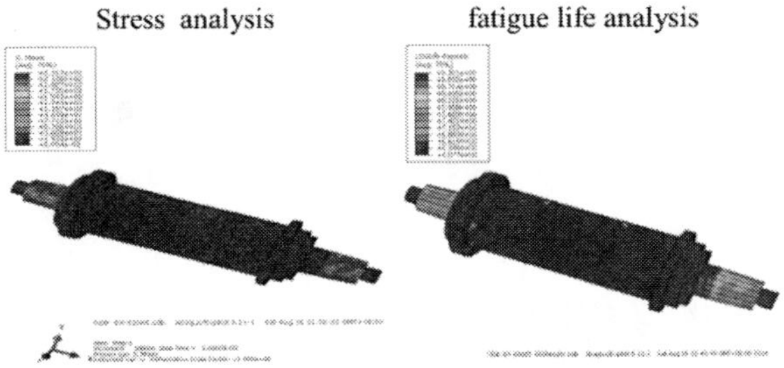

Figure 3. Fatigue analysis of spindle.

5 Topology optimization design

In the further research, topology optimization design may use to reduce the net weight of grinding machine in the case of without derogation strength. The process of optimization analysis need to make some restrictions in objective function, volume constraint and geometry, the structural symmetry factors also should be considered together. The element after topology optimization design should do the same structural strength analysis and compared with the original element. The results can be used to determine whether the topology optimization design to achieve the desired function.

This study makes first choice of outer wheel holder for topology optimization design. After the foregoing related settings of optimization analysis to reduce the weight of element, and execute related analysis of structural strength afterwards. The figure 4 showed the difference of stress and figure 5 shown the change of displacement. In brief the effect of topology optimization design, material was removed 15%, stress from 1.514(Map) down to 1.503(Mpa) and displacement from 2.00(μm) increase to 2.033(μm). These results are not too poor. But the outcome of the analysis of a single element can't determine the effect of optimization. The optimization design after the overall consideration of whole grinder structure, the final result is the basis of the feasibility of the topology optimization design.

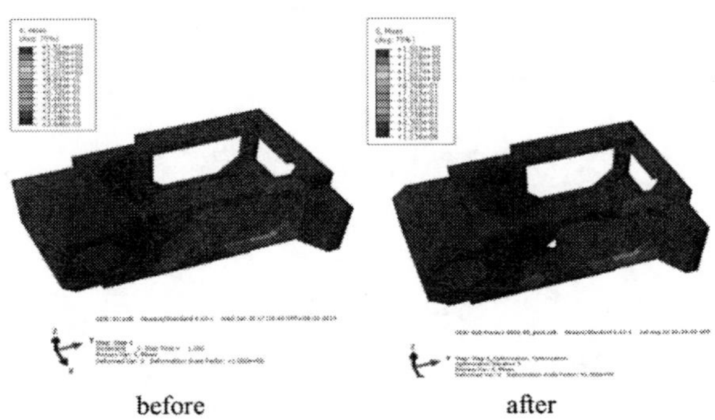

Figure 4. The comparison of stress of outer wheel holder before and after optimization design.

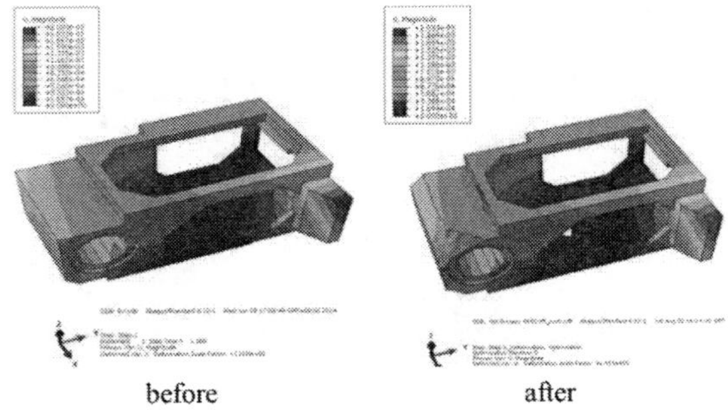

Figure 5. The comparison of displacement of outer wheel holder before and after optimization design.

6 Conclusion

By CAE software to improve the strength of the CNC machine structure, not only to reduce the cost and time of the CNC machine experimental, can be used as the reference of the development of new CNC grind machine structure design.

References

1. Ramezanali, Mahdavinejad, Finite element analysis of machine and workpiece instability in turning. International Journal of Machine Tools & Manufacture, **45**, 753-760 (2005)
2. J. A. Ewing, Humfrey, J. C., The Fracture of Metals under Repeated Alterations of Stress, Philosophical Transaction of the Royal Society, London, **A200**, 241-250 (1903)
3. J. C. Newman, Jr., A Finite Element Analysis of Crack Growth under Monotonic and Cyclic Loading, ASTM STP **637**, 56-80 (1977)
4. N. A. Fleck, Finite Element Analysis of Plasticity-Induced Crack Closure under Plane Strain Conditions, Engineering Fracture Mechanics, **25**, 441-449 (1986)

Design and production of toilet seat lifting mechanism

Tsui-Er Lee[1,a], Rong-Mao Lee[2] and Pao-Ting Liu[2]

[1]*Asia University, Office of Physical Education, 41354 Taichung City, Taiwan*
[2]*National Chin-Yi University of Technology, Department of Mechanical Engineering, 41170 Taichung City, Taiwan*

Abstract. Different physiological structure between genders reflects on daily life, in which "going to toilet" is mostly concerned. Going to toilet is an important triviality every day but is closely related to personal health. Since toilet seats directly contact with the skin, the cleanness presents critical influence on physiological health and psychological comfort. It is common to share a set of sanitary equipment in the family, where females reveal several times higher probability on contacting with toilet seats. Aiming at home toilet seats being easily polluted, an improvement program is proposed in this study to reduce gender conflict with a simple mechanism design, which is improved through experiments. Keywords: female population advantage, toilet seat, female right, convenient design of toilet, use efficiency of toilet.

1 Motivation and purpose

Female population (11.684133 million) in Taiwan for the first time exceeded male population (11.683187 million) by November, 2013 [1]. The problem of gender rights has become more important following the advance of society and gender equality issues. Current research on gender difference and gender conflict focuses more on public places. However, different physiological structure between genders inducing daily convenience conflict and health problems has the trivial "going to toilet" be often concerned. According to the report of Taiwan Health Foundation [2], females in Taiwan appear 8 times higher probability than males suffering from urinary tract inflammation. The factor in such a big difference could be found from male and female physiological structure. The upper part of urinary tract for both genders is the same, i.e. similar length from kidney, ureter to bladder, while the lower part of urinary tract (urethra) appears obvious difference, i.e. males about 10~12 cm and females merely 3~4 cm. Once the environment is not clean or the urination habit is not good, bacteria could easily invade and induce female urinary tract infection. For this reason, maintaining the cleanness of toilet seats is critical. The simple toilet seat lifting could easily induce conflict, and males and females often disagree with each other on who should lift the seats [3]. Females therefore have to suffer from health risks caused by dirty seats. Aiming at home toilet seats, an easy engineering improvement is proposed in this study. Focusing on "plug-in installation and simple operation", the engineering design is expected to create the female-friendly home sanitary environment.

2 Design of seat lifting mechanism

The device is a pure mechanism design, which is operated manually without consuming electricity. The design follows the following requirements.

* Simple mechanism operation.
* Being able to solve gender conflict in toilet.
* Being able to improve toilet cleanness. (Adjusting the standing position of male urination.)
* Not inducing new problems.
* Easy production and assembly.

The mechanism operation process is shown in Figure 1. Both toilet seat lifting and rebound require buffer to avoid damage and noise.

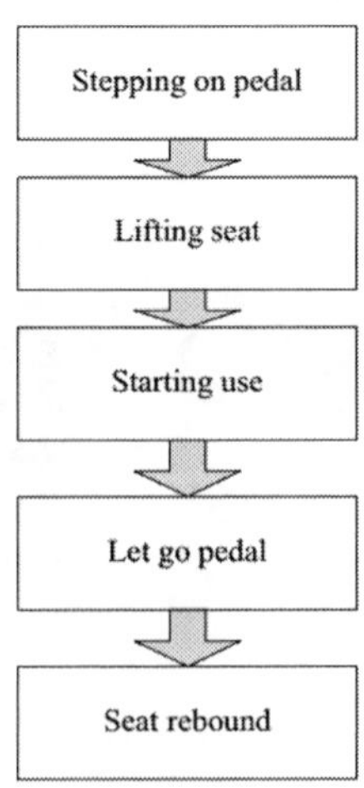

Figure 1. Seat lifting mechanism operation process.

[a] Corresponding author: vivian@asia.edu.tw

DOI: 10.1051/matecconf/20167104010

2.1 Mechanism component

Figure 2 and Figure 3 demonstrate the mechanism, which is expected to contain the elements of seat lifting bar, fixing bar, pedal, and damper. The fixing bar connects the toilet seat base, providing the mechanism fixing force. The pedal presents the functions of (1)leading males standing on the proper position to maintain the toilet cleanness (close to the toilet and aiming at the toilet center) and (2)driving the seat lifting bar to lift the toilet seat upwards and backwards. In order to have the toilet seat automatically "cover" the toilet, after use, by gravity (for the consideration for females), the lifting range appears in 75°~85°. Finally, the toilet seat up and down would induce noise and even result in damage to the toilet body or the seat. A buffer damper is therefore equipped so as to slowly decline the toilet seat after males going to the toilet to control the noise and reduce damage.

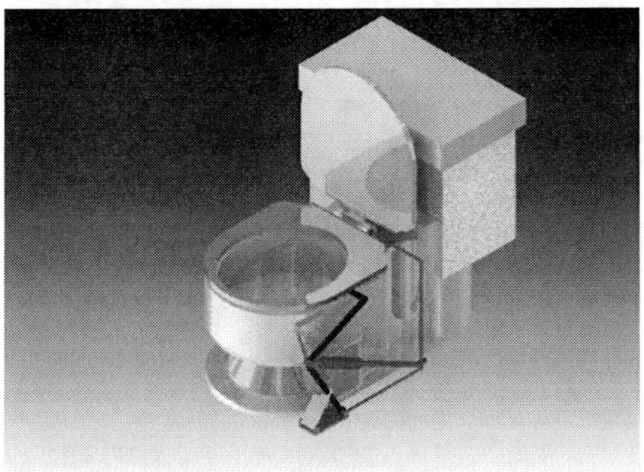

Figure 2. Demonstration of toilet seat lifting mechanism. (static)

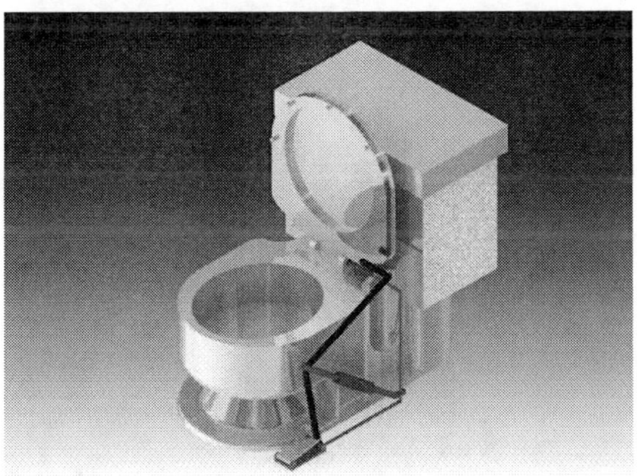

Figure 3. Demonstration of toilet seat lifting mechanism. (lifted)

2.2 Design of mechanism size

Because of distinct toilet specifications and patterns as well as sole sizes, the following items need to be calculated with statistics in order to determine the size and specifications of various mechanism parts.

2.2.1 Common toilet size

The toilet geometric shape and the sizes of various parts would affect the mechanism bar length design. The statistical survey of toilet hardware could enhance the universality of seat lifting mechanism. As HCG toilet shows the highest market share in Taiwan [5], the specification is used as the reference of the basic design (Figure 4 [4]). From Figure 2, the toilet height is important for the seat lifting mechanism size design. The common toilet height is about 360~400mm [6], and the height of 385mm reveals the highest repetition. The mechanism proposed in this study therefore is based on the height of 385mm and offers certain range of adjustable height.

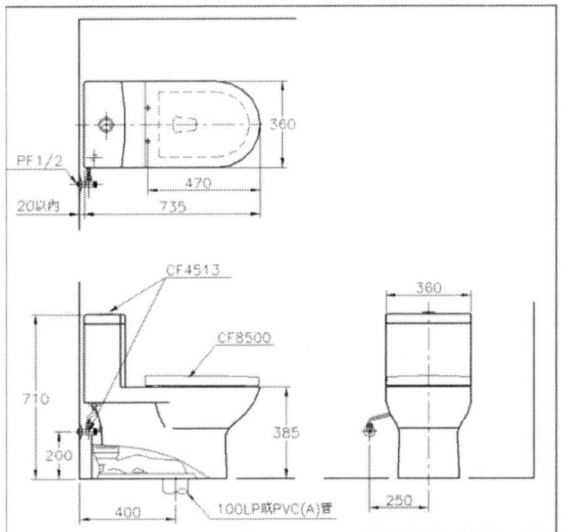

Figure 4. Demonstration of toilet specification [4].

2.2.2 Pedal width and length

The mechanism design aims to allow males conveniently maintaining the toilet cleanness (related to female health) without "bending down to lift the seat". For convenient operation, the average length of the maximum left foot and right foot shapes of domestic males [7] is applied to the standard pedal length (Table 1 and Table 2).

Table 1. Average left sole length of domestic males [7]. (cm)

Age \ Region	Northern Taiwan	Middle Taiwan	Southern Taiwan	East Taiwan	Total
18-50	24.3544	24.244	23.8063	27.8186	24.294
51-60	27.9	23.621	28	-	23.862
61-70	-	26.154	29.74	-	26.552
71~	-	25.7767	-	-	25.7767
Total	24.3837	24.2411	24.1228	27.8186	24.311

Table 2. Average right sole length of domestic males [7]. (cm)

Age \ Region	Northern Taiwan	Middle Taiwan	Southern Taiwan	East Taiwan	Total
18-50	22.7686	21.803	23.4663	25.62	22.186
51-60	24.2	20.54	21.4	-	20.67
61-70	-	22.661	23.44	-	22.197
71~	-	20.5433	-	-	20.5433
Total	22.7804	21.7015	23.4009	25.62	22.0869

2.3 Mechanism design program

According to above requirements and statistics, four types of seat lifting structure are designed the operation functions.

2.3.1 Mechanism A

Following the demonstrated structure (Figure 2), the fixing bar is integrated with the rebound device, Figure 5.

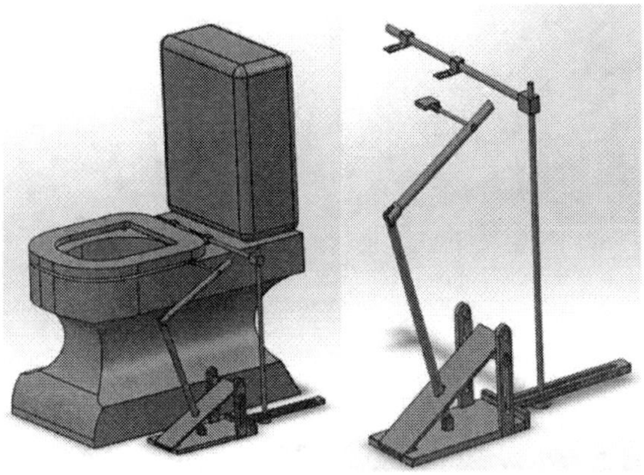

Figure 5. Mechanism A

2.3.2 Mechanism B

The pedal and bars in Mechanism A are easily stocked because of dead center that the contact point is changed into the sliding runner structure (Figure 6).

2.3.3 Mechanism C

In consideration of the pedal stock in Mechanism A, a toggle mechanism [8], Figure 7, is applied to design Mechanism C, in addition to the sliding runner structure in Mechanism B, Figure 8.

2.3.4 Mechanism D

In spite that both Mechanism B and Mechanism C allow smoothly lifting seats, the seat lifting range of such two mechanisms cannot achieve the requirement (75°~85°). Referring to the idea of trash can lid lifting mechanism (Figure 9 [5]), the functional improvement is proceeded in mechanism D, Figure 10, in order to enhance the seat lifting range.

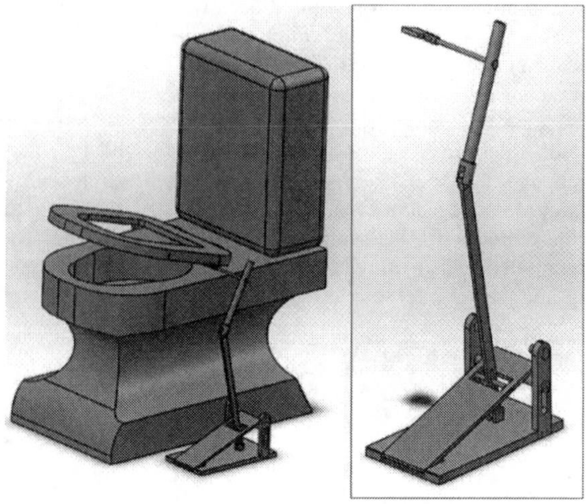

Figure 6. Mechanism B

Figure 7. Toggle mechanism [8].

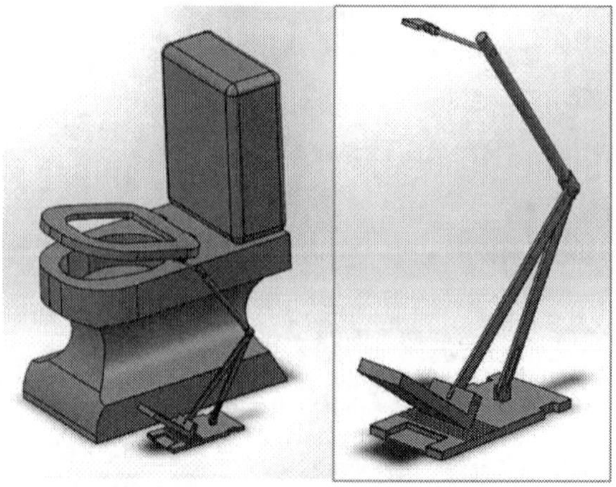

Figure 8. Mechanism C

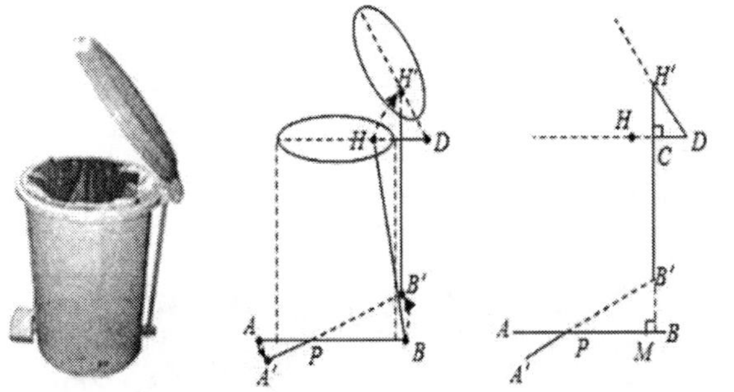

Figure 9. Trash can mechanism [5]

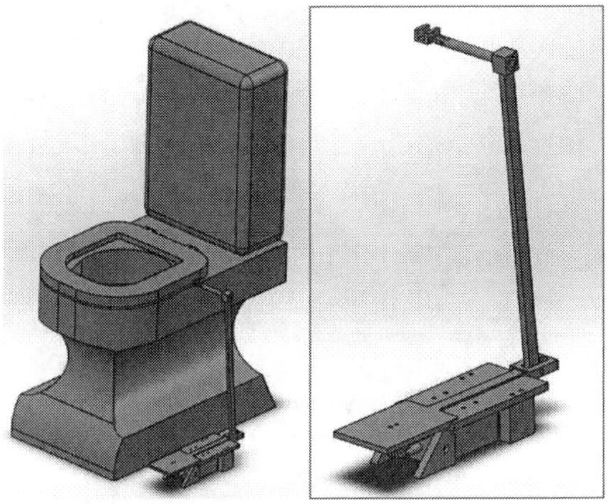

Figure 10. Mechanism D

Table 3 shows the comparison among four types of mechanism design; and, the simulation with Solidworks Motion presents that merely Mechanism D could completely lift toilet seats to 75°~85° (Figure 11) as well as conform to the functional requirement and performance requirement. Mechanism D (Figure 9) is therefore referred to the production of the engineering prototype.

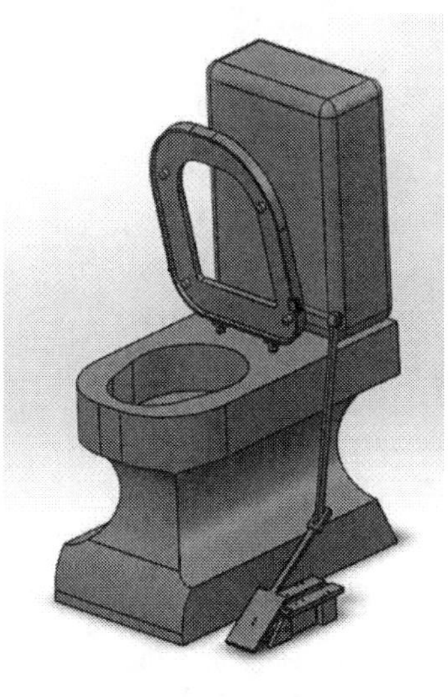

Figure 11. Mechanism D (seat lifting)

Table 3. Comparison of mechanism designs

	Mechanism motion	Seat lifting range
Mechanism A	Unusual (dead point)	Insufficient
Mechanism B	Normal	Insufficient
Mechanism C	Normal	Insufficient
Mechanism D	Normal	Sufficient

3 Seat lifting mechanism prototype production and experiment

The major elements of the mechanism cover base, damper buffer device, structure fixing board, pedal, and seat lifting bar set, as following description.

3.1 Base

The base, Figure 12, presents 4 functions.

* To offer basic height for the pedal operation
* To fix the mechanism for not sliding
* To fix the damper buffer device
* To fix the pedal rotation axis

Figure 12. Base

3.2 Damper buffer device

The pen-shape spring cylinder, Figure 13, is used as the mechanism buffer for smoothly lifting and declining seats. The buffer range is adjustable according user needs (The red square in the figure is the adjuster).

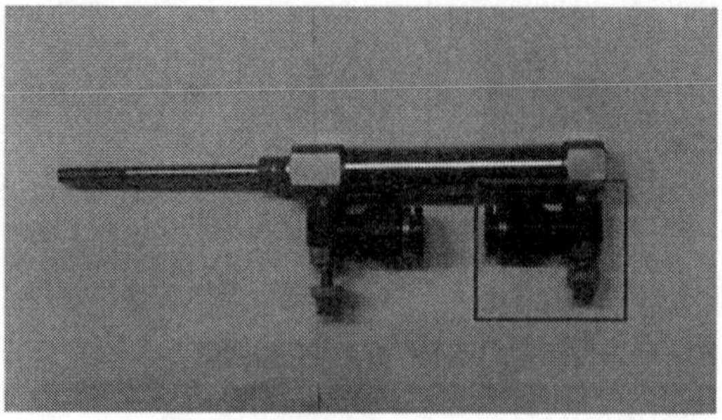

Figure 13. Damper buffer device

3.3 Structure fixing board

This board (Figure 14) is placed between the base and the pedal to provide the moving bars (for pedal and seat lifting bar set) with basic supporting force and to fix the pedal position.

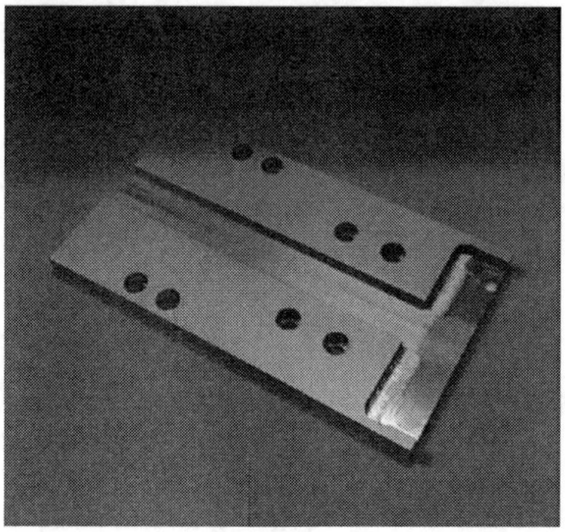

Figure 14. Structure fixing board

3.4 Pedal

The left of the pedal (Figure 15) is connected to the seat lifting bar set, and the base is the rotation center of the right in order to lift the seat lifting bar set to complete the seat lifting.

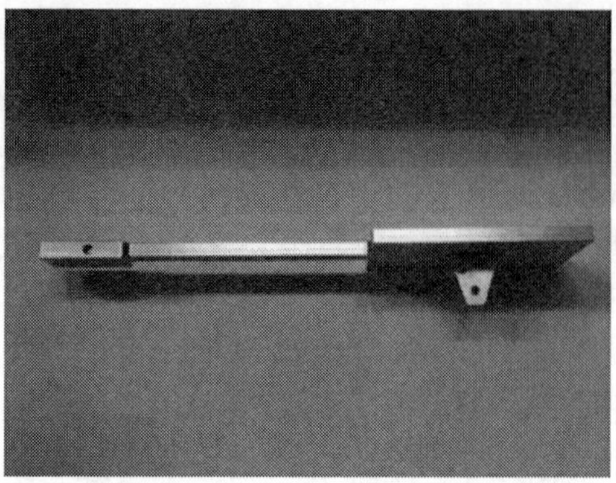

Figure 15. Pedal

3.5 Seat lifting bar set

Bar A (Figure 16) is connected with the pedal, responsible for delivering the upward force. Bar B (Figure 17) is clipped on the toilet seat, and the adjustable clip is designed for the application to toilet seats with different styles and thickness.

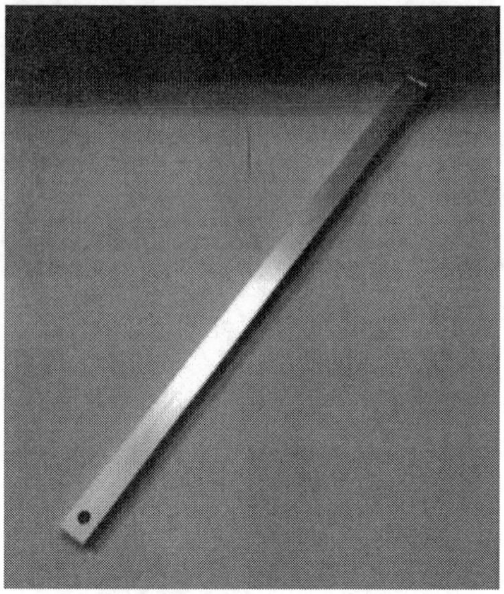

Figure 16. Seat lifting bar set A

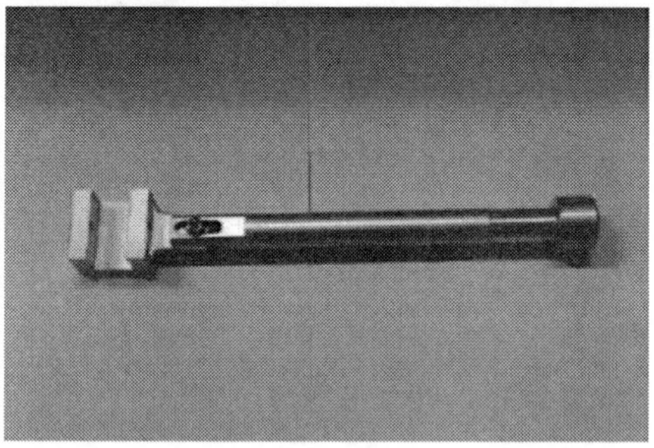

Figure 17. Seat lifting bar set B

The mechanism prototype is shown in Figure 18, and the real installation on a toilet is shown in Figure 19 and Figure 20. The preliminary test reveals the following situations.

1. Uneasy buffer range adjustment

The seat lifting speed is generally different from the declining speed that it requires a period of use time to adjust the damper buffer range.

2. Abnormal sound of bars

The collision of metal bars in the mechanism operation would induce little noise. Such a drawback is taken into consideration of future improvement. It is expected to include some plastic elements or add a rubber pad to the junction for the improvement.

DOI: 10.1051/matecconf/20167104010

3. Mechanism stability

Figure 20 shows that the mechanism can easily shake in the stepping process because of uneven tile surface. The base bottom will be added a soft pad to slightly adjust the ground factor.

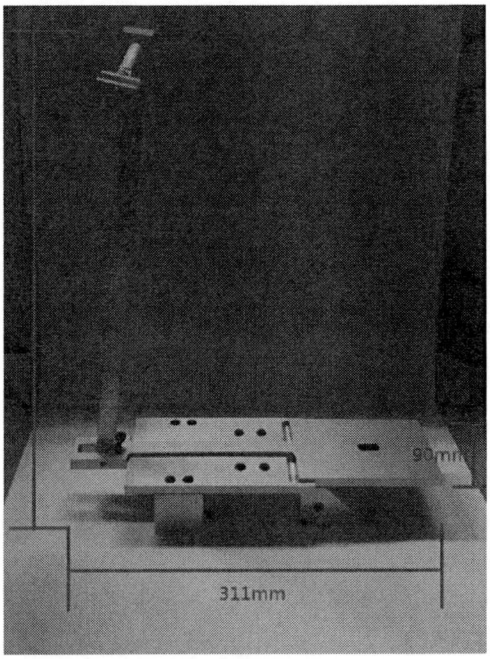

Figure 18. Complete mechanism assembly

Figure 19. Real mechanism assembly

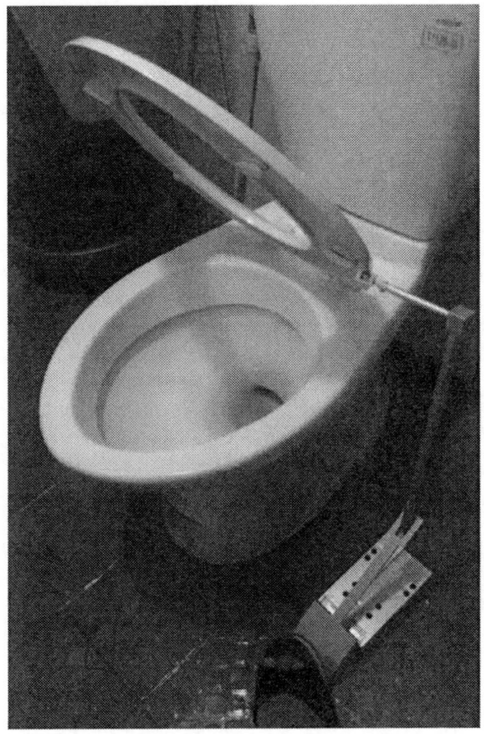

Figure 20. Seat lifting operation

4. Conclusion

Aiming at gender conflict in toilet, an engineering solution is proposed in this study to satisfy the basic requirement of humans through the mechanism design. With the suitable concept proposed from the aspect of requirement and the comparison among several mechanisms, the practicability of the engineering prototype is preliminarily proven. In addition to the improvement of the mechanism design, the public is invited for a longer time experience so as to objectively evaluate the mechanism efficiency for gender equality as well as to collect the shortcoming and the improvement direction of the mechanism design. It is expected to manifest the small problem-big trouble between genders in daily life. The mechanism design includes the requirement of humans that the performance is not the primary consideration. It therefore becomes a bridge to integrate research in different fields.

Acknowledgement

We are grateful for the project funding support of Ministry of Science and Technology (MOST 104-2629-E-468-002) and would like to show our appreciation.

References

1. http://news.ltn.com.tw/news/focus/paper/736358, Last Browse Dates: 2016.03.28.

2. http://www.twhealth.org.tw/index.php?option=com_zoo&task=item&item_id=595&Itemid=19, Last Browse Dates: 2016.03.28.
3. http://blog.udn.com/ubhuang/5243877, Last Browse Dates: 2016.03.28.
4. http://www.hcg.com.tw/Product/Detail/14715749-2b25-46af-8527-34d6df9360b6, Last Browse Dates: 2016.03.28.
5. https://www.google.com.tw/search?q=%E5%9E%83%E5%9C%BE%E6%A1%B6%E6%A9%9F%E6%A7%8B&source=lnms&tbm=isch&sa=X&ved=0ahUKEwitur6ly-LLAhUHlJQKHSHkA_AQ_AUICCgC&biw=1920&bih=955#imgrc=_, Last Browse Dates: 2016.03.28.
6. Construction and Planning Agency, Ministry of the Interior, Taiwan, Public Buildings Hygienic Equipment Design Manual, 2010. (In Chinese)
7. http://www.hcg.com.tw/Bath, Last Browse Dates: 2016.03.28.
8. Footwear & Recreation Technology Research Institute, A Data Analysis Report of Foot Shape Measurement of Adult Male in Taiwan (2011)
9. David H. Myszka, Machines & Mechanisms: Applied Kinematic Analysis (4th Edition), Pearson, London, UK (2011)

MATEC Web of Conferences **71**, 05001 (2016) DOI: 10.1051/matecconf/20167105001
ICCPE 2015

Robust H_∞ Observer-Based Control of Uncertain Neutral Systems with Mixed Delays

Jenq-Der Chen[1,a], Chin-Tan Lee[2], Chao-Tsung Kuo[2], I-Te Wu[3], Ruey-Shin Chen[1]

[1]*Department of Computer Science and Information Engineering, National Quemoy University, Kinmen, Taiwan, R.O.C.*
[2]*Department of Electronic Engineering, National Quemoy University, Kinmen, Taiwan, R.O.C.*
[3]*Department of Life Science-National Taiwan Normal University, and Department of Sports and Leisure National Quemoy University, Kinmen, Taiwan, R.O.C.*

Abstract. In this paper, the robust In this paper, the robust H_∞ control problem of output dynamic observer-based control for a class of uncertain neutral systems with discrete and distributed time delays is considered. Linear matrix inequality (LMI) optimization approach is used to design the new H_∞ output dynamic controls. The minimal H_∞-norm bound and the maximal perturbed bound are given. Based on the result of this paper, the constraint of matrix equality is not necessary for designing the H_∞ observer-based controls. Finally, a numerical example is given to illustrate the efficiency of the proposed approach.

1 Introduction

The existence of the time-delay phenomena in a dynamic system may cause instability or bad performances in dynamic systems [1-3]. In some practical systems, the system models can be described by functional differential equation of neutral type, which the models depend on the state delay but also depend on the state derivatives. Physical examples for neutral system have distributed networks, population ecology, process including steam, heat exchanges, lossless transmission line. Hence the stability and stabilization problems for neutral time-delay system have received some attenuation. By increasing in the equation number of summands and simultaneously decreasing the differences between neighbouring argument values, one naturally arrives at equations with distributed (or continuous) and mixed (both distributed and discrete) delay arguments [2]. In view of Fridman [2], the distributed delays, play an important role about the stability of system.

In the many real-world sysyems, state feedback control will fail to guarantee the stabilizability when some of system states are not measurable. In the observer-based control, output dynamic feedback control is provided and the system state can be estimated from the process. The observer-based controls are often be utilized to stabilize unstable systems or improve the system proformance. Hence, the observer-based control for systems has been an interesting topic in control theory. Lyapunov stability theory is used to design the nonlinear state observers for linear time varying systems [4]. In [5], the LMI approach was introduced to design the observer-based controls for uncertain systems.

On the other hand, the H_∞ control concept was proposed to reduce the effect of the disturbance input on the regulated output to within a prescribed level. The state feedback H_∞ controls for uncertain neutral time-delay systems had been considered in [6-7]. The output H_∞ filtering design for neutral system without uncertainties had been proposed in [8]. To the best of the knowledge of author, the robust H_∞ observer-based control for neutral systems with discrete and distributed time delays has never been considered in the past. In this paper, we will adopt this useful methodology to the design of the robust H_∞ observer-based controls for a class of uncertain neutral time-delay systems. The three classes of H_∞ observer-based controls with known and unknown (uncertain) time-delay values will be considered. Moreover, the minimal H_∞-norm bound and the maximal perturbed bound for the observer-based controls are provided.

2 Problem formulations

Consider the following uncertain neutral system with discrete and distributed time delays:

$$\dot{x}(t)=\left(A_0+\Delta A_0(t)\right)x(t)+\left(A_1+\Delta A_1(t)\right)x(t-h)+\left(A_2+\Delta A_2(t)\right)\dot{x}(t-\tau)$$
$$+\left(A_3+\Delta A_3(t)\right)\int_{t-\eta}^{t}x(s)ds+\left(B+\Delta B(t)\right)u(t)+\left(B_w+\Delta B_w(t)\right)w(t), \quad (1a)$$

[a] Corresponding author: *tdchen@nqu.edu.tw*.

$$y(t)=C_1x(t)+D_1u(t), \quad (1b)$$

$$z(t)=C_2x(t)+D_2u(t)+D_3w(t), \quad (1c)$$

$$x(t)=\phi(t),\ t\in[-H,0], \quad (1d)$$

where $x\in\Re^n$, x_t is the state at time t defined by $x_t(s):=x(t+s)$, $\forall s\in[-H,0]$, $H=\max\{h,\tau,\eta\}>0$, $u\in\Re^m$ is the input, $w\in\Re^l$ is the disturbance input, $y\in\Re^p$ is the measured output, $z\in\Re^q$ is the regulated output, A_0, A_1, A_2, $A_3\in\Re^{n\times n}$, $B\in\Re^{n\times m}$, $B_w\in\Re^{n\times l}$, $C_1\in\Re^{p\times n}$, $D_1\in\Re^{p\times m}$, $C_2\in\Re^{q\times n}$, $D_2\in\Re^{q\times m}$, and $D_3\in\Re^{q\times l}$ are constant matrices, the pairs (A_0,B) and (A_0,C_1) are stabilizable and detectable, respectively. The matrices $\Delta A_0(t)$, $\Delta A_1(t)$, $\Delta A_2(t)$, $\Delta A_3(t)$, $\Delta B(t)$, and $\Delta B_w(t)$, are some perturbations with appropriate dimensions, and the initial vector ϕ is a continuous function. The following assumptions are made on system (1).

Assumption 1. Suppose that the matrix C_1 has full row rank (i.e., $rank(C_1)=p$).

For convenience of discussed, the singular value decomposition of C_1 as follows:

$$C_1=U[S\quad 0]V^T,$$

where $S\in\Re^{p\times p}$ is a diagonal matrix with nonnegative diagonal elements in decreasing order, $0\in\Re^{p\times(n-p)}$ is a zero matrix, and $U\in\Re^{p\times p}$ and $V\in\Re^{n\times n}$ are unitary matrices.

Assumption 2. The perturbed matrices $\Delta A_0(t)$, $\Delta A_1(t)$, $\Delta A_2(t)$, $\Delta A_3(t)$, $\Delta B(t)$, and $\Delta B_w(t)$ satisfy

$$\begin{aligned}&[\Delta A_0(t)\ \ \Delta A_1(t)\ \ \Delta A_2(t)\ \ \Delta A_3(t)\ \ \Delta B(t)\ \ \Delta B_w(t)]\\&=M\cdot F(t)\cdot[N_0\ \ N_1\ \ N_2\ \ N_3\ \ N_4\ \ N_5]\end{aligned},$$

where matrices M, N_0, N_1, N_2, N_3, N_4, and N_5 are constant with appropriate dimensions and the uncertain parameter $F(t)$ satisfies $F^T(t)F(t)\le\nu\cdot I$ with $\nu>0$.

Remark 1. The parameter perturbations $\Delta A_0(t)$, $\Delta A_1(t)$, $\Delta A_2(t)$, $\Delta A_3(t)$, $\Delta B(t)$, and $\Delta B_w(t)$, that appear in system (1) will represent the impossibility for exact mathematical model of a dynamic system due to the system complexity. The uncertainty has been widely used in many practical systems which can be either exactly modeled or overbounded by the condition $F^T(t)F(t)\le\nu\cdot I$. The ν is the bounding parameter on the uncertain perturbation $F(t)$, the matrix $F(t)$ contains the uncertain parameters, and constant matrices M and N_i, $i=0,1,2,3,4,5$, specify how the uncertain parameter $F(t)$ affect the nominal matrix of system (1).

We wish to design the following modified observer-based control with the known time-delay values h, τ and η for system (1):

$$\begin{aligned}\dot{\hat{x}}(t)&=A_C\hat{x}(t)+A_D\hat{x}(t-h)+A_E\dot{\hat{x}}(t-\tau)\\&+A_F\int_{t-\eta}^{t}\hat{x}(s)ds+Bu(t)+L[y(t)-\hat{y}(t)]\end{aligned}, \quad (2a)$$

$$\hat{y}(t)=C_1\hat{x}(t)+D_1u(t), \quad (2b)$$

$$u(t)=K\hat{x}(t), \quad (2c)$$

where $\hat{x}\in\Re^n$ is the estimation of x, $\hat{y}\in\Re^p$ is the observer output, $K\in\Re^{m\times n}$ is the controller gain, $L\in\Re^{n\times p}$ is the observer gain, and A_C, A_D, A_E, $A_F\in\Re^{n\times n}$ are matrices to be determined.

Definition 1. Consider the system (1) with observer-based control (2) and the following conditions are satisfied:

(i) With $w(t)=0$, the closed-loop system (1) with (2) is asymptotically stable.

(ii) With zero initial condition (i.e. $x_0=\hat{x}_0=0$), the signals $w(t)$ and $z(t)$ are bounded by

$$\int_0^\infty\|z(t)\|^2dt\le\gamma^2\cdot\int_0^\infty\|w(t)\|^2dt\ \text{(i.e., }\|z\|_2^2\le\gamma^2\cdot\|w\|_2^2\text{)},$$

$\forall w\in L_2[0,\infty)$, $w\ne0$, for some $\gamma>0$. In this condition, the system (1) is said to be stabilizable with disturbance attenuation γ and degree ν by observer-based control (2), and the control law (2) is said to be an H_∞ observer-based control for system (1). The parameter γ is said to be the H_∞-norm bound for the H_∞ observer-based control, and the parameter ν is said to be the perturbed bound for the H_∞ observer-based control.

Lemma 1.[9] For a given $C_1\in\Re^{p\times n}$ with $rank(C_1)=p$, assume that $X\in\Re^{n\times n}$ is a symmetric matrix, then there exists a matrix $\hat{X}\in\Re^{p\times p}$ such that $C_1X=\hat{X}C_1$ if and only if

$$X=V\begin{bmatrix}\hat{X}_{11}&0\\0&\hat{X}_{22}\end{bmatrix}V^T,$$

where $\hat{X}_{11}\in\Re^{p\times p}$ and $\hat{X}_{22}\in\Re^{(n-p)\times(n-p)}$.

Lemma 2. [10] Let $\Omega_0(x)$ and $\Omega_1(x)$ be two arbitrary quadratic forms over $\Re^n$, then $\Omega_0(x)<0$ for all $x\in\Re^n-\{0\}$ satisfying $\Omega_1(x)\le0$ if and only if there exist $\varepsilon\ge0$ such that

$\Omega_0(x)-\varepsilon\cdot\Omega_1(x)<0$, $\forall x\in\Re^n-\{0\}$.

3 Robust H_∞ control

By (2), system (1) and (2) can be rewritten as

$$\begin{aligned}\begin{bmatrix}\dot{\hat{x}}(t)\\\dot{e}(t)\end{bmatrix}&=\begin{bmatrix}A_C+BK&LC_1\\A_0-A_C&A_0-LC_1\end{bmatrix}\begin{bmatrix}\hat{x}(t)\\e(t)\end{bmatrix}+\begin{bmatrix}A_D&0\\A_1-A_D&A_1\end{bmatrix}\begin{bmatrix}\hat{x}(t-h)\\e(t-h)\end{bmatrix}\\&+\begin{bmatrix}A_E&0\\A_2-A_E&A_2\end{bmatrix}\begin{bmatrix}\dot{\hat{x}}(t-\tau)\\\dot{e}(t-\tau)\end{bmatrix}+\begin{bmatrix}A_F&0\\A_3-A_F&A_3\end{bmatrix}\int_{t-\eta}^{t}\begin{bmatrix}\hat{x}(s)\\e(s)\end{bmatrix}ds\\&+\begin{bmatrix}0\\B_w\end{bmatrix}w(t)+\begin{bmatrix}0\\M\end{bmatrix}\tilde{\Delta}(t)\end{aligned}, \quad (3a)$$

where the signal $e(t)=x(t)-\hat{x}(t)$ is defined as the estimated error of system, $\tilde{\Delta}(t)=F(t)(N_0+N_4K)\hat{x}(t)+N_0e(t)+N_1\hat{x}(t-h)+N_1e(t-h)$

 DOI: 10.1051/matecconf/20167105001

$+N_2\dot{\hat{x}}(t-\tau)+N_2\dot{e}(t-\tau)+N_3\int_{t-\eta}^{t}\hat{x}(s)ds+N_3\int_{t-\eta}^{t}e(s)ds+N_5w(t))$ and the uncertainty $\widetilde{\Delta}(t)$ satisfy the following quadratic inequality:

$$\widetilde{\Delta}^T(t)\cdot\widetilde{\Delta}(t)\le v\cdot\begin{bmatrix}\hat{x}(t)\\e(t)\\\hat{x}(t-h)\\e(t-h)\\\dot{\hat{x}}(t-\tau)\\\dot{e}(t-\tau)\\\int_{t-\eta}^{t}\hat{x}(s)ds\\\int_{t-\eta}^{t}e(s)ds\\w(t)\end{bmatrix}^T\begin{bmatrix}(N_0+N_4K)^T\\N_0^T\\N_1^T\\N_1^T\\N_2^T\\N_2^T\\N_3^T\\N_3^T\\N_5^T\end{bmatrix}\begin{bmatrix}(N_0+N_4K)^T\\N_0^T\\N_1^T\\N_1^T\\N_2^T\\N_2^T\\N_3^T\\N_3^T\\N_5^T\end{bmatrix}^T\begin{bmatrix}\hat{x}(t)\\e(t)\\\hat{x}(t-h)\\e(t-h)\\\dot{\hat{x}}(t-\tau)\\\dot{e}(t-\tau)\\\int_{t-\eta}^{t}\hat{x}(s)ds\\\int_{t-\eta}^{t}e(s)ds\\w(t)\end{bmatrix}. \quad (3b)$$

The minimal disturbance attenuation γ, the allowable maximal bound v, and the H_∞ observer-based control (2) could be solved simultanously from the following result.

Theorem 1. Consider the system (1) with the observer-based control (2). Suppose

$\left\|\begin{bmatrix}A_E & 0\\A_2-A_E & A_2\end{bmatrix}\right\|+\sqrt{2}\cdot\|M\|\cdot\|N_2\|<1$ and the following optimization problem:

$$\min_{\kappa,\rho,X_1,\hat{X}_{11},\hat{X}_{22},X_3,X_4,X_5,X_6,X_7,X_8,W_1,W_2,W_3,W_4,W_5,W_6}\kappa+\rho, \quad (4a)$$

subject to

$$\begin{bmatrix}
\Phi_{11} & \Phi_{12} & W_4 & 0 & W_5 & 0 & W_6 & 0 & 0 & 0 & X & 0 & \eta\cdot X & 0 & \Phi_{115} & \Phi_{116} & \Psi & \Phi_{118}\\
* & \Phi_{22} & \Phi_{23} & A_1X_4 & \Phi_{25} & A_2X_6 & \Phi_{27} & A_3X_8 & B_w & M & 0 & \Phi & 0 & \eta\cdot\Phi & \Phi_{215} & \Phi_{216} & \Phi C_2^T & \Phi_{218}\\
* & * & -X_3 & 0 & 0 & 0 & 0 & 0 & 0 & 0 & 0 & 0 & 0 & 0 & \Phi_{315} & \Phi_{316} & 0 & \Phi_{318}\\
0 & * & 0 & -X_4 & 0 & 0 & 0 & 0 & 0 & 0 & 0 & 0 & 0 & 0 & 0 & \Phi_{416} & 0 & \Phi_{418}\\
* & * & 0 & 0 & -X_5 & 0 & 0 & 0 & 0 & 0 & 0 & 0 & 0 & 0 & \Phi_{515} & \Phi_{516} & 0 & \Phi_{518}\\
0 & * & 0 & 0 & 0 & -X_6 & 0 & 0 & 0 & 0 & 0 & 0 & 0 & 0 & 0 & \Phi_{616} & 0 & \Phi_{618}\\
* & * & 0 & 0 & 0 & 0 & -X_7 & 0 & 0 & 0 & 0 & 0 & 0 & 0 & \Phi_{715} & \Phi_{716} & 0 & \Phi_{718}\\
0 & * & 0 & 0 & 0 & 0 & 0 & -X_8 & 0 & 0 & 0 & 0 & 0 & 0 & 0 & \Phi_{816} & 0 & \Phi_{818}\\
0 & * & 0 & 0 & 0 & 0 & 0 & 0 & -\kappa\cdot I & 0 & 0 & 0 & 0 & 0 & 0 & B_w^T & D_1^T & N_5^T\\
0 & * & 0 & 0 & 0 & 0 & 0 & 0 & 0 & -I & 0 & 0 & 0 & 0 & 0 & M^T & 0 & 0\\
* & 0 & 0 & 0 & 0 & 0 & 0 & 0 & 0 & 0 & -X_3 & 0 & 0 & 0 & 0 & 0 & 0 & 0\\
0 & * & 0 & 0 & 0 & 0 & 0 & 0 & 0 & 0 & 0 & -X_4 & 0 & 0 & 0 & 0 & 0 & 0\\
* & 0 & 0 & 0 & 0 & 0 & 0 & 0 & 0 & 0 & 0 & 0 & -X_7 & 0 & 0 & 0 & 0 & 0\\
0 & * & 0 & 0 & 0 & 0 & 0 & 0 & 0 & 0 & 0 & 0 & 0 & -X_8 & 0 & 0 & 0 & 0\\
* & * & * & 0 & * & 0 & * & 0 & 0 & 0 & 0 & 0 & 0 & 0 & -X_5 & 0 & 0 & 0\\
* & * & * & * & * & * & * & * & * & * & 0 & 0 & 0 & 0 & 0 & -X_6 & 0 & 0\\
* & * & 0 & 0 & 0 & 0 & 0 & 0 & * & 0 & 0 & 0 & 0 & 0 & 0 & 0 & -I & 0\\
* & * & * & * & * & * & * & * & * & 0 & 0 & 0 & 0 & 0 & 0 & 0 & 0 & -\rho\cdot I
\end{bmatrix}<0, \quad (4b)$$

has a solution $\kappa>0$, $\rho>0$, $X_1\in\Re^{n\times n}>0$, $\hat{X}_{11}\in\Re^{p\times p}>0$, $\hat{X}_{22}\in\Re^{(n-p)\times(n-p)}>0$, $X_3\in\Re^{n\times n}>0$, $X_4\in\Re^{n\times n}>0$, $X_5\in\Re^{n\times n}>0$, $X_6\in\Re^{n\times n}>0$, $X_7\in\Re^{n\times n}>0$, $X_8\in\Re^{n\times n}>0$, $W_1\in\Re^{n\times n}$, $W_2\in\Re^{m\times n}$, $W_3\in\Re^{n\times p}$, $W_4\in\Re^{n\times n}$, $W_5\in\Re^{n\times n}$, and $W_6\in\Re^{n\times n}$, where

$$\Phi=V\begin{bmatrix}\hat{X}_{11} & 0\\0 & \hat{X}_{22}\end{bmatrix}V^T,\ \Psi=(C_2X_1+D_2W_2)^T,$$

$\Phi_{11}=W_1+W_1^T+BW_2+W_2^TB^T$, $\Phi_{12}=W_3C_1+X_1A_0^T-W_1^T$, $\Phi_{22}=A_0\Phi+\Phi A_0^T-W_3C_1-C_1^TW_3^T$, $\Phi_{23}=A_1X_3-W_4$, $\Phi_{25}=A_2X_5-W_5$, $\Phi_{27}=A_3X_7-W_6$, $\Phi_{115}=(W_1+BW_2)^T$, $\Phi_{215}=C_1^TW_3^T$, $\Phi_{315}=W_4^T$, $\Phi_{515}=W_5^T$, $\Phi_{715}=W_6^T$, $\Phi_{116}=X_1A_0^T-W_1^T$, $\Phi_{216}=\Phi A_0^T-C_1^TW_3^T$, $\Phi_{316}=X_3A_1^T-W_4^T$, $\Phi_{416}=X_4A_1^T$, $\Phi_{516}=X_5A_2^T-W_5^T$, $\Phi_{616}=X_6A_2^T$, $\Phi_{118}=(N_0X_1+N_4W_2)^T$, $\Phi_{218}=\Phi N_0^T$, $\Phi_{314}=X_3N_1^T$, $\Phi_{418}=X_4N_1^T$, $\Phi_{518}=X_5N_2^T$, and $\Phi_{618}=X_6N_2^T$. Then the system (1) is robustly stabilizable with disturbance attenuation $\gamma=\sqrt{\kappa}$ and degree $\upsilon=\rho^{-1}$ by the H_∞ observer-based control (2). The matrices $A_C=W_1X_1^{-1}$, $A_D=W_4X_3^{-1}$, $A_E=W_5X_5^{-1}$, $A_F=W_6X_7^{-1}$, $K=W_2X_1^{-1}$, and $L=W_3US\hat{X}_{11}^{-1}S^{-1}U^T$ are obtained, respectively.

Proof: Due to the limitation of page, it is omitted.

4 Example

Consider the neutral system (1) with the parameters:

$$A_0=\begin{bmatrix}1 & 0.5 & 1\\-0.5 & -2 & 1\\1 & -1 & -1.1\end{bmatrix},\quad A_1=\begin{bmatrix}-0.7 & 0.6 & -0.6\\-0.3 & -0.2 & 0.7\\0.2 & 0.5 & -0.3\end{bmatrix},$$

$$A_2=\begin{bmatrix}-0.1 & 0.1 & -0.2\\-0.3 & -0.2 & 0.1\\0 & 0 & -0.3\end{bmatrix},\quad A_3=\begin{bmatrix}0.1 & 0 & 0.2\\0.1 & 0.2 & 0.1\\0.1 & 0 & 0\end{bmatrix},$$

$$B=\begin{bmatrix}-1 & 0\\1 & -1\\0 & -1\end{bmatrix},\quad B_w=\begin{bmatrix}0.1 & 0\\0 & 0\\0.1 & 0\end{bmatrix},\quad C_1=\begin{bmatrix}0.1 & 0.2 & 0.1\\0.1 & 0.5 & 0\end{bmatrix},$$

$$D=\begin{bmatrix}0.1 & 0.1 & 0\end{bmatrix},\quad D_2=\begin{bmatrix}0.1 & 0\end{bmatrix},\quad D_3=\begin{bmatrix}0 & 0.1\end{bmatrix},\quad M=\begin{bmatrix}0.2\\0.3\\0.4\end{bmatrix},$$

$N_0=\begin{bmatrix}0 & 0.5 & 0\end{bmatrix}$, $N_1=\begin{bmatrix}0.2 & 0 & 0\end{bmatrix}$, $N_2=\begin{bmatrix}0.1 & 0.1 & 0\end{bmatrix}$, $N_3=\begin{bmatrix}0.1 & 0 & 0.1\end{bmatrix}$, $N_4=\begin{bmatrix}0.2 & 0.1\end{bmatrix}$, $N_5=\begin{bmatrix}0.1 & 0.1\end{bmatrix}$, $\eta=0.1$. (4)

By Theorem 1 with (A1)-(A2), and the optimization problem (4), the dynamic feedback gains in (2) are given by

$$A_C=\begin{bmatrix}0.4654 & 0.2771 & 0.8436\\-2.541 & -2.8088 & 0.4225\\2.8507 & -0.2738 & -0.5808\end{bmatrix},\quad A_D=\begin{bmatrix}-0.8729 & 0.6307 & -0.6077\\-0.445 & -0.1854 & 0.6961\\0.0142 & 0.5494 & -0.3123\end{bmatrix}$$

$$A_E=\begin{bmatrix}-0.1792 & 0.0368 & -0.21\\-0.3692 & -0.2617 & 0.0952\\-0.0813 & -0.0554 & -0.3161\end{bmatrix},\quad A_F=\begin{bmatrix}2.9477 & -0.1598 & 2.9731\\-0.1598 & 0.114 & -0.0912\\2.9731 & -0.0912 & 3.0539\end{bmatrix},$$

$$K=\begin{bmatrix}-15.9336 & -9.5613 & -5.2457\\89.6026 & 37.1001 & 26.8625\end{bmatrix},\quad L=\begin{bmatrix}510.453 & -240.4732\\130.0944 & -53.6957\\524.9408 & -255.4172\end{bmatrix}.$$

The system (1) is stabilizable by the H_∞ observer-based control (2) with (18), disturbance attenuation $\gamma=\sqrt{\kappa}=0.5453$, and degree $v=\rho^{-1}=0.9206$.

In this situation, the states of the system (1) with (4) could not be measurable or have no any practical sense. Hence, the state feedback control schemes in [7] cannot be applied to design the H_∞ control of system. Obviously, the design scheme derived in this paper is more efficient and flexible.

5 Conclusions

In this paper, the problem of H_∞ control design for a class of uncertain systems has been considered. LMI optimization approach has been developed to construct the output H_∞ dynamic observer-based feedback control. A numerical example has been given to demonstrate the merits of the proposed results. Finally, we have noticed that the useful results on the robust observer-based control problem for linear systems with perturbations [11]. Based on the LMI approach, the robust observer-based control has been derived. However, the conditions including the constraint of matrix equality are not the classic LMI feasible form. Their results can not use the efficient software MATLAB to solve the LMI problem with equality constraint. Our obtained results can be performed directly by LMI toolbox of MATLAB.

References

1. J. K. Hale, S. M. Verduyn Lunel, *Introduction to Functional Differential Equations* (Springer Verlag, New York, NY 1993).
2. V. B. Kolmanoskii, A. Myshkis, *Applied Theory of Functional Differential Equations* (Kluwer Academic Publishers, Dordrecht, Netherlands 1992).
3. C. Y. Lu, J. S. H. Tsai, G. J. Jong, T. J. Su, An LMI-Based Approach for Robust Stabilization of Uncertain Stochastic Systems with Time-Varying Delays, IEEE Transactions on Automatic Control **48** 286-289 (2003).
4. D. Salamon, Control and Observation of Neutral Systems, Pitman Advanced Publishing Program, Boston, Massachusetts (1984).
5. J. H. Park, On the Design of Observer-Based Controller of Linear Neutral Delay-Differential Systems, Applied Mathematics and Computation, **150** 195-202, (2004).
6. J. D. Chen, An LMI-Based Robust H_∞ Control of Uncertain Neutral Systems with State and Input Delays, Journal of Optimization Theory and Applications, **126** 553-570 (2005).
7. E. Fridman, U. Shaked, Delay-Dependent Stability and H_∞ Control: Constant and Time-Varying Delays, International Journal of Control, **76** 48-60 (2003).
8. J. H. Park, O. Kwon, S. Won, LMI Approach to Robust H_∞ Filtering for Neutral Delay Differential Systems, Applied Mathematics and Computation, **150** 235-244 (2004).
9. D. W. C. Ho, G. Lu, Robust stabilization for a class of discrete-time non-linear systems via output feedback: the unified LMI approach, International Journal of Control, **76** 105-115 (2003).
10. V. A. Yakubovich, S-Procedure in Nonlinear Control Theory, Vestnick Leningrad Univ. Math., **4** 73-93 (1977).
11. C. H. Lien, An Efficient Method to Design Robust Observer-Based Control of Uncertain Linear Systems, Applied Mathematics and Computation, **158** 29-44 (2004).

MATEC Web of Conferences 71, 05002 (2016)
ICCPE 2015 DOI: 10.1051/matecconf/20167105002

An FPGA-Based Multiple-Axis Velocity Controller and Stepping Motors Drives Design

Chiu-Keng Lai[1,*], *Wei-Nan* Chien[1], *and Yaw-Ting* Tsao[1]

[1]Department of Electrical Engineering, National Chin-Yi University of Technology
No. 57, Sec. 2, Zhongshan Rd, Taiping Dist., Taichung 41170, Taiwan

Abstract. A Field Programmable Gate Array based system is a great hardware platform to support the implementation of hardware controllers such as PID controller and fuzzy controller. It is also programmed as hardware accelerator to speed up the mathematic calculation and greatly enhance the performance as applied to motor drive and motion control. Furthermore, the open structure of FPGA-based system is suitable for those designs with the ability of parallel processing or soft code processor embedded. In this paper, we apply the FPGA to a multi-axis velocity controller design. The developed system integrated three functions inside the FPGA chip, which are respectively the stepping motor drive, the multi-axis motion controller and the motion planning. Furthermore, an embedded controller with a soft code processor compatible to 8051 micro-control unit (MCU) is built to handle the data transfer between the FPGA board and host PC. The MCU is also used to initialize the motion control and run the interpolator. The designed system is practically applied to a XYZ motion platform which is driven by stepping motors to verify its performance.

1 Introduction

A Field Programmable Gate Array (FPGA) based system is a great hardware platform to support the implementation of controllers such as PID controller, fuzzy controller, adaptive controller, optimum controller, FIR filter and even neuro network system [1-4]. It could also be programmed as an ASIC chip, which combines many functions into one FPGA chip, such as PID controller, velocity profile generation, interpolator, and data conversion [5]. It is also used to speed up the mathematic calculation and greatly enhance the performance of the motion planning and drive design [6-8], or used on the other relative field where hardware calculation is important. In addition, due to the properties of open structure, parallel processing design with FPGA is possible and easily designed [7, 9]. Besides, some IPs of soft code processor for system on a chip (SOC) design are provided, such as the MCU NIOS II, Arduino, or ARM DSP [2-4, 10-13], which could be adopted to easily realize the design of embedded system, or design in co-processor operation. With those IPs, the users could easily program their peripheral circuits connecting to the embedded soft code processor to reach their desired functions and performances.

In this paper, we investigate the feasibility of a motion planning and motion control implemented by FPGA. The proposed motion control chip is based on a soft code processor 8051 and as an embedded system to realize the multi-axis motion control where the velocity control is default to be constrained by the trapezoidal velocity profile. The designed system also realizes the interpolator for multi-axis synchronous operation by adopting the Digital Differential Analyzer (DDA) algorithm [13-16]. Furthermore, the designed hardware system is connected to a host PC through the universal asynchronous receiver transmitter (UART) interface to receive the pulse commands dominated by the human machine interface (HMI) on PC. Such a soft code MCU-based embedded system has several advantages. Since it is a dedicated and open structure platform, it can be optimized to run the motion planning algorithms in the most efficient way. It may also be attached to the host computer to reduce the computation time by sharing the PC resources. The cost of such hardware system or processor would be a fraction of a general purpose CPU, hence hundreds of soft code MCU can be deployed in parallel to further increase the computational capabilities as compared with those controlled by PC only, or have the competitiveness as compared to such systems of PC plus axis motion control cards due to their flexibility in design. Most important of all, by implementing such a processor and system on reconfigurable hardware, and utilizing high level abstraction of the motion planning algorithms, the motion system is easily realized and totally integrated into the embedded system.

The paper is organized in the following manner. In Section 2, the architecture of the proposed system is presented. After that, the designed modules for motion control and velocity profile implementation are shown in Section 3. In Section 4, the stepping motor drive design used to the experiment is shown. In Section 5, the

* Corresponding author: chiukl@ncut.edu.tw

experimental results followed the proposed algorithms are shown by two-axis synchronous operation. Finally, the conclusions are given in Section 6.

2 The Architecture of the Integrated System

The block diagram of the developed system is shown in Fig. 1. It includes a Human Machine Interface (HMI) on PC, the motion controller executed by embedded 8051 MCU, the interpolator, the stepping motor drive system. The developed control system could be applied to three-axis motion system driven by stepping motor, it could also be used to the servo systems which are set in position mode and driven by the A/B phases or pulse/direction commands.

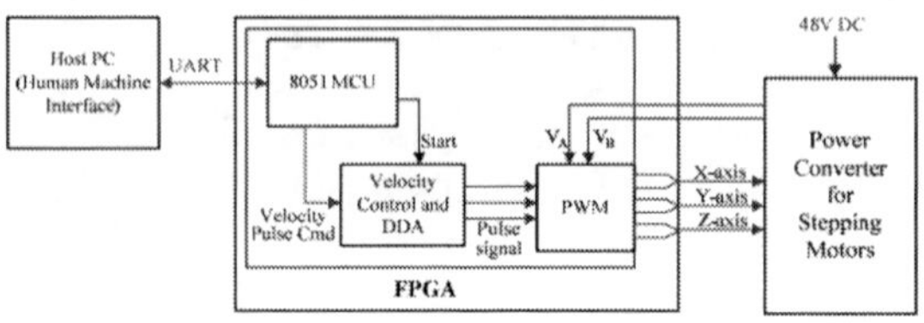

Fig. 1. The architecture of multi-axis motion control system.

The displacement commands for each of the motors are from the HMI on the host PC, and are transferred to the motion controller through the UART interface to generate the stepping pulses under the considerations of trapezoidal velocity profile and multi-axis synchronous operation. After that, the pulse commands are sent to the DDA modules to equally output the stepping pulses to the stepping motor drive system. Finally, the duty cycle controls are added to the stepping commands by passing them through the Pulse Width Modulation (PWM) modules with the current feedbacks, V_A and V_B [17], the motor currents are then regulated on the desired level. As stated above, the overall system integrates the motion planning, velocity control and the stepping motor drive in one chip, which has largely reduced the required complicated structure and the area of PCB.

3 The Motion Control Module Design using FPGA

3.1 The Velocity Programming Module

This module generates the desired acceleration and deceleration performances, and makes the run speed to the permitted maximum feed rate ($V_{\max}$ in pps). The acceleration and deceleration are constrained by the trapezoidal velocity profile as shown in Fig. 2. The sampling time for the testing motion platform is set as $T = 6.5$ ms. Thus, according to the profile shown in Fig. 2, the system maximum feed rate is limited to 180 rpm for X, Y and Z axes. In Fig. 2, the parameters of a is the acceleration and T is the sampling time. To let the system have the same displacements in the acceleration and deceleration durations as Fig. 2 shows, the motors are ideally accelerated to $V_{\max}$ from standstill within duration time 4T, and the duration time is also suitable for the deceleration process. The procedures to generate the velocity command with the trapezoidal velocity profile are as in the following three steps:

(a) In the acceleration duration, the motor is accelerated from the initial speed V_0 with constant acceleration until the motor reaches to the maximum feed rate $V_{\max}$. The amount of pulse in each sampling time is increased by one, which leads to a constant acceleration.

(b) In the constant speed duration, a constant amount of pulse in each sampling time period is generated. Furthermore, the total displacement (in pulse number) and the remaining displacement off the destination are used to decide the time instant for system switching to the deceleration duration. The decisions above-stated are executed by 8051 MCU in FPGA.

(c) In the deceleration duration, the motor speed is decelerated from the $V_{\max}$ to V_0 with constant deceleration rate. In here, deceleration procedure decreases one pulse for each of the sampling time. However, the remaining distance is another consideration when in deceleration duration, and is discussed in subsection 3.3.

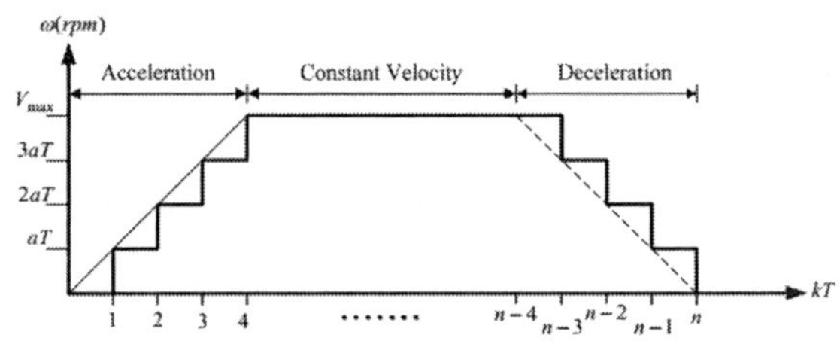

Fig. 2. The trapezoidal velocity profile.

3.2 The Interpolator by Hardware DDA

The parallel processing capability of FPGA makes the multi-axis operation realizable [9] within one chip system. Also, for each of the drive system, its interpolator in here is accomplished by the DDA algorithm. The hardware DDA for one axis, which is shown in Fig. 3, is mainly made by two 8-bit registers and one 8-bit adder. The motion control system accesses the pulse commands processed in last sub-section, and equally sends the stepping pulse to the motor drive system controlled by DDA. In Fig. 3, the output port $\Delta Z[n] = 1$ stands for one stepping signal is occurred.

3.3 The Two-axis Velocity Command Generation

In Fig. 4, it demonstrates the algorithm for two axes, X-axis and Y-axis, motion synchronously, and assumes that the displacement of X-axis (long axis) is greater than Y-axis (short axis). The algorithms are described as follows.

3.3.1 The X-axis Motion Control

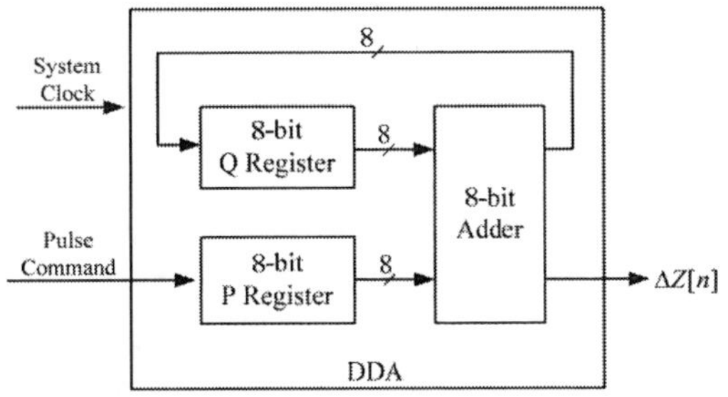

Fig. 3. The block diagram of DDA module.

There are five steps given to demonstrate the generation of stepping pulse for long axis.

Step 1. The pulse commands are stored in the host PC, and the displacements for the two axes are loaded ahead from the host PC to 8051 through the UART interface.

Step 2. The motor dynamics and feed rate are determined by the following equations represented in continuous system,

$$v_X = V_0 + at \tag{1}$$

$$S_X = V_0 t + \frac{1}{2}at^2 \tag{2}$$

$$v_X^2 = V_0^2 + 2aS_X \tag{3}$$

where v_X is the X-axis speed for each of the durations, a is the corresponding acceleration, S_X is the displacement in X-axis and t is the duration time. Motion control by digital controller is usually represented in discrete type. Thus, all the variables shown in Equations (1)~(3) could be represented in discrete type, such as

$$v_X[n] = V_0 + na[n] \tag{4}$$

where $n = 0, 1, 2, \cdots$.

According to the motion control by trapezoidal velocity profile and the dynamics of Equations (1) ~ (3), the motor speed at the three durations shown in Fig. 2 can be decided as follows:

(a) Acceleration duration:

Drive system is accelerated under constant acceleration and the drive speed is represented as

$$v_a[n] = V_0 + na[n] \tag{5}$$

(b) Run duration (Constant velocity):

If the point-to-point displacement is long enough, then the system in run duration may existent, and maximum feed rate is asserted.

$$v_c = V_{\max} \tag{6}$$

(c) Deceleration duration:

Deceleration is made by decreasing the amount of the generated pulse monotonously.

$$v_d[n] = V_{\max} - na[n] \tag{7}$$

In Equations (5)~(7), v_a, v_c and v_d are the velocities for the three durations, and are represented as pulse per sampling duration.

Assuming the displacements of acceleration and deceleration durations are the same, and the constant velocity duration exists, then the total displacements, S_X, could be represented as

$$\begin{aligned} S_X &= S_a + S_c + S_d \\ &\overset{\Delta}{=} L_{des\_X} \end{aligned} \tag{8}$$

where S_a, S_c and S_d are respectively the displacement of acceleration duration, constant velocity duration and deceleration duration, and L_{des_X} is the desired displacement of X-axis for each of the command line. If the desired displacement is long enough, then $V_{\max}$ can be the permitted maximum speed. On the other hand, for a short displacement, $(S_a + S_d) >> S_c$, the motion with trapezoidal velocity profile is modified as triangle velocity profile as shown in Fig. 5. From Eq. (3), the maximum feed speed $V_{\max}$ is thus set as

$$V_{\max} = \sqrt{aL_{des\_X}} \tag{9}$$

Since the displacements have been converted into the amount of pulse, and the stepping motor drive systems are set on position mode, the displacement must be represented in integer. We thus have the maximum velocity in integer as follows:

$$V_{\max} = \left\lfloor \sqrt{aL_{des\_X}} \right\rfloor \tag{10}$$

Step 3. According to Equations (5) and (9), the speed in acceleration duration can be realized and calculated in advance. At the duration, the controller monitors the motor speed, and checks whether the speed is greater than the $V_{\max}$ or not. Once it is true, the motion system is switched to the next duration, constant velocity or deceleration.

Step 4. In the constant velocity duration, the motor runs with the constant speed v_c, and simultaneously 8051 MCU checks the total displacement whether the remaining displacements are smaller than the displacement in the deceleration duration or not, i.e.,

$$\mathrm{Rem}_X = L_{des\_X} - S_a - S_c \le S_d \tag{11}$$

where Rem_X is the remaining displacement to the destination in X-axis. Then, the system goes into deceleration duration.

Step 5. Applying Eq. (7) to implement the deceleration control until the total displacement is reached.

3.3.2 The Y-axis Interpolation

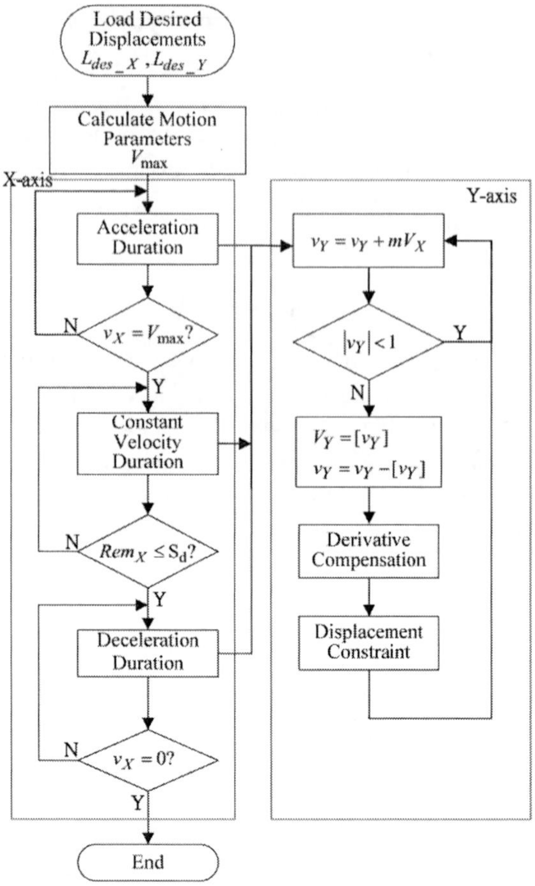

Fig. 4. The algorithm for the two-axis synchronous operation.

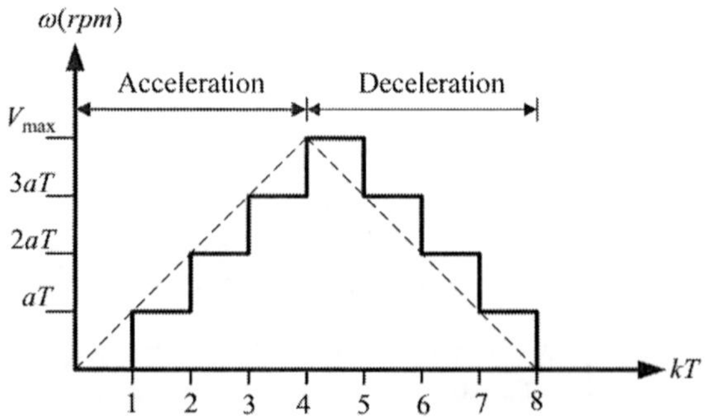

Fig. 5. The triangle velocity profile.

The implementation of Y-axis interpolation is according to the required total displacements, S_Y, and is defined as

$$S_Y = \sum v_Y[n] \tag{12}$$

where $v_Y[n]$ is the speed of Y-axis in the amount of stepping at the sampling instant nT. Define the slope m as

$$m = \frac{L_{des\_Y}}{L_{des\_X}} \tag{13}$$

where L_{des_Y} is the desired Y-axis total displacement. Eq. (13) is used to determine the desired fraction speed of Y-axis as comparing to X-axis. The flow chart is shown in Fig. 4. Since the velocity and position are represented as the amount of pulse, and the remaining displacements are subtracted from the desired position in pps. To avoid the condition of over speed, the algorithm is modified as follow.

Since the motion system is realized by hardware circuit, and finite precision of the data width may cause undesired situations such that the error of displacement may exist. Thus, when we implement the motion control, constraint or compensation is necessary [15]. In the designed system, two conditions may exist during the motion command generation and interpolation. The following is used to compensate those drawbacks.

One of the examples with over speed change on the short-axis (Y-axis) is shown in Fig. 6, where the maximum speed is set to 3 pulses per sampling time. Fig. 6 shows a sudden speed change on Y-axis between the sampling instants 3 and 4. A big sudden speed change may result to the missing of the stepping command owing to the command exceeds the permitted maximum acceleration or deceleration rate.

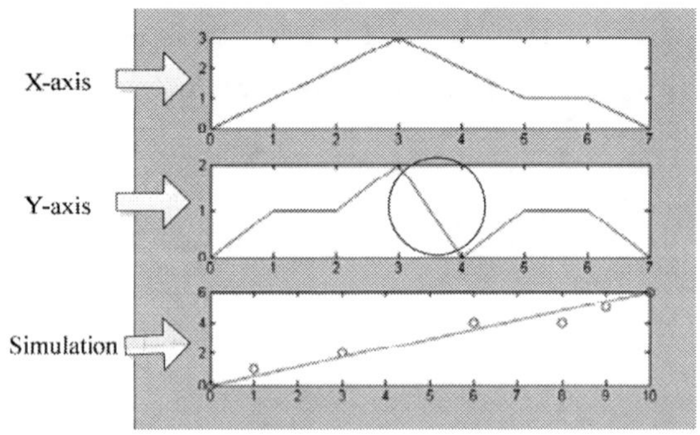

Fig. 6. The simulation results by Matlab without compensation.

To avoid the situation of suddenly large speed change, we add the derivative compensation as shown in Eqs. (14) and (15).

If $\frac{dv_Y}{dt} > a$, then

$$V_Y[n+1] = V_Y[n] + an \tag{14}$$

else

$$V_Y[n+1] = V_Y[n] + (\frac{dv_Y}{dt})n \tag{15}$$

Eq. (14) shows that if the steps 3, 4 and 5 in Fig. 6 lead to the derivative of speed be greater than the permitted maximum acceleration a, the next speed is decided by Eq. (14). Otherwise, the next speed is as the

DOI: 10.1051/matecconf/20167105002

one of Eq. (15). The simulated example shown in Fig. 6 is modified according to Eqs. (14) and (15), and the results are shown in Fig. 7, where the over speed change is avoided to prevent from missing stepping command.

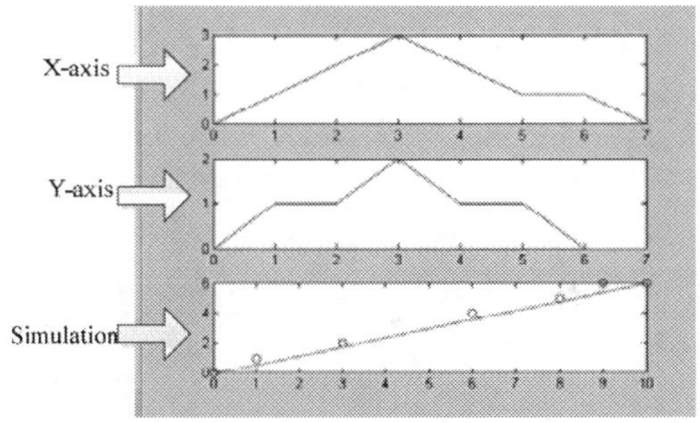

Fig. 7. The simulation results by matlab with acceleration compensation.

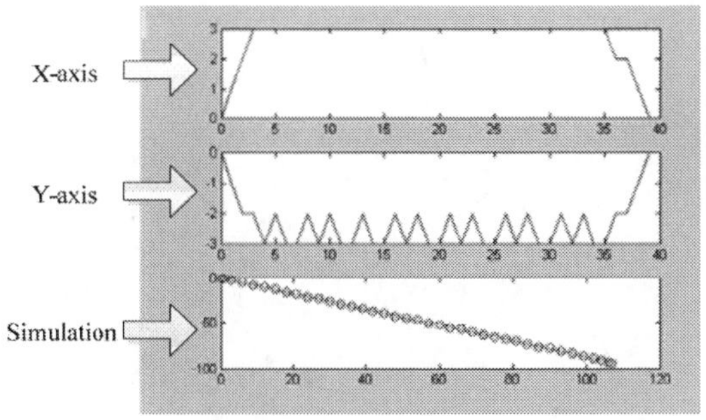

Fig. 8. The simulated results with displacements 107 and -93 on X-axis and Y-axis respectively. It is over interpolation on Y-axis with total displacement -94.

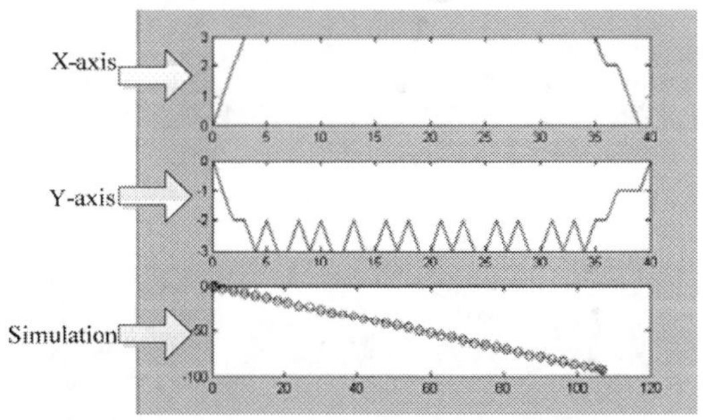

Fig. 9. The simulated results with compensation for the displacements of 107 and -93 on X-axis and Y-axis respectively.

Although the over speed interpolation is reduced with the derivative compensation, the over interpolation may also exist as the simulated results shown in Fig. 8, where the displacement commands on X- and Y-axis are respectively the step numbers of 107 and -93. Thus, if we just consider the over speed change, it is not enough to obtain an exact displacement because of the occurrence of over interpolation.

To avoid the situation of over pulse interpolation, we add the constraints of displacement on the Y-axis as follows:

If $\sqrt{\text{Rem}_Y} > V_Y[n]$, then

$$v_Y = V_Y[n] \tag{16}$$

else

$$v_Y = [\sqrt{\text{Rem}_Y}\,] \tag{17}$$

where Rem_Y is the remaining distance in the amount of pulse to the desired position, V_Y is the speed after the derivative compensation, and v_Y is the true interpolation speed. One of the algorithm used to compensate the problem is shown in Fig. 9 where the over interpolator is avoided since the decision of interpolation step is substituted by the remaining displacement of Y-axis.

4 The FPGA-based Control Board and Stepping Motor Drive based Motion Platform

Fig. 10(a) is the developed FPGA control board, which is mainly designed by Altera cyclone III FPGA, and an USB connector is used to communicate with host PC by UART protocol to download the processed G-code commands to the soft code processor 8051 on control board. The baud rate is 19200 bps. In addition, the control system outputs the stepping signals from the FPGA interface to the stepping motor drive system to drive the motors. The overall system setup is shown in Fig. 10(b).

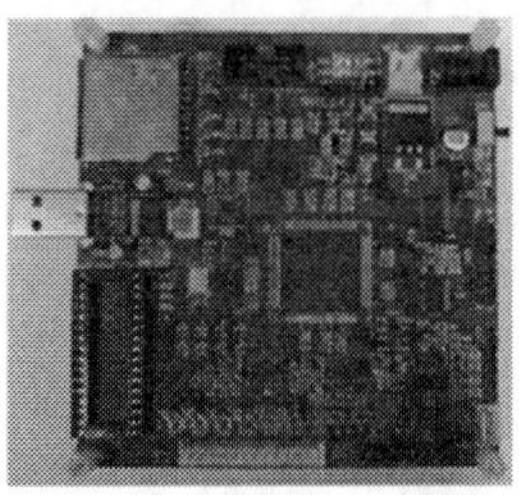

(a)

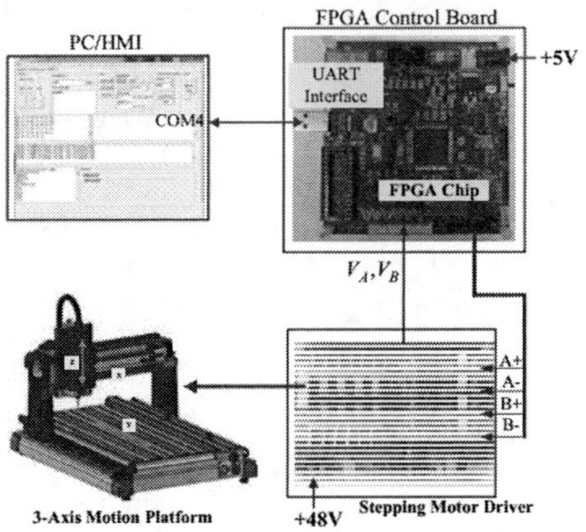

(b)

Fig. 10. (a) Control board, and (b) experimental setup.

DOI: 10.1051/matecconf/20167105002

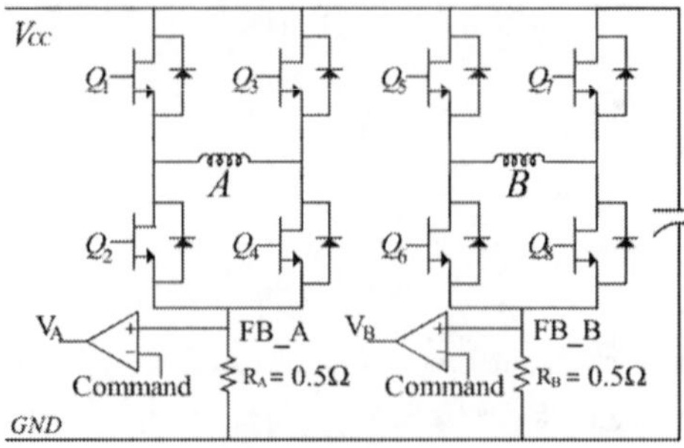

Fig. 11. The power converter of hybrid stepping motor.

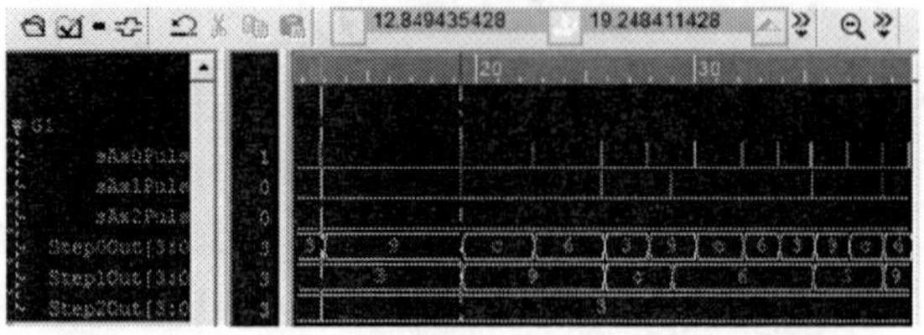

Fig. 12. Stepping signals for stepping motor drive.

The circuit of power converter for hybrid stepping motor drive is shown in Fig. 11. The DC bus voltage is 48 V and current command is set as 1 A. The stepping motors are operated in full-step mode and winding current detection is executed by the two resistors, R_A and R_B, as shown in Fig. 11. The winding current is compared by linear comparators, and their outputs (V_A and V_B) are used to decide the duty cycle of stepping motor drives. The PWM modules are operated in 20 kHz sampling frequency. The detail operations could be found in [17].

5 Simulations and Experimental Results

We apply those considerations and hardware setup stated above to the experiment of two-axis control. They are run under the conditions of six straight lines with different slops to evaluate whether the algorithms set well or not. The six lines are with the following coordinates: {(0,0), (5,5)}, {(1,0), (5,5)}, {(2,0), (5,5)}, {(3,0), (5,5)}; {(4,0), (5,5)}; {(5,0), (5,5)}. The stepping signals for stepping motors are simulated in Fig. 12, where two-axis operation with DDA is shown. The DDA output ports are sAx0Puls and sAx1Puls, and the stepping signals Step0Out[3:0] and Step1Out[3:0] are the signals for the hybrid stepping motor in full step moving. Note that in Fig. 12, the stepping signal for axis 3 isn't activated. The source G-codes are simulated as shown in Fig. 13.

Regarding to the practical experiment by the caving machine as the motion control platform, the system is designed to send the velocity commands from the host PC to the motion controller on the FPGA board through the UART interface. Two conditions are demonstrated by adopting the DDA and without it. Fig. 14 is the one which just executes the trapezoidal velocity motion control and works at two-axis operation, but the DDA modules isn't included. While in Fig. 15, the DDA modules are respectively added to X-axis and Y-axis. The experimental results show that the six lines are straight than those without the DDA. The motion control, interpolation and stepping motor drive are integrated and work well. The experimental results are as expected.

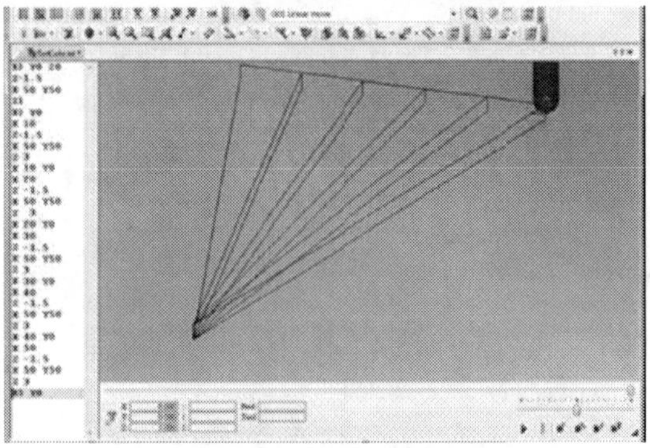

Fig. 13. The simulated results.

Fig. 14. The experimental result without DDA.

Fig. 15. The experimental result with DDA.

6 Conclusions

In the paper, we design the multi-axis velocity control system using the FPGA. The system includes a soft code processor 8051, the UART interface, the motion planning and interpolation, and the stepping motor drive. Simulations and experiments are demonstrated on a two-axis operation. The simulation and experimental results show that the developed multi-axis motion control system has reached the desired performance about the motion control and two axes operation synchronously. To improve the processing performance, a fast baud rate of communication is necessary. We are going to add the USB communication protocol to the system to enhance the performance of data transfer. As regarding to the embedded MCU 8051, it has the most instructions been run in one clock machine cycle. For further requirement on the processing performance, another MCU such as ARM could be adopted to have a better the performance.

References

1. N.K. Quang, Y.S. Kung, Q.P. Ha, "FPGA-based control architecture Integration for Multi-axis Tracking Motion System", IEEE/SICE 2011, pp. 591-596 (2011).
2. R. Nithya, K.R. Sarath Chandran, V.P. Chandramani, "Run-Time Reconfiguration of Processing Elements Through Soft-Core Processor", 2014 International Conference on Communications and Signal Processing (ICCSP 2014), pp. 813-817 (2014).
3. H.P. Sekaran, M.H. Liyanage, N. Krouglicof, "Optimization of Controller Gains for FPGA-based Multi Variable Motion Controller using Response Surface Methodology", IEEE 28th Canadian Conference on Electrical and Computer Engineering (CCECE 2015), pp. 1307-1312 (2015).
4. Z. Szadkowski, E.D. Fraenkel, A.M. Van Den Berg, "FPGA/NIOS Implementation of an Adaptive FIR Filter using Linear Prediction to Reduce Narrow Band RFI for Radio Detection of Cosmic Rays", IEEE Transactions on Nuclear Science, 60(5), pp. 3483-3490 (2013).
5. J.U. Cho, Q.N. Le, J.W. Jeon, "A FPGA-based Multiple-Axis Motion Control Chip", IEEE/IE, 56(3), pp. 856-870 (2009).
6. N. Atay, B. Bayazit, "A Motion Planning Processor on Reconfigurable Hardware", Proceedings of the 2006 IEEE International Conference on Robotics and Automation, pp. 125-132 (2006).
7. A. Ghasemazar, M. Goli, A. Afzali-Kusha, "Embedded Complex Floating Point Hardware Acceleration", 27th International Conference on VLSI Design and 2014 13th International Conference on Embedded Systems, pp. 318-323 (2014).
8. W. Wang, K. Guo, M. Gu, Y. Ma, Y. Wang, "A Universal FPGA-based Floating-point Matrix Processor for Mobile Systems", 2014 International Conference on Field-Programmable Technoloy (FPT2014), pp. 139-146 (2014).
9. C.K. Lai, K.L. Ho, "To Develop a Parallel Processing System with the FPGA for Multi-Axis Motion Control Platform", Applied Mechanics and Materials, 284-287, pp. 2396-2401 (2012).
10. J. Zhang, X. Wang, "Position Loop Method in Multi-axis Motion Controller using Extended DDA Circuit", IEEE International Conference on Automation and Logistics (ICAL '09), pp. 692-696 (2009).
11. X. Yang, Y. Song, P. Wang, "Research and Implementation of a Portable Character Recognition Device", International Conference on Computational and Information Sciences (ICCIS 2013), pp. 1815-1818 (2013).
12. H. Sun, Y. Zhang, J.J. Xue, Z. Wu, "The Remote Control System of the Manipulator", 33rd Chinese Control Conference (CCC 2014), pp. 8283-8286 (2014).
13. P. Desai Dev, "Designing and Improving the Efficiency of Hardware Interpolator", International Conference on Computing, Communication and Automation (ICCCA 2015), pp. 959-962 (2015).
14. C.Y. Chen, P.S. Liao, C.C. Cheng, G.F. Jong, "Design and implementation of integrated non-uniform rational B-spline and digital differential analyzer interpolators for computerized numerical control servo-controllers", Proceedings of the Institution of Mechanical Engineers, Part C: Journal of Mechanical Engineering Science, 221(9), pp. 1075-1087 (2007).
15. K.H. Su, C.K. Hu, M.Y. Cheng, "Design and Implementation of an FPGA-based Motion Command Generation Chip", IEEE International Conference on Systems, Man and Cybernetics (SMC '06), 6, pp. 5030-5035 (2006).
16. W.H. Jiang, H. Wang, "Design of Three-axis Servo System Based on Linear Motor", 2010 International Conference on Computer Design and Applications (ICCDA 2010), 3, pp. V3-182 - V3-185 (2010).
17. C.K. Lai, W.N. Chen; S.L. Tsai, Y.T. Tsao, "Simulation and Hardware Implementation of Stepping Motor Current Regulator by Matlab Simulink and FPGA", International Conference on Electrical and Electronics (EEE2014), Hong Kong, 25-27 April (2014).

Digital Dynamometer and Goniometer in Analyzing Isometric Capacities and Tennis Serving

Chia-Che Liu[1,2] , Chih-Chien Yang[2] , Liang-Ting Tsai[3], Bi-Fon Chang[4] Hsiao-Yun Chang[5], and Chi-Fan Liu[1,*]

[1]*Office of Physical Education and Sport Affairs, Feng Chia University, Taichung, Taiwan, R.O.C.*
[2]*Cognitive Neuro Metrics Laboratory, Graduate Institute of Educational Measurement & Statistics National Taichung University of Education, Taichung, Taiwan, R.O.C.*
[3]*Taiwan Marine Education Center, National Taiwan Ocean University, Keenlung, Taiwan, R.O.C.*
[4]*Department of Physical Education, National Taichung University of Education , Taichung, Taiwan, R.O.C.*
[5]*School of Physical Therapy, Chung Shan Medical University Rehabilitation, Taichung, Taiwan, R.O.C.*

Abstract. The study was to investigate relation between tennis serving techniques and shoulder joint muscles in various skilled players in Taiwan. The study used Micro FET3 (Digital Manual Muscle Dynamometer and Goniometer) which appraised the isometric capacity of the shoulder joint and perimeter muscle. The research consists of 13 collegiate male tennis players; 6 players were in elite groups and the other 7 players were in general groups. The tennis players performed 30 servers testing the range to appraise serve technique reflect. The application of *t*-test and Linear Regression analyzed the serve technique reflect interrelated with the shoulder joint muscle. The muscle strength and the serve accuracy predominate interrelated place where middle trapezius, shoulder internal rotator, and shoulder external rotator of favor hands were obvious. The muscle strength and the serve speed predominated interrelated place where shoulder internal rotator and shoulder external rotator of favor hands were obvious. This study also discovered triceps of favor hands, shoulder extensor, shoulder abductor, triceps, and middle trapezius of non- favor hands muscle strength didn't have any outstanding reference with tennis league table. The research confirmed that the shoulder joint muscle discipline increased shoulder internal rotator, shoulder external rotator, and middle trapezius that help the serve technique reflect. .

1 Introduction

The tennis players have tonicity muscle that can exert powerful serve technique and avoid sport injuries [1]. The shoulder joint sport injuries are frequent performed power sport of the upper limb movement repeat, such as the serve of the tennis, volleyball, and the flinger of baseball pitch movement [2]. All those sports injuries are more often occurred. Because of the patterns of the sport movement resemblance with one another, the sport injuries in shoulder joints are very similar. Therefore, if a tennis player does not have strong upper limb muscle, it is easy for him/her to get sport injuries. Strength testing is one of the cornerstones of physical examination and an increase in strength is the aim of many rehabilitation programs [3].

Micro FET3 Digital Manual Muscle Dynamometer and Goniometer by Fabrication Enterprises is a unique product used to help in diagnosing and treating muscular disorders. In addition, it allows you to accurately and scientifically measure Muscle Testing (MT) and Range of Motion (ROM). For MT, Micro FET3 offers a true measure of actual force produced. It measures force from multiple directions simultaneously, so even the slightest tip of the transducer won't affect test results. As for ROM, Micro FET3 design uses a single gauge inclinometer that has been used for years in performing ROM tests.

Micro FET3 has the capability of using the latest wireless radio frequency system which eliminates the need for cables, wires or cords, making it completely portable. Weighing less than 1 lb, Micro FET3 provides quick, reliable and accurate digital results for muscle and range of motion testing. Besides, it delivers muscle testing accuracy within 2%, and range of motion accuracy within 1 degree, which makes it a useful alternative in sport science research.

2 Tennis serve technique

Being capable of serving in high speed doesn't guarantee a tennis player can save the serve game. In other words, though not being able to serve in high speed, a tennis player can raise the odds of saving the serve game by controlling the ball precisely, altering the direction of serving once in a while, serving toward wide, his rival's bodies or center line, which makes it difficult for receivers to predict ball placements[4].

The tennis serve movement drift is divided into five stages- preparative serve movement, back swing movement, toss movement, slash movement, and shot-making follow -through movement. Serve fundamental stroke technique enact very importance role in the competition. A perfection serve technique can handily win the serve game during the competition. It is basically not failing. This is a tendency of power tennis. The big

* Corresponding author: fanchiliu@yahoo.com

league players need possess one of requisite [5]. The tennis competition starts prologue from serve and received. Whereas, the power tennis leads the serve is not just active attack, but also is one of the cage convenience [6]. Therefore, the serve is very importance technique in the tennis sports, also is a singularity grip by oneself, and did not influence by adversary. A lots of male and female tennis players serve constantly created new high speed and time speed over 200 kilometer in the professional tennis now. Nowadays, the professional tennis mechanism tends to be power tennis. Winning linchpin have to be shot-making morality and speed [7]. Consequently, in a professional sport that requires speed and explosive power, a tennis player can boost the odds on winning by making good use of his/her technique. Performing shot-making movement that need in-form grip to retain racket stability from instant hitting, and penetrate body rotate and kinetic chain impact. Ranking of joint is from the leg along the basin, the torso, and the upper limb which speed and power will lastly deliver to the racket and comprise the speed of the racket [8]. The upper limb power movement expression is very important pointer for excellence players. Therein, the upper limb joint movements have ultimate affect on the speed of the ball [9]. Serve is the most important technique in the tennis competition, besides, it is the easiest to incurrence sport injuries.

The serve technique, the ball speed and the angle cannot be entirely concomitant. If the serve has a big angle, than the ball needs aggrandize rotation. The tennis player might increase the rotation of the serve to comprise the height hop and the big angle serve. Thus, it can decrease molarity of the serve, the received backstroke and enhance proximate attack opportunity. This can enhance the percentage of winning. Therefore, the tennis player possess bionic serve accuracy and serve speed. Then the serve technique level will be promoted to great success.

Both the speed of the serve and the elbow extension were related [10]. The different speeds of the serve on the female tennis players. The muscle denomination weight of the elbow extension movement would predominate the dissimilitude [11]. The research proved that the tennis players who could serve fast had elbow joint extensive muscles that were in better condition. This research was to treat the shoulder joint muscle to see whether it was influenced with the serve speed or the serve accuracy.

3 Method

The research targets 13 collegiate tennis players and divides them into 2 groups- elite group and general group on the basis of their national ranking of Chinese Taipei Tennis Association. The 6 players whose level ranking are within 32 are labeled elite, while the others are labeled general. All of the players have no severe upper limb injuries within a year and participate in the test voluntarily.

The shoulder muscle strength evaluation is to gauge the isometric strength of shoulder joints and their ambient muscles, including shoulder extensor, shoulder flexor, shoulder abductor, biceps, triceps, serratus anterior, middle trapezius, lower trapezius, shoulder internal rotator (shoulder IR), and shoulder external rotator (shoulder ER) with MicroFET3. All of the test receivers use both their favor hands and non-favor hands in the tests.

The evaluation of serve technique performance is to gauge serve speed with Stalker SPORT Radar Gun (Applied Concepts, Inc., TX). In the research, the definition of "efficient serve" is to serve the ball into designated placements. Figure 2 explains the test of serve speed and accuracy, we divide the sever placements into three blocks- A, B and C.

Every test receiver has to serve 10 balls into each block randomly. If the ball falls in to the designated block, it's called "an efficient ball." The accuracy criterion is based on the percentage of "efficient balls" in every 30 serve. If the sever is "efficient", the searchers will record its speed. If it's inefficient, they won't [12].

This research analyzes the data with the computer statistical software- SPSS for windows 12.0. First, on the basis of independent sample t-test, it assays the difference of shoulder joint muscle strength in favor hands and non-favor hands between elite group players and general group players. And then, with Linear Regression, it analyzes the relativity of shoulder joint muscle strength in both hands and serve speed, serve accuracy. In the research, we set the significance level as $\alpha = .05$.

Table 1. test receivers' personal information

	age	height	weight	experience
Elite (N=6)	8.83±0.9	173.60±9.1	65.66±11.9	10±1.6
General (N=7)	21±1.1	175±4.4	69.20±5.7	9.85±1.0

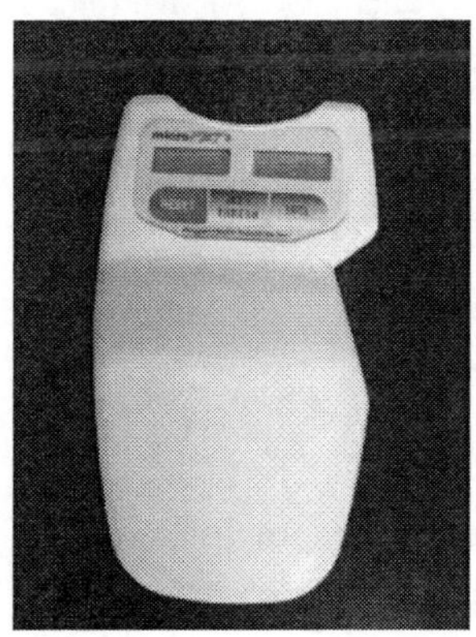
(a)

(b)

Fig. 1. (a) Micro FET3 (Digital Manual Muscle Dynamometer and Goniometer) (b) Stalker SPORT Radar Gun

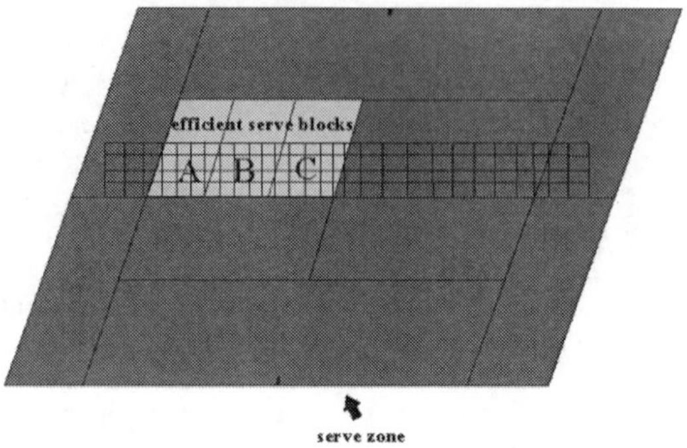

Fig. 2. test on serving techniques performance

4 Results

The comparison of shoulder joint muscle strength between the elite players and general ones is shown as Table 2 and 3; the relevance of the players' shoulder joint muscle strength and his serve speed, accuracy is presented as Table 4 and Table 5. On average, the serve accuracy of the elite players is 33 ± 6.89 %; the serve speed is 160 ± 10.08 km/hr. As for the general players, the serve accuracy is 23 ± 11.35 %; the serve speed is 161 ± 9.58 km/hr. Besides, in the research, we get the ratio of shoulder internal rotation to external rotation. The ratio of shoulder internal rotation to external rotation of elite players is 1.07 ± 0.15, while that of general players is 1.16 ± 0.08.

According to Table 2, as we compare the elite players' shoulder joint muscle strength of favor hands to that of the general players, we find only "triceps" reaches the outstanding standard ($p < .05$). The elite players' muscle strength of triceps is 33.57 ± 5.35 pounds, while that of the general players is 45.27 ± 7.93 pounds. The data shows that the elite players' muscle strength of triceps is inferior to that of the general players. As for the other items on the table, they don't show any significant distinction between the groups.

Table 2. elite and general players shoulder joint muscle strength in favor hands in *t*-test

variable	groups	mean	SD	T
Shoulder extensor	elite	36.55	7.95	-1.93
	general	43.67	5.33	
Shoulder flexor	elite	36.12	6.86	-2.02
	general	42.61	4.69	
Shoulder abductor	elite	36.52	10.10	-1.17
	general	41.39	4.26	
Biceps	elite	55.52	10.25	-1.83
	general	63.73	5.68	
Triceps	elite	33.57	5.35	-3.06*
	general	45.27	7.93	
Serratus anterior	elite	34.30	12.05	-1.42
	general	43.10	10.37	
Middle trapezius	elite	29.67	6.29	-1.60
	general	34.01	3.31	
Lower trapezius	elite	14.58	3.35	-2.15
	general	19.56	4.73	
Shoulder IR	elite	29.60	8.30	-0.56
	general	31.59	3.04	
Shoulder ER	elite	27.60	6.59	0.16
	general	27.13	3.04	

* p-value <.05

According to Table 3, the elite players and the general ones' muscle strength of shoulder extensor, shoulder abductor, triceps and trap middle in favor hands all reach the outstanding standard ($p < .05$). The elite players' muscle strength of shoulder extensor is 35.85 ± 9.15 pounds, while that of the general players is 45.20 ± 6.36 pounds. The data shows that the elite players' muscle strength of shoulder extensor is inferior to that of the general players.

The elite players' muscle strength of shoulder abductor is 34.75 ± 6.53 pounds, which shows that the elite players' muscle strength of shoulder abductor is inferior to that of the general players. The elite players' muscle strength of triceps is 33.10 ± 4.73 pounds, while that of the general players is 42.20 ± 7.67 pounds. The data shows that the elite players' muscle strength of triceps is inferior to that of the general players.

The elite players' muscle strength of middle trap is 25.31 ± 3.33 pounds, while that of the general players is 31.96 ± 2.73 pounds. The data shows that the elite players' muscle strength of middle trap is inferior to that of the general players. As for the other muscular factors on Table 3, they don't show any significant distinction between the two groups.

Table 3. elite and general players shoulder joint muscle strength in non- favor hands in *t*-test

variable	groups	mean	SD	T
Shoulder extensor	elite	35.85	10.03	-1.99*
	general	45.20	6.87	
Shoulder flexor	elite	34.73	10.12	-1.30
	general	40.03	3.62	
Shoulder abductor	elite	34.75	7.16	-2.26*
	general	42.01	4.30	
Biceps	elite	48.25	7.64	-2.18
	general	56.20	5.48	
Triceps	elite	33.10	5.18	-2.32*

	general	42.20	8.28	
Serratus anterior	elite	35.58	12.71	-1.38
	general	43.77	8.59	
Middle trapezius	elite	25.31	3.65	-3.64*
	general	31.96	2.95	
Lower trapezius	elite	13.77	3.76	-1.34
	general	15.93	1.91	
Shoulder IR	elite	26.25	9.52	0.32
	general	25.03	3.41	
Shoulder ER	elite	26.98	6.21	-0.94
	general	29.96	5.16	

* p-value <.05

Table 4. elite and general players shoulder joint muscle strength in favor hands in Linear Regression

	serve accuracy			
variable	β	SE	T	p-value
intercept	-81.51	44.98	-1.81	.21
Shoulder extensor	-3.49	1.21	-2.88	.10
Shoulder flexor	-0.43	0.50	-0.85	.49
Shoulder abductor	0.25	0.78	0.32	.78
Biceps	3.59	1.33	2.71	.11
Triceps	0.16	0.34	0.47	.69
Serratus anterior	-3.24	0.97	-3.35	.08
Middle trapezius	8.18	1.84	4.44	.04*
Lower trapezius	-6.31	1.61	-3.91	.06
Shoulder IR	4.41	0.73	6.07	.03*
Shoulder ER	-4.61	0.88	-5.25	.03*
	serve speed			
variable	β	SE	T	p-value
intercept	77.12	28.49	2.71	.11
Shoulder extensor	-0.02	0.78	-0.03	.99
Shoulder flexor	0.56	0.32	1.75	.22
Shoulder abductor	-0.23	0.49	-.46	.69
Biceps	1.50	0.84	1.78	.22
Triceps	0.77	0.21	3.63	.07
Serratus anterior	-0.88	0.61	-1.43	.29
Midddle trapezius	0.25	1.17	0.21	.85
Lower trapezius	-1.88	1.02	-1.84	.21
Shoulder IR	-2.36	0.46	-5.14	.04*
Shoulder ER	2.99	0.56	5.37	.03*

* p-value <.05

Table 4 and Table 5 are the Linear Regression of shoulder joint muscle strength. The parameters stand for the estimated values of variables and intercepts.

In Linear Regression on serve accuracy, the muscle strength of middle trapezius, shoulder IR, and shoulder ER in favor hands all shows remarkable distinction ($p < .05$). As for serve speed, the the muscle strength of shoulder IR and shoulder ER in favor hands both shows remarkable distinction ($p < .05$). Moreover, in non-favor hands, none of the variables in Linear Regression on serve accuracy and speed shows significant distinction.

Table 5. elite and general players shoulder joint muscle strength in non- favor hands in Linear Regression

	serve accuracy			
variable	β	SE	T	p-value
intercept	96.75	33.12	2.92	.10
Shoulder extensor	0.49	0.99	0.50	.67
Shoulder flexor	1.97	1.96	1.01	.42
Shoulder abductor	-2.19	1.39	-1.58	.26
Biceps	0.32	1.54	0.21	.85
Triceps	-1.40	1.37	-1.02	.42
Serratus anterior	0.95	0.78	1.22	.35
Middle trapezius	-0.43	1.69	-0.25	.83
Lower trapezius	-0.30	1.55	-0.19	.86
Shoulder IR	-1.30	1.41	-0.92	.45
Shoulder ER	-1.06	1.64	-0.65	.58
	serve speed			
variable	β	SE	T	p-value
intercept	142.88	24.39	5.86	.03*
Shoulder extensor	-0.16	0.73	-0.22	.85
Shoulder flexor	-0.47	1.44	-0.32	.78
Shoulder abductor	-1.55	1.02	-1.52	.27
Biceps	-0.20	1.13	-0.18	.87
Triceps	1.82	1.01	1.80	.21
Serratus anterior	-0.78	0.58	-1.35	.31
Middle trapezius	-0.64	1.24	-0.51	.66
Lower trapezius	-0.39	1.14	-0.34	.77
Shoulder IR	-0.56	1.04	-0.54	.64
Shoulder ER	3.94	1.21	3.26	.08

* p-value <.05

5 Conclusion

In general, elite tennis players have high-speed and efficient serve technique. And their shoulder muscle strength is highly related to their serve movements, ground strokes, net interception, etc. Hence, to give full play of one's muscle strength in high speed, especially when s/he accelerates a racket, high-energy power in shoulder joints is required, which is an important index in racket sports [13].

According to the research, the strength of triceps of favor hands, shoulder extensor, shoulder abductor, triceps and middle trapezius of non-favor hands of elite players is remarkably minor to that of general players, while statistics on other muscles doesn't show much distinction. The results of this research are different from the prior studies [14][15][16].

The researchers ascribe the outcome to the fact that tennis is a sport requiring whole-body coordination and pliability, not just focusing on the training of a few parts. Besides, the researchers presume the decline of muscle strength in elite players is because they attain more training and attend more tournaments, which causes fatigue. We suggest more research be done on the training of elite tennis players and their counterparts.

The shoulder joint muscle affects the serve reflect of the tennis players. The consequences of this investigation figures out the serve accuracy domain were predominately interrelated when overused hand, middle trapezius, shoulder internal rotator and shoulder external rotator. The serve speed domain were predominately interrelated when overused hand, shoulder internal rotator and shoulder external rotator, the shoulder internal rotators were a very important linchpin with the control of the serve falling point [15]; because trapezius at the launch process attach tranquilize muscle and can decrease scapula drag ahead.

This manifest trapezius at the launch movement keeps the shoulder joint steady which was very important [17]. Other research results were similar to these results. This research result shows that the shoulder joint muscle of the tennis players in the elite group's serve techniques, had an accuracy average of 33 ± 6.89% better than the tennis players in the general group's serve which averaged 23 ± 11.35%. This manifest the tennis player's serve accuracy and the shoulder joint muscle are not exactly interrelated. However, the tennis serve technique requires height mechanism movement, also needs to consider the integer profluent movement and the time of the body center gravity displacement. This has to cooperate with the release integer swing arm.

The serve will amplify efficiency culmination. The shoulder joint rotator can contribute to the ball speed 71% and can generate important effort on the ball [18]. To improve serving technique, tennis players should enhance their whole-body physical coordination and aptly increase the training on middle trapezius, shoulder internal rotator and shoulder external rotator, etc. In conclusion, in current power tennis age, tennis players pay more attention on the accuracy of serve technique. They not only care about the speed, quality and quantity of ball serving but the ball placements, angles, and twirling. Current professional tennis players have better figure and skills; however, they won't be able to control the game or their opponents without accurate ball placements when they serve.

According to the research, we find high relativity between enhancing the training of shoulder internal rotator and shoulder external rotator and the speed and accuracy of serving, which pinpoints via full-scale muscle training, improving serve technique and reducing the probability of getting hurt are crucial to professional tennis players.

References

1. B. C. Elliott, "Biomechanics of the Serve in Tennis. A Biomedical Perspective. Sports Medicine, **6**, 285-294 (1988)
2. H. W. Lee, "Mechanisms of Neck and Shoulder Injuries in Tennis Player. The Journal of Orthopaedic and Sports Physical Therapy, **21**, 28-37 (1995)
3. A. S. Falrey, K. S. Courneya, and C. J. Field, "Physical Exercise and immune system function in Cancer survivors. Cancer. **94**(2), 539-551 (2002)
4. H. S. Wang, W. Y. Lin, and H. H. Chao, "The role of serve placement in Tennis matches- An example study of the 2004 US Open Men's Singles. Physical Education Journal, **38**(4),109-120 (2005)
5. Y. H. Lien, "A qualitative study on modern man Power Tennis. Physical Education Journal, **37**,113-131 (2004)
6. B. F. Chang, C. J. Chen, "Outcome Analysis of Skills Serve and Receive Among Professional Tennis Players. Sports Coaching Science, **16**, 83-94 (2009)
7. I. T. Wang, "Sports Biomechanical Analysis of the Flat Service in Tennis. Taipei Physical Education College Institute of Sports and Exercise Science Master's Thesis (2007)
8. B. Elliott, "Biomechanics and tennis. British Journal of Sports Medicine, **40**, 392-396, (2006)
9. C. M. Chen, W. T. Tang, "The Correlation Data of Upper Lime Muscle for Baseball Pitcher Exerted discipline. The University Physical Education & Sports. **75**, 61-67 (2004)
10. M. W. Huang, L. C. Liang, and C. J. Chen, "The Correlation of the Serve to the Shoulder and Elbow Joint Muscle for Female Tennis Player. Physical Education, Fitness and Inactive Periodical of Nation Chiayi University, **9**(1), 127-132 (2010)
11. L. C. Liang, "Upper Limb Isokinetic Strength for Tennis Players with Different Serve Velocities. Sports & Exercise Research, **13**(2), 181-187 (2011)
12. B. F. Chang, "Relationship between balance and serve skill for collegiate tennis players. Physical Education Journal. **44**(1),.45-56 (2011)
13. L. H. Wang, "The Application of Physiology to Tennis. The University Physical Education & Sports, **55**,.127-132 (2001)
14. S. L. Huang, "Isokinetic analysis of shoulder muscles in three levels of women tennis players. Chinese Culture University Institute of sports coaching science Master's Thesis (1996)
15. M. W. Huang, "The Differences of Upper Isokinetic Strength Profile and Specific Performance between Elite and General Female Tennis Players. National Hsinchu University of Education Institute of Physical Education Master's Thesis (2009)
16. A. Stanley, R. McGann, J. Hall, L. Mckenna, and N. K. Briffa, "Shoulder strength and range of motion in female amateur-league tennis players. The Journal of Orthopaedic & Sports Physical Therapy, **34**,.402-409 (2004)

DOI: 10.1051/matecconf/20167105003

17. Y. C. Liu, S. H. Chen, and H. Y. Chen, "The Electromyography Analysis of Grenade Throwing. Journal of Physical Education and Sports, **16**(3), 205-218 (2005)
18. A. S. Clements, K. A. Ginn, and E. Henley. "Correlation between muscle strength and throwing speed in adolescent baseball players. Physical Therapy in Sport, **2**.123-131 (2001)

MATEC Web of Conferences 71, 05004 (2016) DOI: 10.1051/matecconf/20167105004
ICCPE 2015

Development of Intelligent Auxiliary System for Customized Physical Fitness and Healthcare

Chung-Chi Huang[1,a], Hsiao-Man Liu[2] and Yen-Ting Ke[3]

[1]*Department of Automation and Control Engineering, Far East University, Tainan City, 744, Taiwan, ROC*
[2]*Department of Leisure and Sports Management, Far East University, Tainan City, 744, Taiwan, ROC*
[2] *Department of Mechanical Engineering, Far East University, Tainan City, 744, Taiwan, ROC*

Abstract. With the advent of global high-tech industry and commerce era, the sedentary reduces opportunities of physical activity. And physical fitness and health of people is getting worse and worse. At present, the shortage of physical fitness instructors greatly affected the effectiveness of health promotion. Therefore, it is necessary to develop an auxiliary system which can reduce the workload of instructors and enhance physical fitness and health for people. But current general physical fitness and healthcare system is hard to meet individualized needs. The main purpose of this research is to develop an intelligent auxiliary system for customized physical fitness and healthcare. It records all processes of physical fitness and healthcare system by wireless sensors network. The results of intelligent auxiliary systems for customized physical fitness and healthcare will be generated by fuzzy logic Inference. It will improve individualized physical fitness and healthcare. Finally, we will demonstrate the advantages of the intelligent auxiliary system for customized physical fitness and healthcare.

1 Introduction

With the sedentary reduces opportunities of physical activity, the hypo kinetic disease will lead to the chronic diseases. In recent years, developing of customized physical fitness and healthcare system is more and more flourishing. A PDA-Based healthcare system was proposed (C.C.Yang 2002). And users can make inquiry about the information of the nutritional value, calories of a food, and the calories consumed by exercise. Based on his own personal needs, the individual can design an appropriate diet and exercise with the information. It not only help people to manage personal healthcare no matter in sport, diet or medicine issues, but also help doctors diagnose diseases and enhance the efficiency of medical administration and quality. Considering that the most of the instruments price was too high to popularize at present, a simple design and a fair price instrument focused on fitness testing of sit-up was proposed. (J.C. Su et al. 2009). It combined logic electric cable with an inducted machine that matched with an organization of machinery for sit-up testing and passed through the score indication. The whole process was accomplished by logic electric cable and an inducted machine instead of testing by manual operation that could reduce labour costs, avoid mistakes, also enhanced accuracy. And then a web-based physical fitness system was proposed. (J.C. Su et al. 2008). It focus on simple assessment of personal health risk factor, detection and condition assessment of physical fitness, and information of group exercise prescription...etc. A physical fitness system by using the fuzzy theory is proposed (H.C. Wu 2013). It can increase the learning effects and improve the accuracy of training selection. So that lower training costs can be achieved by using the fuzzy theory to estimate the effective assessment for physical fitness test. An automated interactive exercise coaching system using the Microsoft Kinect was developed (F. Ofli et al. 2015). The system guides users through a series of video exercises, tracks and measures their movements, provides real-time feedback, and records their performance over time. The system consists of exercises to improve flexibility, balance, endurance, and strength, with the aim of improving performance of daily activities and reducing fall risk. A homecare sensory system that assesses the elder's physical fitness through quantifying their home rehabilitation exercises was proposed (C.Y. Chiang et al. 2011). The sensory system used a tri-axial accelerometer to collect the motion acceleration and transmitted the data through wireless personal area networks in the home environment. Through the quantification method for the senior adults, the proposed system attempted to provide appropriate exercise advises for better health administration. To sum up the above, it is very important to develop a customized auxiliary system which can enhance physical fitness and health for people.

2 Intelligent customized physical fitness and healthcare system

[a] Corresponding author: cchuang@cc.feu.edu.tw

In this paper, an intelligent assessment and prediction system was proposed to construct an intelligent auxiliary system for complete physical fitness platform system. It constructs a complete physical fitness platform which uses fuzzy logic method to infer customized individual prescriptions according the age, body, fitness and strength. It contains two parts: (1) Assessment for customized physical fitness and healthcare (2) Prediction for customized physical fitness and healthcare. And it uses the architecture of cloud-dust based intelligent system to realize intelligent auxiliary system for customized physical fitness and healthcare. It showed as figure 1.

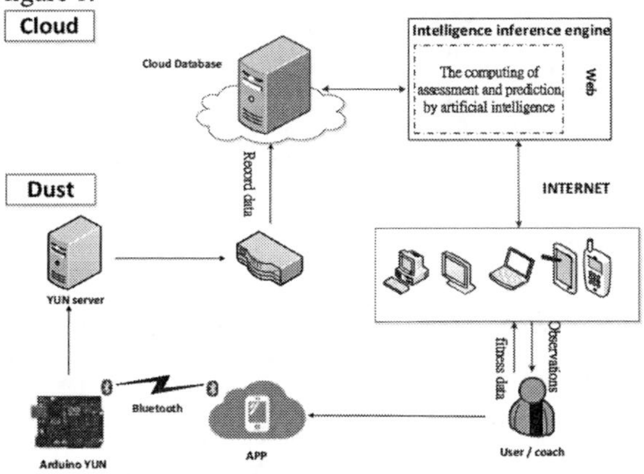

Figure 1. The architecture of cloud-dust based intelligent auxiliary system for customized physical fitness and healthcare

3 Records and surveillance for customized physical fitness and healthcare

The App of mobile devices is designed by Zigbee, WiFi, and Bluetooth. The maximum heart rate, the effective heartbeat, the current heart rate and average heart rate will be calculated by using microprocessor to sense data of heartbeats per 5 seconds. It was shown as figure 2.

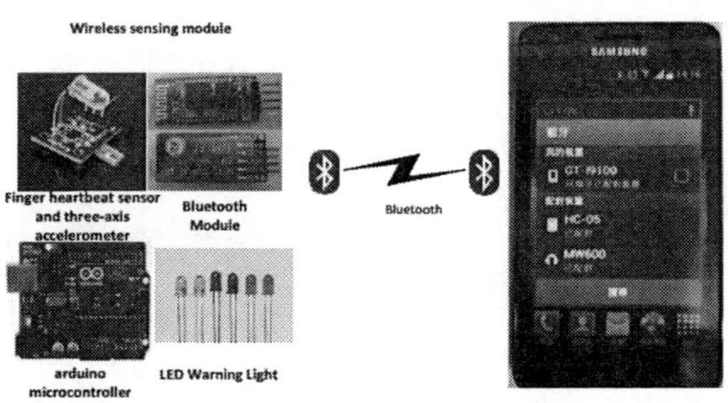

Figure 2. Records and surveillance of heartbeats for customized physical fitness and healthcare

As figure 3 shown, the LED will show the alerts. The green light is showed if the heart rate of the user is in the range of 0-65%. The yellow light is showed in the range of 65-85% and the red light is showed in the range of 85-100%. It will provide a voice to alert the user and the current state of the coach to keep the sports safety if the red light shows for three times simultaneously. The flowchart of records and surveillance for customized physical fitness and healthcare was shown as figure.

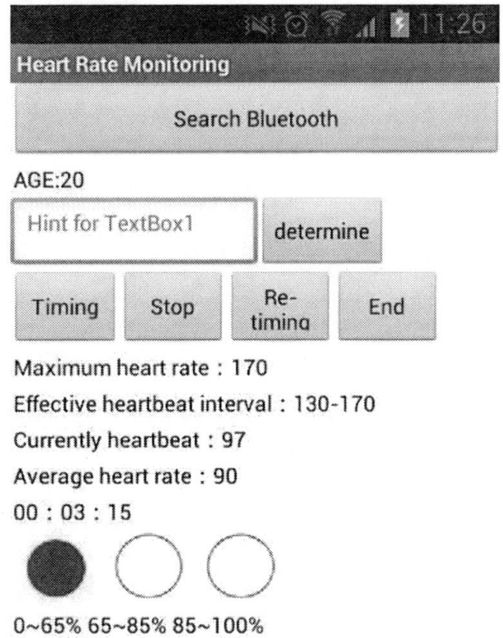

Figure 3. The alert lights of heartbeat states in the system.

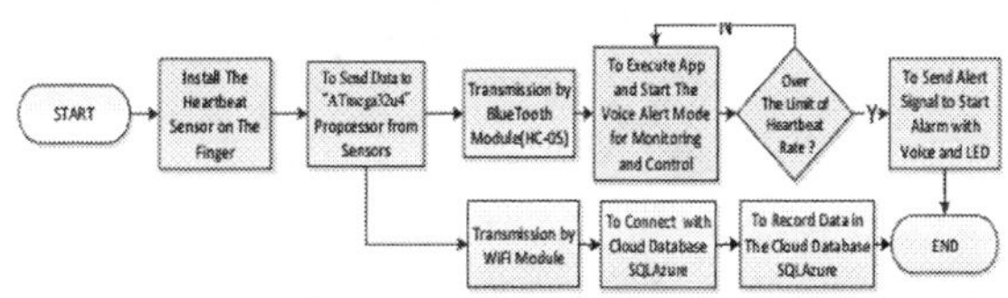

Figure 4. The flowchart of records and surveillance for customized physical fitness and healthcare.

4 Intelligent assessment for customized physical fitness and healthcare system

Considering the input parameters as muscular endurance, flexibility and the 3-minute step test, use fuzzy inference to generate the results of assessment for customized physical fitness and healthcare system. The flowchart of assessment using fuzzy inference for customized physical fitness and healthcare system is shown as figure 5.

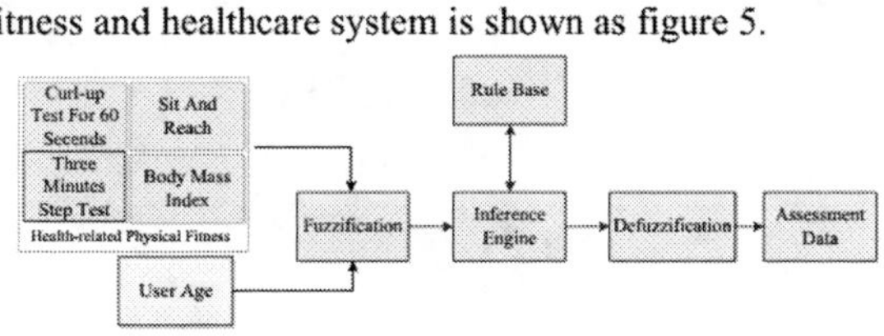

Figure 5. The flowchart of assessment using fuzzy inference for customized physical fitness and healthcare system

The input parameters as muscular endurance, flexibility and the 3-minute step test will be sent to cloud database.

And then the results of assessment by the computing of cloud fuzzy engine will be generated and showed in web-based human-machine-interface. The architecture of assessment using fuzzy inference for customized physical fitness and healthcare system is shown as figure 6.

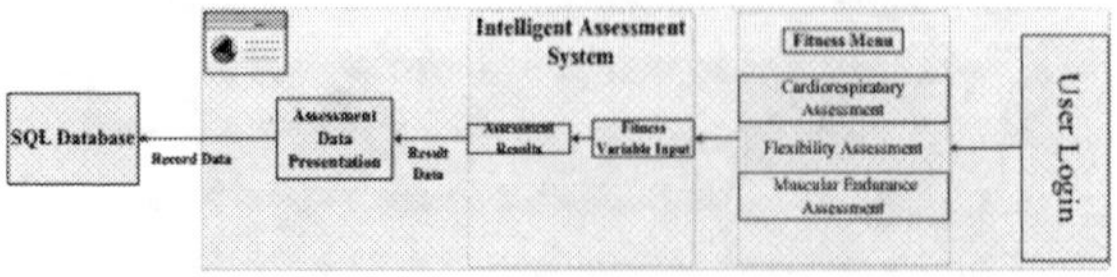

Figure 6. The architecture of assessment using fuzzy inference for customized physical fitness and healthcare system.

The process of fuzzy inference for Health-related Physical Fitness (HRPF) is as the followings:

Step 1. To build up the model of customized physical fitness and healthcare system as figure. 7. And then set Cardiorespiratory-Fitness, Muscular-Endurance, Flexibility and BMI as input parameters and HRPF as output parameter.

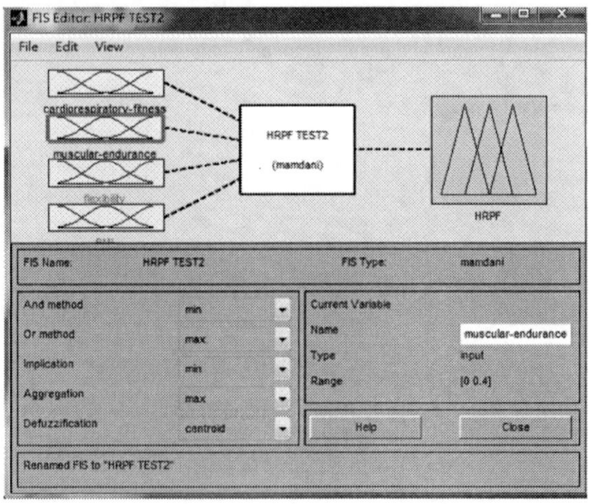

Figure 7. The model of customized physical fitness and healthcare system.

Step 2. To set the parameters of cardiorespiratory-fitness as 0-0.75 and the membership as S(0,0,0.54), M(0.5, 0.575,0.65), L(0.62,0.75,0.75). It is showed as figure 8.

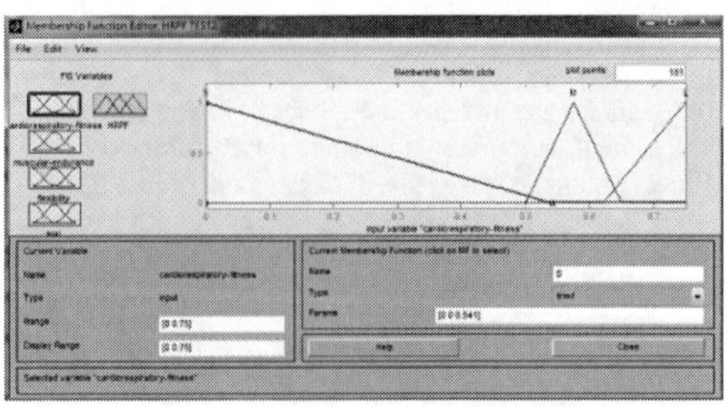

Figure 8. Membership setting for cardiorespiratory fitness.

Step 3. To set the parameters of muscular endurance as 0-0.4 and the membership as S(0, 0, 0.2), M(0.1, 0.2, 0.3), L(0.25,0.4,0.4). It is showed as figure 9

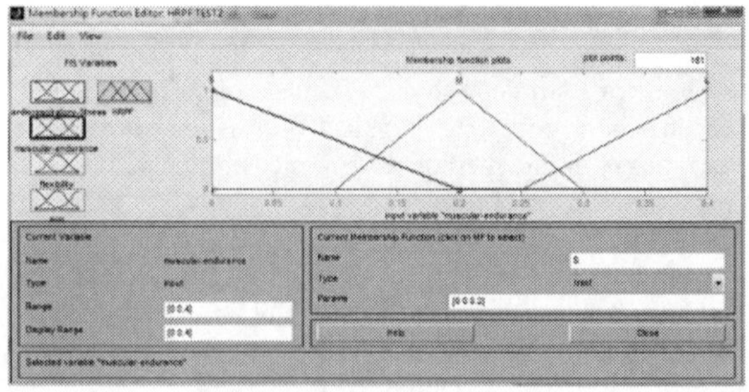

Figure 9. Membership setting for muscular endurance.

Step 4. To set the parameters of flexibility as 0-0.1 and the membership as S(0.01, 0.01, 0.05), M(0.02, 0.055, 0.09), L(0.06,0.1,0.1). It is showed as figure 10.

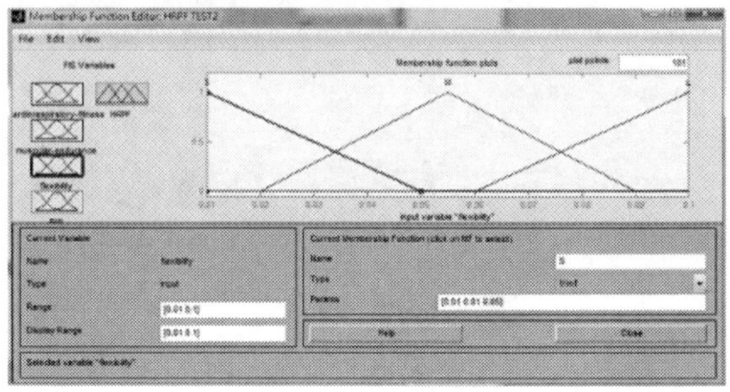

Figure 10. Membership setting for flexibility.

Step 5. To set the parameters of BMI as 0-0.4 and the membership as S(0, 0, 0.2), M(0.18, 0.24, 0.3), L(0.27, 0.4, 0.4). It is showed as figure 11.

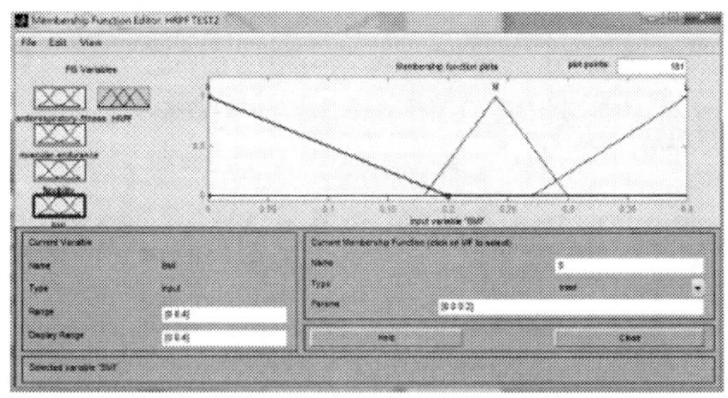

Figure 11. Membership setting for BMI.

Step 6. To set the parameters of HRPF as 0-1 and the membership as VS(0, 0, 0.25), S(0, 0.25, 0.5), M(0.25, 0.5, 0.75), L(0.5, 0.75, 1), ML(0.5, 0.75, 1). It is showed as figure 12.

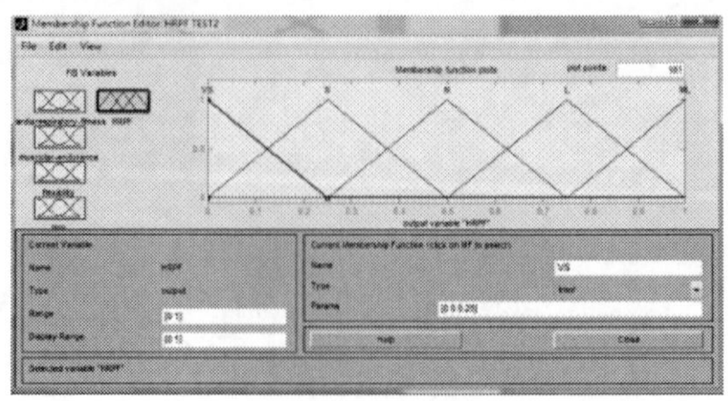

Figure 12. Membership setting for HRPF.

 DOI: 10.1051/matecconf/20167105004

Step 7. To set 81 rules for HRPF. It is showed in figure 13. Cardiorespiratory fitness is a main item for HRPF.

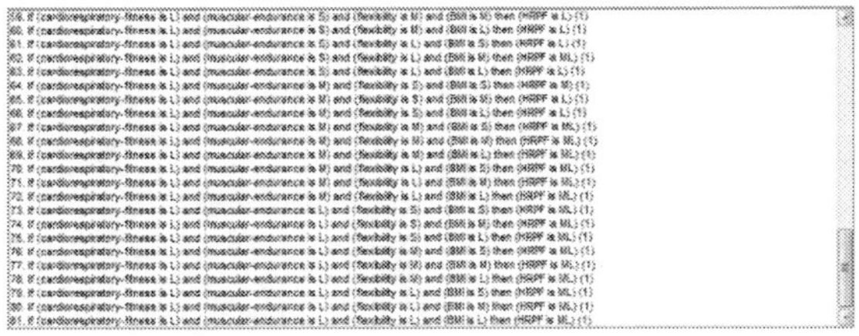

Figure 13. Rules setting for HRPF.

Step 8. The results of fuzzy inference for Health-related Physical Fitness (HRPF) by inputting Cardiorespiratory-Fitness, Muscular-Endurance, Flexibility and BMI. If Cardiorespiratory-Fitness=0.405 (S Area), Muscular-Endurance=0.182 (M Area), Flexibility=0.074 (M Area and L Area), BMI=0.238 (M Area), the system inferred HRPF = 0.438 (M Area). It means the status of user's Physical Fitness is "Middle". It is showed as figure 14.

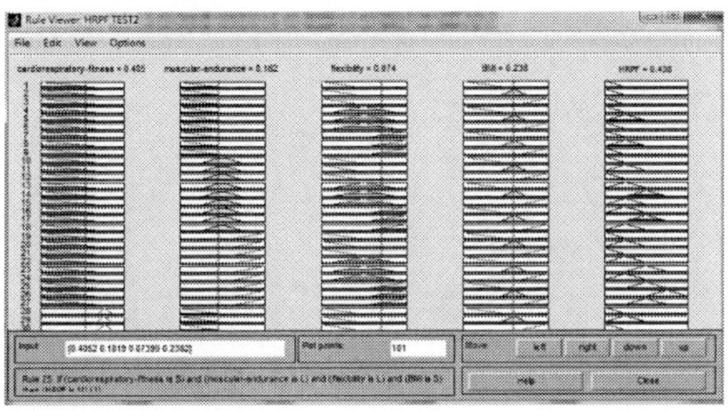

Figure 14. Fuzzy inference for HRPF.

5 Intelligent prediction for customized physical fitness and healthcare

Considering the historical records of muscular endurance, flexibility and the 3-minute step test, use neural network to generate the results of trend prediction for customized physical fitness and healthcare system. The flowchart of trend prediction using neural network method for customized physical fitness and healthcare system is shown as figure 15. And the architecture of intelligent inference using neural network method is shown as figure 16.

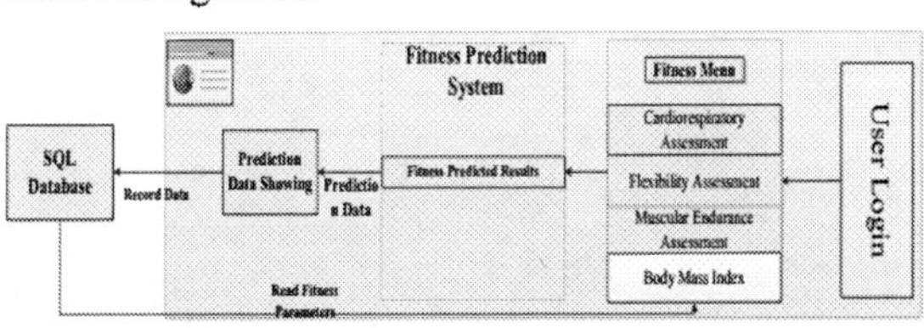

Figure 15. The flowchart of trend prediction using neural network method for customized physical fitness and healthcare.

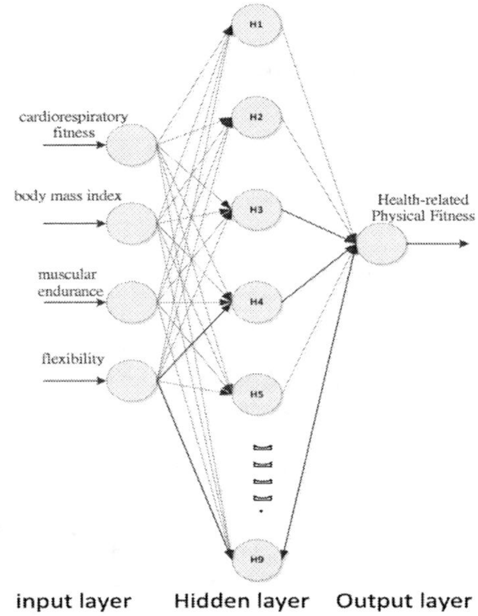

Figure 16. The architecture of intelligent inference using neural network method for trend prediction of HRPF.

Artificial neural network was training by the historical records of muscular endurance, flexibility and the 3-minutes step test for trend prediction of HRPF. Based on cloud-dust architecture, the trend prediction of HRPF will be shown in mobile devices or PCs. It is shown as figure 17.

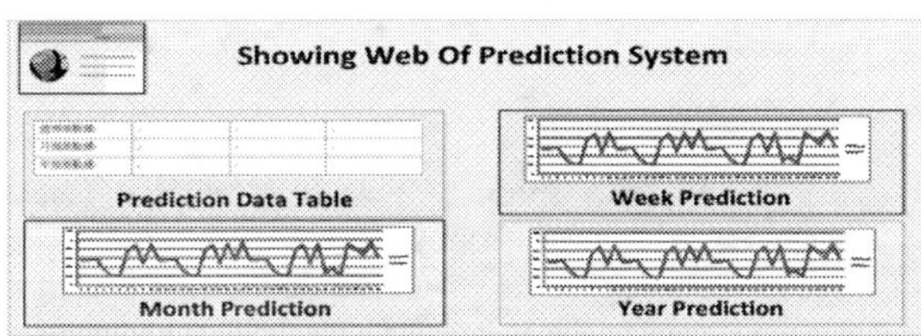

Figure 17. The trend prediction of HRPF based on cloud-dust architecture

6 Results and discussions

We can find the performance of intelligent assessment and prediction for customized physical fitness and healthcare system from the experimental results in figure 18 and figure 19. It shows the system has an excellent performance of assessment and trend prediction for customized physical fitness and healthcare system.

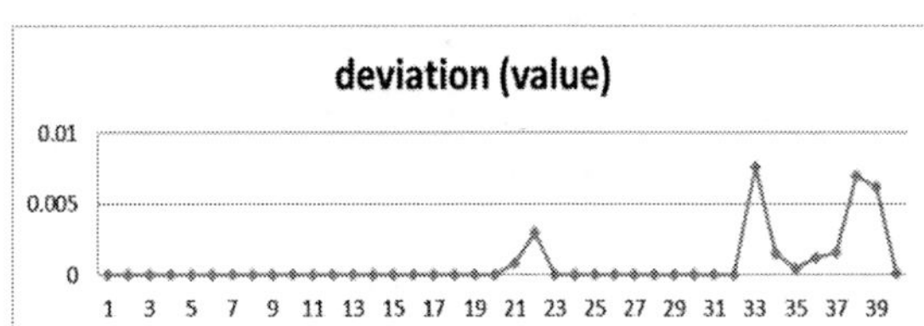

Figure 18. Total deviation of assessment

DOI: 10.1051/matecconf/20167105004

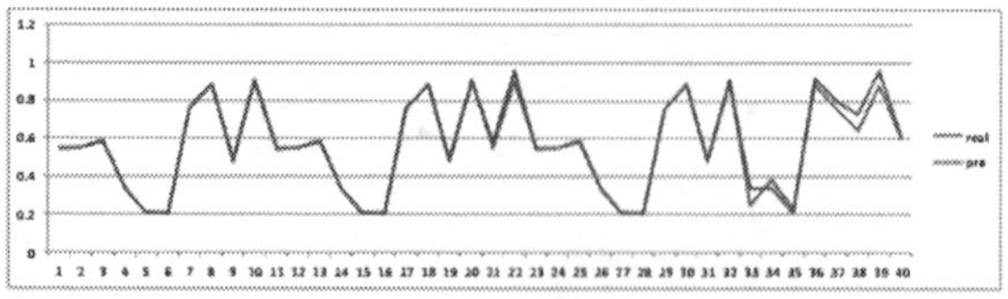

Figure 19. Trend comparison of actual and prediction results

7 Conclusions

The results of assessment and prediction for customized physical fitness and healthcare system will be generated by inference of fuzzy logic and neural network. It will improve individualized physical fitness and healthcare. Finally, we will demonstrate the advantages of the intelligent assessment and prediction for customized physical fitness and healthcare system. In order to reduce labour costs, avoid mistakes, also enhanced accuracy, the intelligent system plays an important role in HRPF field. By the experimental results, we can find it meets the both needs of the efficiency and the lowest cost for intelligent assessment and prediction for customized physical fitness and healthcare system.

Acknowledgment

We would like to thank the Ministry of Science and Technology of the Republic of China (Taiwan) for financial support of this research under contract numbers MOST 104-2622-H-269 -002 -CC3

References

1. C.C. Yang, The Study of Portable Personal Health Managment System (2002)
2. C.Y. Chiang et al., Quantification of Home Rehabilitation Exercise for the Elder's Physical Fitness Monitoring (2011)
3. F. Ofli et al., Design and Evaluation of an Interactive Exercise Coaching System for Older Adults: Lessons Learned (2015)
4. H.C. Wu, Physical Fintness Fuzzy Assessment, (2013)
5. J.C. Su et al., The Research of the Development on Electric Sit-up Testing System (2009)
6. J.C. Su et al., The Development of Fitness Testing Software System (2008)

MATEC Web of Conferences **71**, 05005 (2016) DOI: 10.1051/matecconf/20167105005
ICCPE 2015

The Acceptance of Smart Wearable Devices through Health Cognitive

Fen- Fen Huang [1,a], Yi-Horng Lai [1]

[1] *Department of Health Care Administration, Oriental Institute of Technology, New Taipei City, Taiwan*

Abstract. Many people have become "health consumers" and "health conscious" by seeking products to more effectively manage health-related lifestyle aspects such as fitness activities, healthcare, sports, and rehabilitation. The results revealed that perceived ease of use, perceived usefulness, and health cognitive had a significant impact on attitude to use. To increase individuals' intention to use smart wearable devices, it is critical to encourage a positive attitude toward using the devices to acquire health knowledge. Emphasizing devising effective means to communicate the health utility of the devices to customers should be the highest priority for industry producers.

1 General Instructions

Remote health monitoring has the potential to improve the quality of health services and to reduce the total cost in healthcare by avoiding unnecessary hospitalizations and ensuring that those who need urgent care get it faster. Huge progress occurred in signal processing and mobile telecommunications as well as in human-computer interface technologies and medical knowledge.

Remote health monitoring can significantly contribute to the enhancement of disease prevention, early diagnosis, disease management, treatment and home rehabilitation.

The healthcare and health delivery such as increase of, ageing population, chronic diseases though rehabilitation, monitoring, homecare and so on and by the remarkable progress in sciences and technologies. People become more "health conscious" and "health consumers" asking for better health and life style management, including healthcare, rehabilitation, fitness, sport, etc.

2 Review of the Literature

Variety of portable smart devices started to emerge in 21st century. It allows people to use more advanced computing ability and offer instantaneous connectivity more than the personal computer [1]. In Addition, these smart devices provide a platform where people can easily access to the information that they need [2] and real-time information can be exchanged anywhere [3]. A smart wearable personal device should be light weight and low power consumption, of reasonable price by totally unskilled persons.

Health anxiety usually appears to be associated with increased worry [4], anxiety [5], and fear [6] though overall, during the experience of health anxiety. Most studies [5, 7] have examined cognitions and behaviors associated with health anxiety. Some health anxiety is associated with increased functional incidences of anxiety and depression [8]. Individuals with high health anxiety used more health care and caused higher costs for the health service [9].

Some studies on health anxiety have examined the cognitive processes in maintaining health anxiety [10]. A critical review of studies examining cognitive processes implicated in health anxiety [11] reported that health anxious individuals tend to negatively emotion as indicating an illness, a phenomenon known as amplification.

Behaviors associated with health anxiety are referred to as safety behaviors, which intention to protect one's health. The case with anxiety disorders, seeking behavior played a significant role in the maintenance of health anxiety [5, 12]. This study adopted Moon and Kim's (2001) definition of Perceived Playfulness in this study [13]. According to the authors, Perceived Playfulness can be defined as "the extent to which the individual perceives that the attention is focused on the interaction with the device; and finds the interaction intrinsically enjoyable or interesting."

Based on the theory of Flow [14], they considered playfulness as an intrinsic belief or motive, which is shared by the individual's experience with the environment. Flow represents most interesting experience possible when the person is unconsciously engaged in an activity such that he becomes so engrossed thereby losing the self-sense [15]. Flow experience can occur in structured activities such as sports, ritual events, games, artistic performances, and so on [14]. In other words, enjoyment, experiences of concentration, and curiosity in an activity characterize the experience of flow [16].

Other researchers have studied Perceived Playfulness in the mobile Internet services. In the study of mobile services, Nysveen et al. (2005) found playfulness plays a key role in determining user acceptance of these services

[a] Corresponding author: author@e-mail.org

[17]. Fang et al. (2006) founded mobile tasks that can be performed on handheld devices into some broad types [18]. It found Perceived Playfulness was significant when users perform gaming tasks. The results suggest Perceived Playfulness played a special role in user's favorable acceptance of mobile Internet services.

Technology Acceptance Model (TAM) has become the popular theory of technology use in the information systems literature [19]. Davis et al. (1989) developed the technology acceptance model (TAM) to specify the relationships, related attitude and behavior: perceived ease of use and perceived usefulness, and users' attitudes, behavioral intention and actual behavior [20]. There are two predictors of attitude to use a technology and highly predictive model. The first predictor is perceived usefulness. It refers to one's belief that using a technology will lead to desirable outcomes and usually involving an increase in personal effectiveness. The second predictor is perceived ease of use. It is one's belief that using the technology will be free of effort. Usability dimensions or other ease of use like time efficiency also be relevant in healthcare relative research [21].

3 Method

3.1 Procedure

The sample consisted of the people who attended road running in Taiwan. Roadrunners from six sections of the road running were invited to complete the questionnaire. A total of 163 questionnaires were returned. This study extended Davis's (1989) TAM to explore health anxiety and acceptance of mart wearable devices.

In this study, the model hypothesizes that road-runners perceived health anxiety and health cognitive had a significant influence on their perceived usefulness. The degree to which smart wearable device is easy to use, as perceived by roadrunners' experience, affects both their perception of the usefulness of the technology and their attitude toward using the technology in general.

Attitude is also influenced by the level of the technology's usefulness, as perceived by health anxiety and health cognitive. Finally, the intensity of road-runners' intention to use a technology can be explained and predicted by their attitude toward using the technology and the technology's perceived usefulness.

SPSS 22 with the data also supported most of the individual causal paths postulated by TAM. We intend to provide an integrated theoretical basis to evaluate new generations of TAM and point out the potential problem of user acceptance. Behavior intention is predicted by both attitude and perceived usefulness. Perceived ease of use influences both attitude and perceived usefulness.

4 Results

As to the demand for smart wearable device, 82.8% (135) were willing to purchase the products at present. When asked about an acceptable payment for buying the smart wearable devices, 73.6% (120) of the respondents chose NTD$1,000 or less. The result is consistent with previous studies.

There are significant relationships between health anxiety and the use frequency, devices' number of health information searching. People who are more anxious about their health are more likely to seeking health information frequently and using the smart wearable device. Health information could be used by people with an illness in order to understand their diagnosis and treatment options as well as by healthy people for risk evaluation and disease prevention.

In this study, most of the respondents are healthy persons. And the purpose of using the smart wearable device is to increase their health information. It is supposed that the smart wearable device is helpful for disease prevention and health promotion. The results revealed that perceived ease of use, perceived usefulness, and health anxiety had a significant impact on attitude to use.

Table 1. Description of participants

Variables	N	%
Gender		
Male	96	58%
Female	67	41%
Education		
University	127	77%
High school	36	22%
Marriage		
Yes	24	14%
No	139	85%
Employment		
Yes	42	25%
No	121	74%
Increasing health Information		
Yes	151	92%
No	12	7%

Table 2. Regression Model

Variables	B	t	p
Age	-0.009	-1.691	0.051
Marriage	-0.107	-0.985	0.421
Employment	0.099	1.018	0.315
Perceived usefulness	-0.271	-3.239	0.003**
Perceived ease of use	0.233	3.172	0.006**
Health anxiety	0.36	4.121	0.000**
Increasing health information	0.331	4.026	0.000**
Health cognitive	0.182	2.916	0.010**
Constant	0.216	3.013	0.009**
R		0.533	
R square		0.284	

*p<0.05, **p<0.01

5 Discussion and Conclusion

This study represents research in examining the applicability of TAM to explaining roadrunners' acceptance of telecare technology. The model was evaluated using data collected from roadrunners practicing the smart wearable device in Taiwan. Some implications can be drawn from the findings of the study.

Lack of experience among the general public is one of the barrier to the development of the smart wearable device. This survey demonstrated that roadrunners agree with the concept and the potential benefits of the smart wearable device when they receive information. Another barrier is financial burden. Many roadrunners are not willing to pay for the smart wearable device unless it is inexpensive. The relative industry need to consider how to enhance public awareness of smart wearable devices' benefits and broadcast worthiness of purchasing the smart wearable device.

From a managerial standpoint, in order to raise individual intentions to use a technology, it is important to encourage a positive attitude about health anxiety toward using the technology. Positive perception of the technology's usefulness is crucial, whereas the technology's ease of use may not be equally important for professionals. Emphasize devising effective means to communicate the health cognitive of the technology to roadrunners should be the high priority task for the healthcare industry management.

6 Competing Interests

No conflicting interests can be indicated.

Acknowledgements

This research was supported by grants from the Far Eastern Memorial Hospital for the project of "exploring the hospital applications of smart wearable devices" (RD1040073).

References

1. Chen, J., Yen, D., & Chen, K. The acceptance and diffusion of the innovative smartphone use. Information & Management, 46, 241-248. (2009).
2. Lee, T. The impact of perceptions of interactivity on customer trust and transaction intentions in mobile commerce. Journal of Electronic Commerce Research, 6(3), 165-180. (2005).
3. Hennig-Thurau, T., Malthouse, E. C., Friege, C., Gensler, S., Lobschat, L., Rangaswamy, A., & Skiera, B. The impact of new media on customer relationships, Journal of Service Research, 13(3), 311-330. (2010).
4. Longley P.A., M. F. Goodchild, D. J. Maguire and D. W. Rhind. Geographic Information Systems and Science. Second Edition. John Wiley, Chichester, (2005).
5. Abramowitz, J.S. & Moore, E.L. An experimental analysis of hypochondriasis. Behavior Research and Therapy, 45, 413–424. (2007).
6. Fergus, T.A., Valentiner, D.P., McGrath, P.B., & GierLonsway, S.L. Comparing Two Short-Form Versions of the Social Interaction Anxiety Scale and the Social Phobia Scale in a Clinical Sample. Paper presented at the 46th Annual ABCT Convention. (2012).
7. Gramling, A.A., J.W. Schatzberg, and W.A. Wallace. The role of undergraduate auditing coursework in reducing the expectations gap. Issues in Accounting Education 11 (1): 131-161. (1996).
8. Noyes, R., Happel, R. L., & Yagla, S. J. Correlates of hypochondriasis in a nonclinical population. Psychosomatics, 40, 461–469. (1999).
9. Fink, P., Ørnbøl, E., & Christensen, K.S. The outcome of health anxiety in primary care. A two-year follow up study on health care costs and selfrated health. Plos One, 5, 1–12. (2010).
10. Lees A, Mogg K, Bradley BP. Healthy anxiety, anxiety sensitivity, and attentional biases for pictorial stimuli and linguistic health threat cues. Cognition Emotion; 19(3):453–62. (2005)
11. Williams M. J., McManus F., Muse K., & Williams J. M. G. Mindfulness-based cognitive therapy for severe health anxiety (hypochondriasis): An interpretative phenomenological analysis of patients' experiences. British Journal of Clinical Psychology, 50, 379–397. (2011).
12. Olatunji, B. O., Etzel, E. N., Tomarken, A. J., Ciesielski, B. G., & Deacon, B. J. The effects of safety behaviors on health anxiety: An experimental investigation. Behaviour Research and Therapy, 49, 719-728. (2011).
13. Moon, J. W., & Kim, Y. G. Extending the TAM for a World-Wide-Web context. Information & Management, 38(4), 217–230. (2001).
14. Csikszentmihalyi, M. The future of Flow. In M. Csikszentmihalyi & I. S. Csikszentmihalyi (Eds.), Optimal experience: Psychological studies of Flow in consciousness (pp. 364–383). Cambridge University Press: New York. (1988).
15. Chung, J., & Tan, F. B. Antecedents of Perceived Playfulness: An exploratory study on user acceptance of general information-searching Websites. Information & Management, 41(7), 869–881. (2004).
16. Agarwal, R., & Karahanna, E. Time flies when you're having fun: Cognitive absorbtion and beliefs about information technology usage. MIS Quarterly, 24(4), 665–694. (2000)
17. Nysveen, H., Pedersen, P. E., & Thorbjørnsen, H. Intentions to use mobile services: Antecedents and cross-service comparisons. Journal of the Academy of Marketing Science, 33(3), 330–346. (2005).
18. Fang, X. W., Chan, S., Brzezinski, J., & Xu, S. Moderating effects of task types on wireless technology acceptance. Journal of Management Information Systems, 22(3), 123–157. (2006).
19. Talukder, M., Harris, H., Mapunda, G. Adoption of innovations by individuals within organizations: An Australian study. Asia Pacific Management Review, 13(2), 463-480. (2008)
20. Davis, F.D., Bagozzi, R.P., Warshaw, P.R. User acceptance of computer technology: A comparison of two theoretical models. Management Science, 35(8), 982-1003. (1989)

21. Gagnon, M.P., Godin, G., Gagne, C., Fortin, J.P., Lamothe, L., Reinharz, D., and Cloutier, A. An adaptation of the theory of interpersonal behaviour to the study of telecare adoption by heart disease patients. International Journal of Medical Informatics, 71, 103–115. (2003)

MATEC Web of Conferences **71**, 05006 (2016) DOI: 10.1051/matecconf/20167105006
ICCPE 2015

Test for Thermoelectric Self Cooling

Kuo-Chi Liu [a], Yuen-Shin Chen and Mu-Kai Chen

Department of Mechanical Engineering, Far East University, 49 Chung Hua Rd., Hsin-Shih, Tainan 744, Taiwan

Abstract. Low grade waste heat is easily appears in working devices and raises their temperature. Therefore, forced convection cooling system became a necessity for keeping their performance. For saving energy, the ideal of thermoelectric self cooling that recovers and transforms the waste heat into electricity with thermoelectricity modules for driving the cooling system was proposed. In order to explore the techniques of its application, this study sets up a simulator of thermoelectric self cooling device for experimental study. The effects of the structure of heat sink on the function of thermoelectric self cooling have been investigated.

1 Introduction

The development of energy harvesting technology has received more attention in a variety of application areas. Currently, the environment around us is full of various waste heats which can not be recycled effectively by conventional methods. Therefore, recovering waste heat for electric power generation, especially low grade waste heat, has been regarded as the new research direction. For entire solid-state energy conversion, thermoelectric generation technology appears to have advantages in low grade waste heat recovery [1]. Thermoelectric generation technology can directly converse thermal energy into electricity by using a thermoelectric generator, which consist of p-type and n-type semiconductors, has no moving parts and is compact, quiet, highly reliable and environmentally friendly [2, 3].

Low grade waste heat easily appears in working devices, such as electrical power converters, transformers and control systems. Waste heat not only decreases energy efficiency, also raises working temperature of devices and reduces performance of devices. To ensure performance, forced convection cooling is always putted on devices. An extra electricity demand will be required by the cooling system. From the viewpoint of saving energy, such cooling behavior will consume more energy. Therefore, Martínez et. al. [4] proposed the idea of thermoelectric self cooling for saving electricity. The waste heat from devices was thermoelectrically converted into electricity to drive the forced cooling systems. The validity of thermoelectric self cooling has been experimentally evidenced by Martínez et. al. [4]. This paper would do an extend study to the techniques of its application. The effect of geometry of the heat sink on the dissipation efficiency is experimentally explored. It is also investigated how the rate of waste heat affects the function of generation unit.

2 Experiment works

2.1 Device with thermoelectric self cooling

Figure 1. Photograph of the prototype made in this study.

A thermoelectric self cooling device similar to that in Ref. [4] is set up, as shown in Figure 1. A film heating resistor, which is connected in parallel to an adjustable AC power source, is sandwiched by two pieces of aluminum plate to simulate the cooled device. The power generation unit is composed of four thermoelectric modules (Kryotherm TGM-287-1.0-1.5) in series. It converts a part of the dissipation heat from the cooled device into electricity

[a] Corresponding author: kuochi.liu@msa.hinet.net

DOI: 10.1051/matecconf/20167105006

(a) (b)

Figure 2. Perspective view of the device: (a) heat sink with cube base (situation A) and (b) rectangular pyramidal base (situation B).

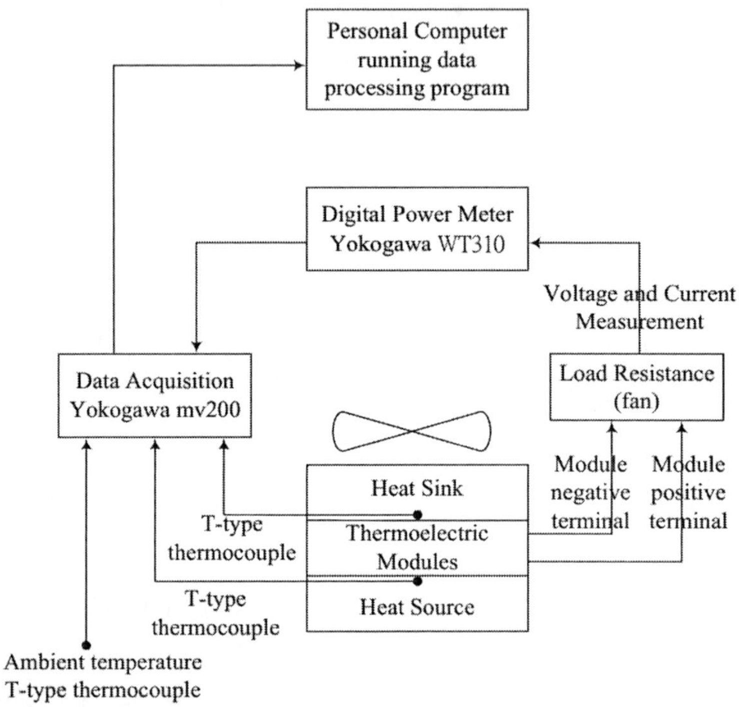

Figure 3. Scheme diagram of the test system.

to run the cooling fan, and the forced convection effect is induced. The bottom side of the cooled device is thermally insulated.

To explore the effect of heat sink, this study designs and manufactures the heat sinks with cube and rectangular pyramidal bases, as shown in Figure 2.

2.2 Data acquisition

This study establishes a test system for the experimental study of thermoelectric self cooling. Figure 3 shows the scheme diagram of the test system. The measurement equipments consist of a personal computer running data processing program, Yokogawa data acquisition WT310, Yokogawa digital power meter mv200, and a number of T type thermal couples. Measuring the temperature of the hot and cold sides of the thermoelectric module is not easy. T-type thermocouples are attached onto two sides of the thermoelectric generation unit to measure the cold temperature, T_c, and the hot temperature, T_h.

The experiments are executed within the environments temperature 25±1 ^{0}C. In order to explore the effect of dissipation heat, the power offered to the device is controlled by the transformer and power meter to be 80W 、120W and 160W respectively. The bottom side of the device is thermally insulated. Temperature, current and voltage are synchronously measured. Multiplying the voltage by the current can obtain the power generation rate. As a result, the effects of the geometry of heat sink and the starting delay of cooling fan on the thermoelectric generation unit can be known from the temperature difference between two sides of the generation unit and the generation power.

3 Results and discussion

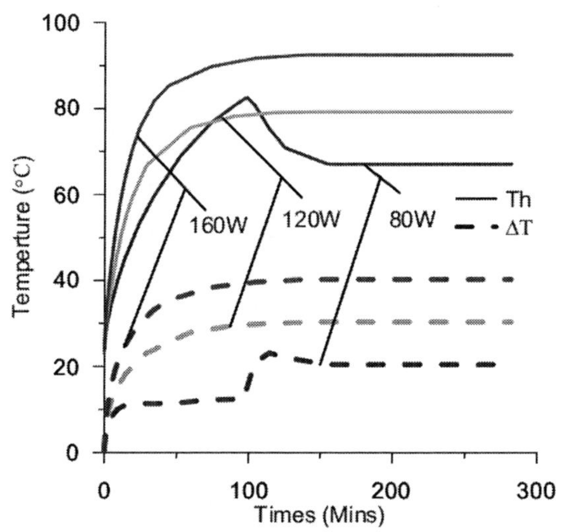

(a) For situation A.

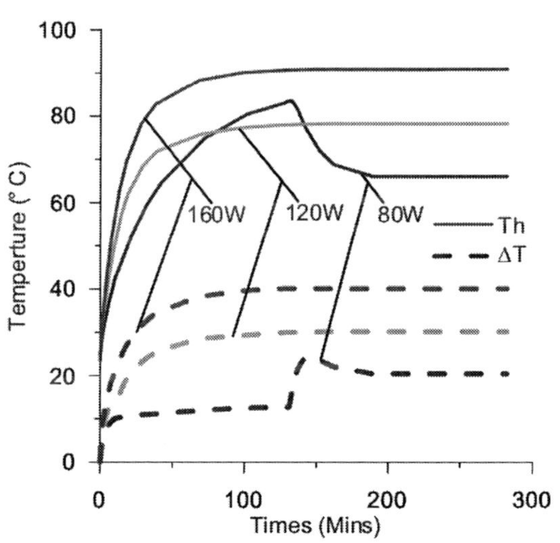

(b) For situation B.

Figure 4. Temperature difference variation between the hot and cold sides of the thermoelectric generation unit for the situations A and B.

In order to know the effect of heat sink on the performance of thermoelectric self cooling, the heat sinks, with a cube base or a rectangular pyramidal base, were used. For convenience of statement, the situation of the heat sink with a cube base and the situation of the heat sink with a rectangular pyramidal base are called as situation A and situation B, respectively.

Figure 4 shows the temperature difference variation between the hot and cold sides of the thermoelectric generation unit for the situations A and B with the input power 80W, 120W, and 160W. As the input power is 120W and 160W, the temperature of hot side of the thermoelectric generation unit is rapidly raised at the beginning of heating. Therefore, the temperature difference between the hot and cold sides of the thermoelectric generation unit makes the generation unit generate an enough power to run the cooling fan. In other words, if the heat emitted from the device is enough, the cooling fan will be run with the power generated by thermoelectric modules in a short time, and the forced cooling is achieved. Relatively, as the input power is 80W, the difference between the hot and cold temperatures can not be enlarged and the generation power can not achieve the level that is able to run the cooling fan. Because the device is under nature convection conditions in a long time, the hot temperature continually increases, and its maximum temperature is over the maximum temperature caused by the power 120W, as shown in Figure 4.

It is observed from Figure 4 that the difference between the hot and cold temperatures rapidly increases for the case of the power 80W after the cooling fan has been started. The hot temperature also drops along. The comparison between the experimental results of situations A and B further shows that, as the input power is 120W and 160W, the variation of the side temperatures of situation A is similar to that of situation B. The chamfered surfaces increases the dissipation area, so the hot temperature is lower in the situation B. Due to the dissipation of the chamfered surfaces, the difference between the hot and cold temperatures get a lag to achieve the level that make the thermoelectric modules generate enough power to drive the fan.

Martínez et. al. [4] considered that the phenomenon of thermal bridge may happen in the gap between the heat sink and the device and reduce the power generation. Therefore, they mounted a cold extender between the power generation unit and the heat sink in order to avoid the phenomenon of thermal bridge. The cold extender is not installed in the present experiment. The experiment results show the effect of thermal bridge is not obvious.

4 Conclusions

The results show that the larger the dissipation heat, the larger the difference between the hot and cold temperatures, and the effect of forced cooling more easily appears. The effect of thermal bridge is not obvious in the present study. The chamfered surfaces of heat sink with a rectangular pyramidal base do not benefit on the function of thermoelectric self cooling, but they increase the amount of heat dissipation in the early times and weaken the phenomenon of overheat.

Acknowledgments

This research was supported by the National Science Council in Taiwan through Grant NSC 101-2221-E-269-006.

References

1. L.E. Bell, Science **321**, 1457 (2008).
2. J. Yang and F.R. Stabler, J. Electron Mater **38**, 1245 (2009).
3. P. Yodovard, J. Khedari, and J. Hirunlabh, Energy Source **23**, 213 (2001).
4. A. Martínez, D. Astrain, and A. Rodríguez, Energy **36**, 5250 (2011).

Total Ionization Radiation Sensor Performance Improvement by Using Si-rich MONOS Device

Hsieh Wen-Ching [1*], Hao-Tien Daniel Lee [2], Fuh-Cheng Jon[3], Shich-Chuan Wu[4] Dawei Heh[5], Tzu-I Tsai[6]

[1]*Minghsin Universuty of Science and Technology, Xinxing Rd, 1, Xinfeng 30401, Taiwan*
[2]*Treasure Giant Technology Inc. 3F-1, 42, Lyushuei Rd. Hsinchu City, 30068 Taiwan*
[3]*Southern Taiwan University of Science and Technology, 1, Nan-Tai Street, Yungkang District, Tainan 710, Taiwan*
[4]*National Nano Device Laboratories, Hsinchu City, 30068 Taiwan*
[5]*Taiwan Semiconductor Manufacturing Company Limited, Hsinchu City, 30068 Taiwan*
[6] *United Microelectronics Corporation, Hsinchu City, 30068 Taiwan*

Abstract. The Si-rich metal -oxide-nitride-oxide-silicon (hereafter Si-MONOS) can be candidates for non-volatile total ionization dose (TID) radiation sensors. In the case of Si-MONOS gamma radiation sensors, the gamma radiation induces a significant decrease of threshold voltage V_T. The change of V_T for Si-MONOS after gamma irradiation has a strong correlation to the TID of gamma ray exposure as well. The Si-MONOS capacitor device in this study has demonstrated the better feasibility for non-volatile TID radiation sensing in the future.

1 Introduction

Total ionizing dose (TID) radiation effect is the major application concern for the operation of electronic devices in advanced X-ray lithography semiconductor manufacture process and outer space application, as well as other harsh environments such as accelerators, where high and low energy particles exist. When a MOS structure is irradiated by gamma rays, the positive charges build-up at Si-SiO2 interface and the interface state occurred [1]. The irradiation effects of a Metal-Nitride-Oxide- Silicon (MNOS) device with stacked insulation layers composed silicon nitride and silicon dioxide has been reported [2]~[5]. However the detailed mechanism of the charged trappings has not been understood yet. In this study, the radiation-induced trapped charges in the insulation layer of Si rich metal - oxide-nitride-oxide-silicon (hereafter Si-MONOS) and metal-oxide-nitride-oxide-silicon (hereafter MONOS) capacitor structures are investigated. The gamma ray radiation induces significant decrease of threshold voltage V_T for both Si-MONOS and MONOS devices. It is considered that this V_T change mostly owing to significant increase of radiation induced positive trapped charges in the Si rich oxide-nitride-oxide (hereafter Si-ONO) and oxide-nitride-oxide (hereafter ONO) gate dielectric after gamma ray irradiation, respectively. In this paper, the change of V_T for both Si-MONOS and MONOS devices after Gamma ray expose were compared; the change of V_T for Si-MONOS is more significant than that of MONOS. This Si-MONOS capacitor device with ONO gate dielectric in this study has demonstrated the better potential for the application of non-volatile gamma ray TID dosimeter.

2 Proposed Mechanism

In cases of Si-MONOS and MONOS capacitor devices, gamma ray radiation and fixed gate voltage was applied on these Si-MONOS and MONOS capacitor devices simultaneously for gamma ray data writing. When these Si-MONOS and MONOS capacitor structures are irradiated by gamma rays, the electron-hole pairs are generated throughout the gate insulation layers. These free carriers are swept by electric field, and some of them are captured by the charge trap centers, especially at the oxide-nitride (O-N) interface and silicon- oxide (S-O) interface. The amount of captured charge depends on the local density and the capture cross section of the trap centers. When a negative gate voltage VG is applied on these Si-MONOS and MONOS capacitor devices, the holes generated by radiation in the insulation layers are swept toward metal gate side by electric field, and some of the holes are captured by the charge trap centers at the O-N interface; however electrons generated by radiation in the insulation layers are swept toward silicon base side by electric field, and some of electrons are captured by the charge trap centers at the O-N interface. As shown in experiment data, the gamma ray radiation induces significant decrease of V_T for both Si-MONOS and MONOS capacitor devices. It is considered that the amount of trapped holes in gate insulator due to irradiation is greater than the amount of trapped electron. It is suggested that the amount of holes trapped at O-N interface is higher than the amount of electrons trapped at

[a] Corresponding author: wchsieh@must.edu.tw

O-N interface .The trapped charges change the electrical characteristic of these Si-MONOS and MONOS capacitor devices in terms of gate diode capacitor, gate V_T, and drain current. Therefore, the change of V_T in this case can be correlated to the amount of trapped charges and the total ionization dose (TID) of gamma radiation as well. These trapped charges due to irradiation are accumulated in gate dielectric layer and can't be destroyed by data reading process. In addition, data in these Si-MONOS and SONOS capacitor devices can be erased to original null state by charges injection mechanism. In this paper, the changes of V_T for Si-MONOS and MONOS after gamma ray expose are compared. As shown in experiment data, the change of V_T for Si-MONOS is more significant than that for MONOS. It is suggested that the amount of positive trapped charges in the Si-ONO gate dielectric layer is much greater than that in the ONO gate dielectric layer; so the decrease of V_T for Si-MONOS is more significant than that for MONOS. This Si-MONOS capacitor device is potentially more suitable for the application of gamma ray dosimeter.

3 Experimental

N-type MONOS & Si-MONOS were used and compared in this paper. N-type MONOS capacitor structures were fabricated on p-type resistivity 15-25ohm-cm Si <100> substrate. We use thermal SiO2 for tunnelling oxide, CVD Si3N4 or Si-rich CVD Si3N4 for trapping layer, CVD TEOS SiO2 for blocking oxide of ONO gate dielectric. In order to manipulate the Si composition of the nitride layer, the SiH2Cl2–NH3 gas flow-rate ratio was set differently for standard nitride and uniform Si-rich nitride, respectively. The SiO2–Si3N4–SiO2 (hereafter ONO) gate stack consists of a 100 Å ~200 Å thick silicon nitride trapping layer and 50 Å ~150 Å bottom and top silicon oxides. To write radiation data on these MONOS & Si-MONOS capacity devices, gamma radiation was exposed on these MONOS & Si-MONOS capacity devices. For the data read, V_T was measured at room temperature by using a HP4156A parameter analyzer. The experimental results of gate capacitance applied with various gate voltages (C-V) were obtained by using a computer-controlled HP4284 parameter analyzer, and the C-V curves were measured by sweeping VG together with zero source-and-drain bias condition at room temperature. Figure 2 shows the charge generation and trapping states of the gate dielectric for the Si-MONOS device after ionization radiation exposure.

4 Results and Discussion

4.1. Threshold voltage shift after ionization radiation exposure

As illustrated in figure 3 it can be seen that the capacity to voltage curves (C-V curve) of MONOS shifted to the left after 10 Mrad of gamma irradiation. This implies that gamma radiation induces a decrease of V_T for MONOS. The amount of decrease of V_T is about 2.0 volts. It is considered that the change is due to increase of net positive trapped charges in the ONO gate dielectric layer after gamma irradiation.

From figure 4, it also can be seen that C-V curve of Si-MONOS shifted far to the left after 10 Mrad of gamma irradiation. The amount of decrease of V_T is up to about 2.5 volts. As shown in the experiment data, the change of V_T for Si-MONOS is more than that for MONOS after 10Mrad TID of gamma irradiation. The dependence of V_T shift on TID for Si-MONOS is more than that for SONOS after gamma irradiation. It can be explained that the amount of net positive trapped charges due to gamma irradiation in the Si-ONO gate dielectric is greater than that in the ONO gate dielectric after gamma irradiation. These radiation-induced shifts in the irradiated device are a combination of two effects; the first effect is a result from the loss of stored charge in the ONO trapping layer and the second effect is due to a build-up of positive charge resulted from asymmetric trapping of electrons and holes in the ONO trapping layer. These two combined effects cause a large V_T shift in the Si-MONOS but create a smaller V_T shift in the MONOS. [5],[6]

4.2. Gate leakage current after 10Mrad gamma ray exposure

As illustrated in figure 5 and figure 6, both can show that the gate oxide leakage current of Si-MONOS and MONOS device did not increase significant after 10Mrad gamma ray exposure. The gate oxide leakage current of Si-MONOS increase more significant than MONOS after 10Mrad gamma ray exposure.

4.3. V_T decay vs. TID

In both figure 7 and figure 8, the decrease of V_T for MONOS and Si-MONOS are plotted against the total Gamma ray radiation dose TID received. The decay of V_T for MONOS and Si-MONOS device increases as a function of Gamma TID is indicated in Figure 7 and figure 8. The effect of gamma TID on the V_T change on the Si-MONOS device is more significant than that of the MONOS device. The decrease of V_T in this case can be correlated to the increase of positive trapped charges in the insulator and the increase of exposure TID for Gamma radiation as well. It also shows that decay of V_T increases more sharply after Gamma irradiation levels up to 100 krad TID. This result is agreed with the previous studies [5, 6].

4.4. V_T stability vs. retention time

As illustrated in Figure 9 and Figure 10, it also can be seen that the V_T decay vs time of Si-MONOS after 10Mrad gamma irradiation are better than that of MONOS. It shows that retention charge loss in Si-MONOS device after 10Mrad gamma irradiation is strongly affected by Si composition ratio and decreases as nitride trapping layer becomes Si rich. This capacity device with Si rich ONO gate dielectric in this study has demonstrated the better feasibility of using Si rich MONOS capacity device for gamma ray TID dosimeter.

5 Conclusions

As shown in experiment data, the change of V_T for Si-MONOS is more significant than that of MONOS. The dependencies of V_T shift on TID in Si-ONO gate dielectric is more than the one in ONO gate dielectric. It can be explained that the amount of positive trapped charges in the Si-ONO gate dielectric layer is greater than the one in the ONO gate dielectric layer; so the decrease of V_T for Si-MONOS is more than the one for MONOS. It is suggested that the amount of hole trapped at O-N interface in Si rich ONO gate dielectric layer is more significant than the one in the ONO gate dielectric layer. This capacity device with Si rich ONO gate dielectric in this study has demonstrated the better feasibility of using Si rich MONOS capacity device for the application of non-volatile gamma ray TID dosimeter.

Acknowledgments

The authors would also like to thank National Nano Device Laboratories (NDL), National Tsing Hua University (NTHU), and National Chiao Tung University (NCTU) to provide the instruments for wafer fabrication and testing. This paper was funded in part by the National Science Council (NSC) sponsor.

References

1. Watanabe, K.; Kato, M.; Okabe, T.; Nagata, M.: IEEE Trans. Nucl. Sci. NS-33, 1216–1222 (1986)
2. Takahashi, Y.; Ohnishi, K.; Fujimaki, T.; Yoshikawa, M.: IEEE Trans. Nucl. Sci. 461578–1585 (1999)
3. Gasperin, A.; Ghidini, G.; Cester, A.; Paccagnella, A. Oxide-nitride-oxide capacitor reliability under heavy-ion irradiation. IEEE Trans. Nucl. Sci. , 54, 1898–1905 (2007)
4. B.; Dockerty, R.; Shaneyfelt, M.; Habermehl, S.; Murray, J. Total dose radiation response of NROM-style SOI non-volatile memory elements. IEEE Trans. Nucl. Sci. 55, 3202–3205 (2008)
5. Qiao, F.Y.; Yu, X.; Pan, L.Y.: In Proceedings of the 19th IEEE International Symposium on the Physical and Failure Analysis of Integrated Circuits (IPFA), Singapore,(2–6 July 2012)
6. Hsieh, W.-C.; Lee, H.-T.D.; Jong, F.-C.: Sensors 14 14553-14566 (2014)

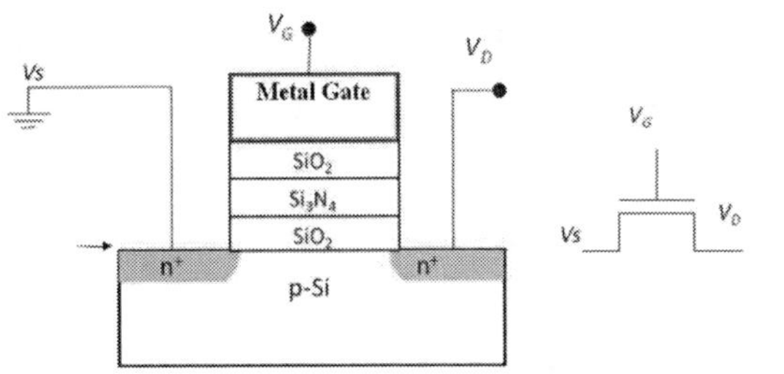

Figure 1. Cross-section view of MONOS devices.

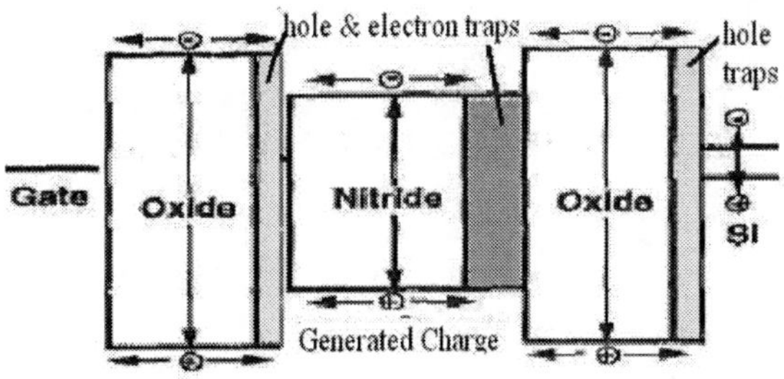

Figure 2. Charges generation and trapping in the MONOS device after Gamma exposure.

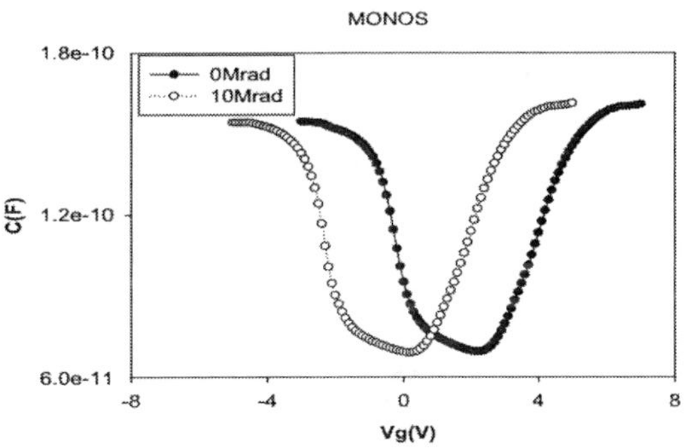

Figure 3. The C-V curve of MONOS device after 10 Mrad Gamma ray irradiation.

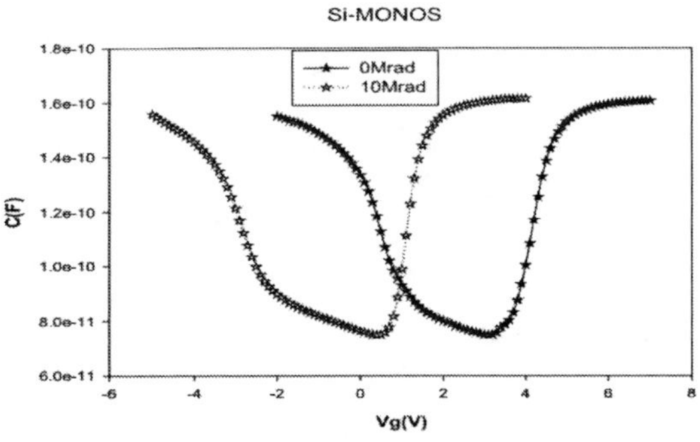

Figure 4. The C-V curve of Si-MONOS device after 10 Mrad Gamma ray irradiation.

DOI: 10.1051/matecconf/20167105007

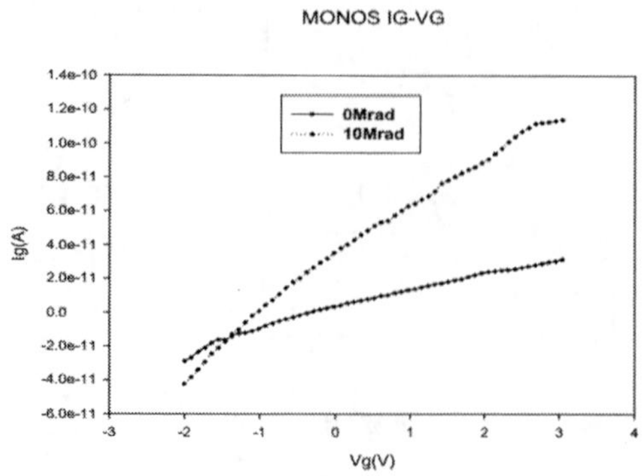

Figure 5. The gate current of MONOS device after 10Mrad gamma exposure.

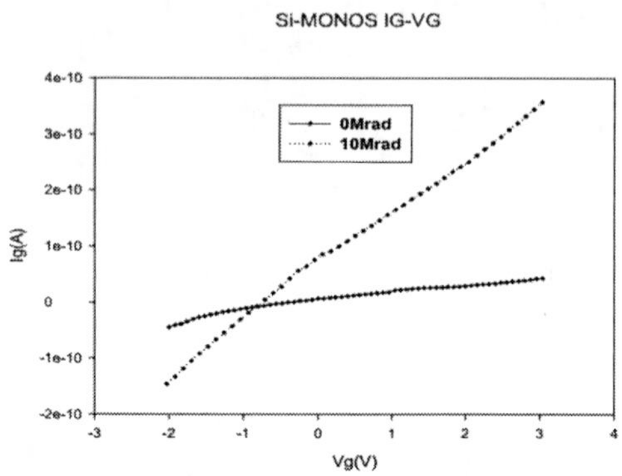

Figure 6. The gate current of Si-MONOS device after 10Mrad gamma ray exposure.

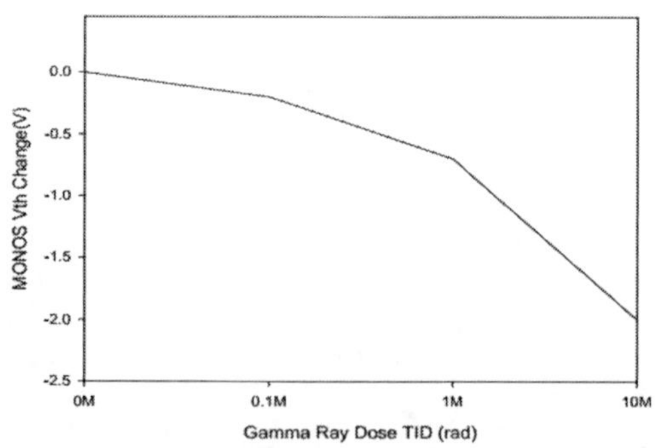

Figure 7. The dependencies of the V_T decrease on gamma radiation TID for MONOS

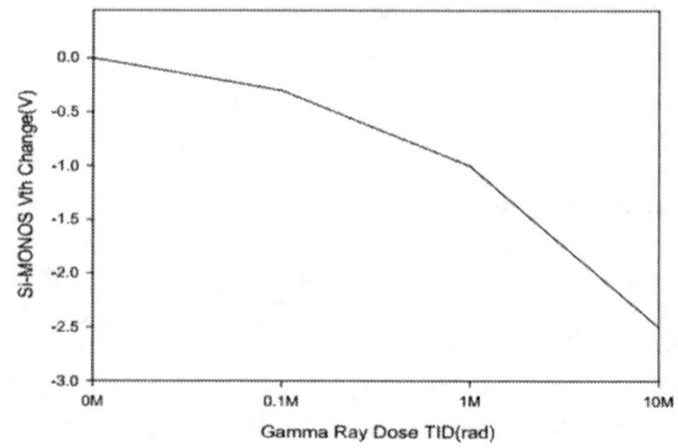

Figure 8. The dependencies of the V_T decrease on gamma radiation TID for Si-MONOS

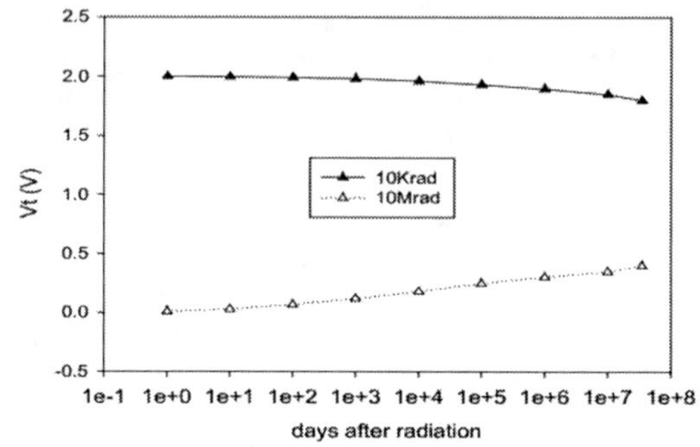

Figure 9. The V_T decay vs time of MONOS after 10Mrad gamma irradiation

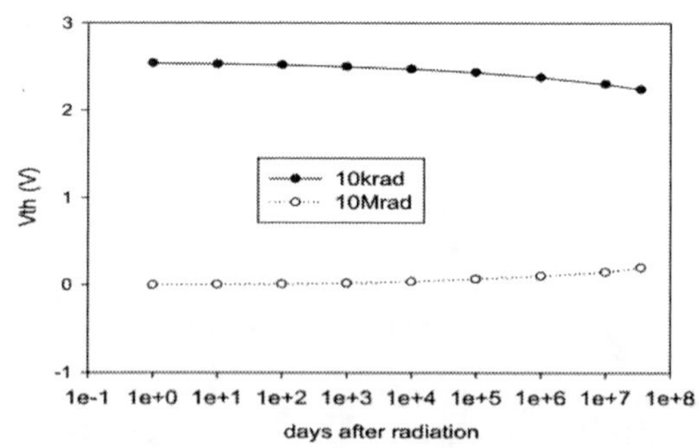

Figure 10. The V_T decay vs time of Si-MONOS after 10Mrad gamma irradiation

AUTHOR INDEX

EDP Sciences – Web of Conferences
17, Avenue du Hoggar
Parc d'Activité de Courtabouf
BP 112
F-91944 Les Ulis Cedex A
France

ISBN 978-1-5108-2748-6

9 781510 827486

CURRAN ASSOCIATES INC.
proceedings.com

EOS Optical Technologies Conference 2019

EPJ Web of Conferences Volume 215 (2019)

Munich, Germany
24 - 26 June 2019

edp sciences